Pharmaceutical Experimental Design

DRUGS AND THE PHARMACEUTICAL SCIENCES

A Series of Textbooks and Monographs

Pharmaceutical Experimental Design

Gareth A. Lewis
Synthélabo Recherche
Chilly Mazarin, France

Didier Mathieu
University of the Mediterranean
Marseille, France

Roger Phan-Tan-Luu
University of Law, Economics, and Science of Aix-Marseille
Marseille, France

MARCEL DEKKER, INC.   NEW YORK · BASEL

ISBN: 0-8247-9860-0

This book is printed on acid-free paper.

Headquarters
Marcel Dekker, Inc.
270 Madison Avenue, New York, NY 10016
tel: 212-696-9000; fax: 212-685-4540

Eastern Hemisphere Distribution
Marcel Dekker AG
Hutgasse 4, Postfach 812, CH-4001 Basel, Switzerland
tel: 44-61-261-8482; fax: 44-61-261-8896

World Wide Web
http://www.dekker.com

The publisher offers discounts on this book when ordered in bulk quantities. For more information, write to Special Sales/Professional Marketing at the headquarters address above.

Current printing (last digit):
10 9 8 7 6 5 4 3

PRINTED IN THE UNITED STATES OF AMERICA

Preface

Recent years have seen an upsurge in the application of statistical experimental design methods to pharmaceutical development, as well as to many other fields. These techniques are no longer the preserve of a few experts and enthusiasts but are becoming standard tools for the experimenter whose utility is recognised by management. Along with, or rather preceding, this interest has been a rapid development of the statistical and computational tools necessary for exploiting the methods. However, their use is not always optimal and the full range of available methods is often unfamiliar.

Our goal in writing this text is, therefore, to present an integrated approach to statistical experimental design, covering all important methods while avoiding mathematical difficulties. The book will be useful not only to the pharmacist or pharmaceutical scientist developing pharmaceutical dosage forms and wanting the best and most efficient methods of planning his experimental work, but also to formulation scientists working in other sectors of the chemical industry. It is not designed primarily for professional statisticians but these too may find the discussion of pharmaceutical applications helpful.

Chapter 1 introduces experimental design, describes the book's plan, and elucidates some key concepts. Chapter 2 describes screening designs for qualitative factors at different levels, and chapter 3 covers the classic two-level factorial design for studying the effects of factors and interactions among them. Chapter 4 summarises some of the mathematical tools required in the rest of the book.

Chapters 5 and 6 deal with designs for predictive models (response surfaces) and their use in optimization and process validation. Optimization methods are given for single and for multiple responses. For this we rely mainly on graphical analysis and desirability, but other methods such as steepest ascent and optimum path are also covered. The general topic of optimization and validation is continued in chapter 7, on treating variability, where the "Taguchi" method for quality assurance and possible approaches in scaling-up and process transfer are discussed.

Chapter 8 covers non-standard designs and the last two chapters are concerned with mixtures. In chapter 9, the standard mixture design and models are described and chapter 10 shows how the methods discussed in the rest of the book may be applied to the more usual pharmaceutical formulation problem where there are severe constraints on the proportions of the different components. In all of these chapters we indicate how to choose the optimal design according to the experimenter's objectives, as those most commonly used — according to the pharmaceutical literature — are by no means optimal in all situations.

However, no book on experimental design of this scope can be considered exhaustive. In particular, discussion of mathematical and statistical analysis has been kept brief. Designs for factor studies at more than two levels are not discussed. We do not describe robust regression methods, nor the analysis of correlations in responses (for example, principle components analysis), nor the use of partial least squares. Our discussion of variability and of the "Taguchi" approach will perhaps be considered insufficiently detailed in a few years. We have confined ourselves to linear (polynomial) models for the most part, but much interest is starting to be expressed in highly non-linear systems and their analysis by means of artificial neural networks. The importance of these topics for pharmaceutical development still remains to be fully assessed.

The words of Qohelet are appropriate here: "And furthermore be admonished, that of making of books there is no end" (*Ecclesiastes* 12:13)! He was perhaps unduly pessimistic as he continued, "and much study is a weariness of the flesh." This was not our experience, and we trust it will not be that of our readers. In fact, the experience of putting together this volume on statistical experimental design has proven to be personally most rewarding. Since we issue from rather different backgrounds, academic and industrial, chemical and pharmaceutical, French and British, we found that our approaches contrasted! Reconciling our differences, learning from each other, and integrating our work, was not always easy, but was invariably stimulating. If we have succeeded in transmitting to the reader something of what we have learned though this process, we will be well content.

We would consider all the topics and methods we have covered here likely to be useful in development, which already takes the book well beyond the two-level factorial designs equated by many with experimental design itself. This makes for a lot of material to be digested, but when it is considered that the tools of statistical design may be applied to such a large part of the development process, we do not feel any apology is called for! However, the beginner should feel reassured that he does not need such a complete knowledge before being able to apply these methods to his own work. He should, at least, read the introductions to each chapter carefully, especially chapters 2, 3, 5, 8 and 9 in order to decide which *situation* he is in when a problem arises, whether screening, factor influence study, modelling, or optimization, and whether there are particular difficulties and constraints involved. Inevitably he will make some mistakes, as we ourselves have done, but he will still find that a less-than-ideal choice of design will still give interpretable results and will be more efficient than the alternative non-design approach.

We wish to express our appreciation to our colleagues, without whose help this book would not have been written because its origin is in the investigation of pharmaceutical problems and product development that we have been carrying out together over many years. In particular we would like to thank Maryvonne Chariot, Jean Montel, Veronique Andrieu, Frederic André, and Gérard Alaux, present and past colleagues of G.A.L. at Synthélabo, and Michelle Sergent from the University of Aix-Marseille.

Gareth A. Lewis
Didier Mathieu
Roger Phan-Tan-Luu

Contents

1

OVERVIEW

The Scope of Experimental Design

I. INTRODUCTION

A. Designing an Experimental Strategy

In developing a formulation, product or process, pharmaceutical or otherwise, the answer is rarely known right from the start. Our own past experience, scientific theory, and the contents of the scientific and technical literature may all be of help, but we will still need to do experiments, whether to answer our questions or to confirm what we already believe to be the case. And before starting the experimentation, we will need to decide what the experiment is actually going to be. We require an experimental strategy.

Any experimentalist will, we hope and trust, go into a project with some kind of plan of attack. That is, he will use an experimental design. It may be a "good" design or it may be a "bad" one. It may even seem to be quite random, or at least non-systematic, but even in these circumstances the experimenter, because of his experience and expertise or pharmaceutical intuition, *may* arrive very quickly at the answer. On the other hand he could well miss the solution entirely and waste weeks or months of valuable development time. The "statistical" design methods described in this book are ways of choosing experiments efficiently and systematically to give reliable and coherent information. They have now been widely employed in the design and development of pharmaceutical formulations. They are invaluable tools, but even so they are in no way substitutes for experience, expertise, or intelligence.

We will therefore try to show where statistical experimental design methods may save time, give a more complete picture of a system, define systems and allow easy and rapid validation. They do not always allow us to find an individual result more quickly than earlier optimization methods, but they will generally do so with a far greater degree of certainty.

Experimental design can be defined as the strategy for setting up experiments in such a manner that the information required is obtained as efficiently and precisely as possible.

B. The Purpose of Statistical Design

Experimentation is carried out to determine the relationship (usually in the form of a mathematical model) between factors acting on the system and the response or properties of the system (the system being a process or a product, or both). The information is then used to achieve, or to further, the aims of the project

We therefore aim to plan and carry out the experiments needed for the project, or part of a project, with maximum efficiency. Thus, use of the budget (that is, the resources of money, equipment, raw material, manpower, time, etc. that are made available) is optimized to reach the objectives as quickly and as surely as possible with the best possible precision, while still respecting the various restrictions that are imposed.

The purpose of using statistical experimental design, and therefore of this book, is not solely to minimize the number of experiments (though this may well happen, and it may be one of the objectives under certain circumstances).

C. A Pharmaceutical Example

1. Screening a large number of factors

We now consider the *extrusion-spheronization* process, which is a widely used method of obtaining multiparticulate dosage forms. The drug substance is mixed with a diluent, a binder (and possibly with other excipients), and water, and kneaded to obtain a wet plastic mass. This is then extruded through small holes to give a mass of narrow pasta-like cylinders. These are then spheronized by rapid rotation on a plate to obtain more or less spherical particles or pellets of the order of 1 mm in diameter.

This method depends on a large number of factors, and there are many examples of the use of experimental design for its study in the pharmaceutical literature. It is treated in more detail in the following two chapters. Some of these factors and possible ranges of study are shown in table 1.1. At the beginning of a project for a process study, the objective was to discover, as economically as was reasonably possible, which of these factors had large effects on the yield of pellets of the right size in order to select ones for further study. We shall see that this involves postulating a simple *additive* or *first-order* model.

Table 1.1 Factors in Extrusion-Spheronization

Factor		Lower limit	Upper limit
% of binder	%	0.5	1
amount of water	%	40	50
granulation time	min	1	2
spheronization load	kg	1	4
spheronization speed	rpm	700	1100
extruder speed	rpm	15	60
spheronization time	min	2	5

Varying one factor at a time

One way of finding out which factors have an effect would be to change them one at a time. We carry out, for example, the experiment with all factors at the lower level, shown in the third column of table 1.1 (experiment 1, say). Then we do further experiments, changing each factor to the upper limit in turn. Then we may see the influence of each of these factors on the yield by calculating the difference in yield between this experiment and that of experiment 1. This is the "one-factor-at-a-time" approach, but it has certain very real disadvantages:

- Eight experiments are required. However the effect of each factor is calculated from the results of only 2 experiments, whatever the total number of experiments carried out. The precision is therefore poor. None of the other experiments will provide additional information on this effect.
- We cannot be sure that the influence of a given factor will be the same, whatever the levels of the other factors. So the effect of increasing the granulation time from 1 to 2 minutes might be quite different for 40% water and 50% water.
- If the result of experiment 1 is wrong, then all conclusions are wrong. (However, experiment 1, with all factors at the lower level could be replicated. Nor does each and every factor need to be examined with respect to this one experiment.)

The one-factor-at-a-time method (which is in itself an experimental design) is inefficient, can give misleading results, and in general it should be avoided. In the vast majority of cases our approach will be to vary all factors together.

Varying all factors together
A simple example of this approach for the above extrusion-spheronization problem, one where each factor again takes only 2 levels, is given in table 1.2.

Table 1.2 Experimental Plan for Extrusion-Spheronization with All Factors Varied Together

Run no.	Binder (%)	Water (%)	Granul. time (min)	Spher. load (kg)	Spher. speed (rpm)	Extrud. speed (rpm)	Spher. time (min)
1	0.5	40	1	2	700	60	5
2	1.0	40	1	1	1100	15	5
3	1.0	50	1	1	700	60	2
4	0.5	50	2	1	700	15	5
5	1.0	40	2	2	700	15	2
6	0.5	50	1	2	1100	15	2
7	0.5	40	2	1	1100	60	2
8	1.0	50	2	2	1100	60	5

Examination of this table shows that no information may be obtained by comparing the results of any 2 experiments in the table. In fact, to find the effect of changing any one of the factors we will need to use the results of *all 8* experiments in the design. We will find in chapter 2 that, employing this design:

- The influence of each factor on the yield of pellets is estimated with a far higher precision than by changing the factors one at a time. In fact, the standard error of estimation is halved. To obtain the same precision by the one-factor-at-a-time method each experiment would have to be done 4 times.
- The result of each experiment enters equally into the calculation of the effects of each factor. Thus, if one experiment is in error, this error is shared evenly over the estimations of the effects, and it will probably not influence the general conclusions.
- The number of experiments is the same as for the one-factor-at-a-time method.

Clearly, this second statistical design is far better than the first. The screening design is adapted to the problem, both to the objectives (that is, *screening* of factors) and to the constraints (7 factors studied between maximum and minimum values). If we had been presented with a different problem, needing for example to compare in addition the effects on the yield of 3 different binders, or those of 3 models of spheronizing apparatus, then the design would have needed to have been quite different. Designs and strategies for all such situations are given in the various chapters.

2. More complex designs

The above design would have allowed average effects to be estimated with maximum efficiency. However the effect of changing the spheronization speed might be quite different depending on whether the extrusion rate is high or low (figure 1.1). For a more complete analysis of the effects, experiments need to be carried out at more combinations of levels. The equivalent to the one-factor-at-a-time approach would involve studying each pair of factors individually. Again we will find that a global solution where all factors are varied together will be the most efficient, requiring fewer experiments and giving more precise and reliable estimations of the effects. One of a number of possible approaches might be to take the original design of table 1.2 and carry out a *complementary* design of the same number of runs, where all the levels of certain columns are inverted. The 16 experiments would give estimations of the 7 effects, and also information on whether there are interactions between factors, and some indications (not without ambiguity) as to what these interactions might be.

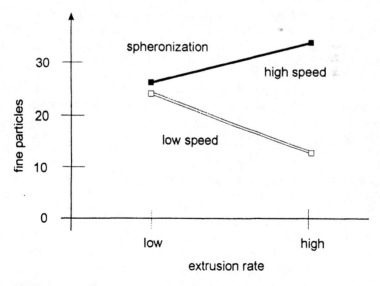

Figure 1.1 Percentage mass of particles below 800 µm: main effect of spheronization speed and interaction with extrusion rate.

In the same way the sequential one-run-at-a-time search for an optimized formulation or process is difficult and often inefficient and unreliable. Here also, a structured design where all factors (less factors than for screening) are changed together, and the data are analysed only at the end, results in a more reliable and precise positioning of the optimum, allowing prediction of what happens when process or formulation parameters are varied about their optimum values, information almost totally lacking when the alternative approach is used. Only experimental design-based approaches to formulation or process development result in a predictive model. These also are highly efficient in terms of information obtained for the number of experimental runs carried out. Optimization on the three most influential factors would probably require between 10 and 18 runs, depending on the design.

II. PLAN OF THE BOOK

It is necessary to say something to the reader about how this book is organized. There are a number of excellent texts on experimental design and the fact that we have sometimes approached matters differently does not indicate that we think our approach better, only that it may be a useful alternative.

A. Stages in Experimentation

The various steps of an experiment using statistical design methodology are typically those given in table 1.3. The many examples in this book will be described for the most part according to such a plan. It is simplified, as there is almost certain to be a certain amount of "coming and going", with revision of previous stages. For example, it might be found that all designs answering the objectives were too large, with respect to the available resources. All stages in the process should be formally documented and verified as they are approached in turn by discussion between the various parties concerned. We will take this opportunity to emphasise the importance of the first two (planning) stages, consisting of a review of the available data, definition of the objectives of the experimentation and of likely subsequent stages, and identification of the situation (screening, factor study, response surface methodology etc.).

     Statistical design of experiments cannot be dissociated from the remainder of the project and it is necessary to associate all the participants in the project - from the project manager to those who actually carry out the experiments. It is important that the "statistician" who sets up the experimental design be integrated in the project team, and in particular that he is fully aware of the stages that preceded it and those that are likely to follow on after. Planning of the experiment is by far the most important stage. For a full discussion of the planning of a "designed" experiment see for example the very interesting article by Coleman and Montgomery (1), the ensuing discussion (1), and an analysis by Stansbury (2).

It is necessary for the entire team to meet together, to define the objectives and review existing data, before going to the next stage. Before attempting to draw up any kind of protocol, it will be necessary to identify the *situation* (such as screening, factor studies or optimization). The book is to a major extent organized around them, as described in the next section. Most improper uses of experimental designs result from errors in or neglect of these planning stages, that is, insufficient analysis of the problem or a wrong identification of the situation or scenario.

Table 1.3 Stages in a Statistically Designed Experiment

1	Description of problem	7	Constraints and limits (definitive)
2	Analysis of existing data	8	Experimental design matrix
3	Identification of situation	9	Experimental plan
4	List of factors (definitive)	10	Experimentation
5	List of experimental responses	11	Analysis of data
6	Mathematical model	12	Conclusions and next stage

There follows a detailed listing of the variable and fixed factors, and the constraints operating, which together make up the domain of experimentation. Then, and only then, can the model and experimental design be chosen, and a protocol (experimental plan) be drawn up. These steps are defined in section II.B of this chapter.

B. Designs and Methods Described

1. The "design situation"

It is essential to recognise, if one is to choose the right design, treatment or approach, in what situation one finds oneself. Do we want to find out *which* factors amongst a large number of factors are significant and influence or may influence the process or formulation? If so the problem is one of *screening* and is covered to a major extent in chapter 2. If we have already identified 4 to 5 factors which have an influence, we may then wish to quantify their influence, and in particular discover how the effect of each factor is influenced by the others. A *factor influence study* is then required. This normally involves a factorial design (chapter 3). If on the other hand we have developed a formulation or process but we wish to *predict the response(s)* within the experimental domain, then we must use an appropriate design for determining mathematical models for the responses. This and the method used (response surface methodology or RSM) are covered in chapter 5. These 3 subjects are closely related to one another and they form a continuous whole.

Designs discussed in detail in these chapters are given in full in the text. Where the rules for constructing them are simple, the designs for a large number of factors may be summarized in a single table. Other designs are tabulated in the appendices.

2. Optimization

Screening, factor studies and RSM are all part of the search for a product or process with certain characteristics. We are likely to require a certain profile or a maximized yield, or to find the best compromise among a large number of sometimes conflicting responses or properties. We show in chapter 6 how to identify the best combination of factors by graphical, algebraic, and numerical methods, normally using models of the various properties of the system obtained by the RSM designs. However, we will also indicate briefly how the sequential simplex optimization method may be integrated with the model-based approach of the rest of the book.

3. Process and formulation validation

Validation has been a key issue in the industry for some time and it covers the whole of development. Since validation is not an activity that is reserved for the end of development, but is part of its very conception, systematic use of statistical design in developing a formulation or process ensures traceability, supports validation, and makes the subsequent confirmatory validation very much easier and more certain. It is discussed in a number of the later chapters, especially in the final section of chapter 6.

4. Quality of products and processes

There is variability in all processes. It is assumed constant under all conditions in the first part of the book, but in chapter 7 we look at the concept in more detail. First we will describe how to use the methods already described under circumstances of non-constant variability. This is followed by the study of variability itself with a view to minimizing it.

This leads to the discussion of how modern statistical methods may be introduced to assure that quality is built into the product. For if experimental design has become a buzz-word, *quality* is another. The early work in this field was done in Japan, but the approach of Taguchi and others in building a quality that is independent of changes in the process variables has been refined as interest has widened. These so-called Japanese methods of *assuring quality* are having a large effect on engineering practice both in Europe and North America. It seems that the effect on the pharmaceutical industry is as yet much less marked – although there is much interest, it is not yet transformed into action. This may happen in time.

In the meanwhile we explore some ways in which Taguchi's ideas may be applied using the experimental design strategy described in the rest of the book, and the kind of problems in pharmaceutical development that they are likely to help solve, in assuring reliable manufacture of pharmaceutical dosage forms.

5. Experimental design quality

The concept of the *quality of an experimental design* is an essential one, and is introduced early on. Most of the current definitions of design quality, optimality, and efficiency are described more formally and systematically in chapter 8, leading to methods for obtaining *optimal experimental designs* for cases where the standard designs are not applicable. The theory of the use of *exchange algorithms* for deriving these designs is discussed in very simple terms. Mathematical derivations are outside the scope of the book and the reader is referred to other textbooks, or to the original papers.

6. Mixtures

The analysis and mathematical modelling of mixtures is significant in pharmaceutical formulation and these present certain particular problems. The final chapters cover methods for optimizing formulations and also treat "mixed" problems where process variables and mixture composition are studied at the same time. In both chapters 9 and 10 we continue to illustrate the graphical and numerical optimization methods described earlier.

7. Some general comments on the contents

The emphasis throughout this book is on those designs and models that are useful, or potentially useful, in the development of a pharmaceutical formulation. This is why such topics as asymmetric factorial designs and mixtures with constraints are discussed, and why we introduce a wide variety of second-order designs for use in optimization. We stress the design of experiments rather than analysis of experimental data, though the two aspects of the problem are intimately connected. And we indicate how different stages are interdependent, and how our choice of design at a given stage depends on how we expect the project to continue, as well as on the present problem, and the knowledge already obtained.

C. Examples in the Text

1. Pharmaceutical examples

Many of the examples, taken either from the literature or from unpublished studies, are concerned with development of solid dosage forms. Particular topics are:
   • drug excipient compatibility screening,
   • dissolution testing,
   • granulation,
   • tablet formulation and process study,
   • formulation of sustained release tablets,
   • dry coating for delayed release dosage forms,
   • extrusion-spheronization,

- solubility,
- nanoparticle synthesis,
- microcapsule synthesis,
- oral solution formulation,
- transdermal drug delivery.

However, the methods used may be applied, with appropriate modifications, to the majority of problems in pharmaceutical and chemical development. Some of the general themes are given below.

2. Alternative approaches to the same problem

There is such a variety of designs, of ways of setting up the experiments for studying a problem, each with its advantages and disadvantages, that we have indicated alternative methods and in some cases described the use of different designs for treating the same problem. For example, a study of the influence of various factors - type of bicarbonate used, amount of diluent, amount of acid, compression - on the formulation of an effervescent paracetamol tablet is described in chapter 3. A complete factorial design of 16 experiments was used, and in many circumstances this would be the best method. However, it would have been possible to investigate the same problem using at least 4 other designs that would each have given similar information about the formulation, information of lesser quality it is true, but also requiring fewer experiments. The experiments of all these other smaller designs were all found in the design actually used, so the results of the 5 methods could be compared using the appropriate portion of the data. We shall see, however, that there is no need for any actual experimental data in order to compare the *quality* of information of the different methods.

The problem of excipient compatibility screening is also discussed and it is shown how, according to the different numbers of diluents, lubricants, disintegrants, glidants requiring testing, we need to set up different kinds of screening designs.

There is usually a variety of possible methods for treating a given problem and no one method is necessarily the best. Our choice will depend in part on the immediate objectives, and the immediate restraints that operate, but it will also be influenced by what we see as the likely next stage in the project. The method which is the most efficient if all goes well, is not always going to be the most efficient if there are problems. In this sense, design is not totally unlike a game of chess!

3. Linking designs

No experimental design exists on its own, but it is influenced by the previous phase of experimentation and the projected future steps. Its choice depends partly on the previous results. The strategy is most effective if statistical design is used in most or all stages of development and not only for screening, or optimizing the formulation or the process. It is sometimes possible to "re-use" experiments from previous studies, integrating them into the design, thus achieving savings of time and material. This may sometimes be anticipated.

For example, certain designs make up a part of other more complex designs, that may be required in the next stage of the project. We will illustrate these links and this continuity between steps as some of the examples given will be referred to in a number of the chapters. For example, the central composite design used in response surface modelling and optimization may be built up by adding to a factorial design. This sequential method is simulated using a literature example, the formulation of a oral solution. A part of the data were given and analysed in chapter 3 in order to demonstrate factor influence studies. Then in chapter 5 all the data are given so as to show how the design might be augmented to enable response-surface modelling. Finally, the estimated models are used to demonstrate graphical optimization in chapter 6.

Another factorial design, used for studying solubility in mixed micelles, introduces and demonstrates multi-linear regression and analysis of variance. It is then extended, also in chapter 5, to a central composite design to illustrate the estimation of predictive models and their validation.

4. Building designs in stages

In addition to the possibility of reusing experiments in going from one stage to another, it is worth noting that many designs may be carried out in steps, or *blocks*. It has already been mentioned that extrusion-spheronization, as a method for producing multiparticulate dosage forms, has been much studied using statistical experimental design. We use it here to introduce methods for choice and elimination of factors (factor influence studies), and at the same time, to demonstrate the sequential approach to design. A factor-influence study is carried out in several stages. The project may therefore be shortened if the first step gives sufficient information. It may be modified or, at the worst, abandoned should the results be poor, or augmented with a second design should the results warrant it.

Quite often we may be unable to justify carrying out a full design at the beginning of a project. Yet with careful planning, the study may be carried out in stages, with all parts being integrated into a coherent whole.

5. Different methods of optimization

When optimizing a formulation or process, there are a number of different methods for tackling the problem and the resulting data may also be analysed in a number of different ways. By demonstrating alternative treatments of the same data, we will show advantages and weaknesses of the various optimization methods and how they complement one other.

For example, the production of pellets by granulation in a high-speed mixer is used to illustrate properties of the uniform shell (Doehlert) design, and it is shown how the design space may be expanded using this kind of design. The resulting mathematical models are also used to demonstrate both the optimum path and canonical analysis methods for optimization. Both graphical and numerical optimization are described and compared for a number of examples: an oral liquid formulation, nanosphere synthesis, a tableting process study, and a placebo tablet formulation.

D. Statistical Background Needed

We aim to give an ordered, consistent approach, providing the pharmaceutical scientist with those experimental design tools that he or she really needs to develop a product. But much of the theory of statistical analysis is omitted, useful background information though it may be. A basic understanding of statistics is assumed – the normal distribution, the central limit theorem, variance and standard deviation, probability distributions and significance tests. The theory of distributions and of significance testing, if required, should be studied elsewhere (3, 4, 5). Other than this, the mathematics needed are elementary. Proofs are not given. Analysis of data is by multi-linear least squares regression and by analysis of variance. For significance testing, the F-test is used. These methods are introduced in chapter 4, together with discussion of the extremely important $X'X$ (information) and $(X'X)^{-1}$ (dispersion) matrices. Thus, although linear regression is used to analyse screening and factorial designs, in general an understanding of the method is not necessary at this point. A brief summary of the matrix algebra needed to understand the text is to be found in appendix I.

III. STARTING OUT IN EXPERIMENTAL DESIGN

A. Some Elementary Definitions

1. Quantitative factors and the factor space

Quantitative factors are those acting on the system that can take numerical values, rate, time, percentage, amount... They are most often *continuous,* in that they may be set at any value within prescribed limits. Examples are: the amount of liquid added to a granulate, the time of an operation such as spheronization or granulation, the drying temperature and the percentage of a certain excipient. Thus, if the minimum granulation time is 1 minute and the maximum is 5 minutes the time may be set at any value between 1 and 5 minutes (figure 1.2).

However, because of practical limitations, a quantitative factor may sometimes be allowed only *discrete* levels if only certain settings are available – for example, the sieve size used for screening a powder or granulate, the speed setting on a mixer-granulator. Unless otherwise stated, a quantitative factor is assumed continuous.

Natural variables for quantitative factors
The natural variable associated with each quantitative factor takes a numerical value corresponding to the factor level. The level of a quantitative factor i, expressed in terms of the units in which it is defined (so many litres, so many minutes ...), will be written as U_i, as in figure 1.2. We do not usually find it necessary to distinguish between the factor and natural variable. However it is possible to define different natural variables for the same factor. The factor temperature would normally be expressed in units K or °C, but could equally well be expressed as K^{-1}.

Associated (coded) variables
With each natural variable, we associate a *non-dimensional*, coded variable X_i. This coding, sometimes called normalization, is obtained by transforming the natural variable, usually so that the level of the central value of the experimental domain (see below) is zero. The limits of the domain in terms of coded variables are usually identical for all variables. The extreme values may be "round numbers", ± 1, though this is by no means always so.

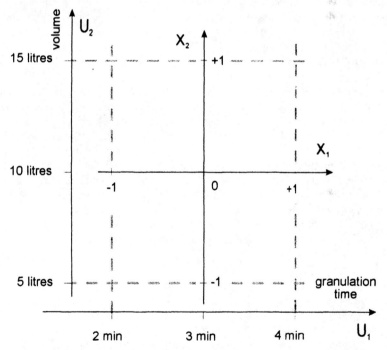

Figure 1.2. Quantitative factors and the factor space. The axes for the natural variables, granulation time and volume are labelled U_1, U_2 and the axes of the corresponding coded variables are labelled X_1, X_2. The factor space covers the page and extends beyond it, whereas the design space is the square enclosed by $X_1 = \pm 1$, $X_2 = \pm 1$).

Factor space

This is the k dimensional space defined by the k coded variables X_i for the continuous quantitative factors being investigated. If we examine only two factors, keeping all other conditions constant, it can be represented as a (two dimensional) plane. Except for the special case of mixtures, factor space is defined in terms of independent variables. If there are 3 factors, the factor space can be represented diagrammatically in three dimensions.

We are only interested in studying quite a small part of the factor space, that which is inside the **experimental domain**. Sometimes called a *region of interest*, it is the part of the factor space enclosed by upper and lower levels of the coded variables, or is within an ellipsoid about a centre of interest.

The **design space** is the factor space within this domain defined in terms of the coded variables X_i.

Factor space and design space for mixtures

For the important class of mixture experimental designs, the variables (proportions or percentages of each constituent of the mixture) are not independent of one another. If we represent the factor space by two-dimensional or pseudo-three-dimensional drawings, the axes for the variables are not at right angles and the factor space that has any real physical meaning is not infinite.

2. Qualitative factors

These take only discrete values. In pharmaceutical development, an example of a qualitative variable might be the nature of an excipient in the formulation, for example "diluent", the levels being "lactose", "cellulose", "mannitol" and "phosphate". Or it might be the general type of machine, such as mixer-granulator, used in a given process, if several models or sizes are being compared. In medicinal chemistry it might be a group on a molecule. So if all the factors studied are qualitative, the *factor space* consists of discrete points, equal in number to the product of the available levels of all the factors, each representing a combination of levels of the different factors, as shown in figure 1.3 for two factors.

Mixed factor spaces, where some factors are qualitative, or quantitative and discrete and other quantitative factors are continuous, are also possible.

Coded variables for qualitative factors

The levels of qualitative factors are sometimes referred to by numerical quantities. If we were comparing three pieces of equipment, the machine might take levels 1, 2, and 3 as in figure 1.3. However, unlike the qualitative continuous variable, no other values are allowed and these numbers do not have any physical significance. Level 3 is not (necessarily) "greater than" level 1. Among the 4 diluents in the same figure, phosphate (level 4) is not greater than cellulose (level 2).

Design space for qualitative factors

This consists of the points representing all possible combinations of the levels that are being studied for the associated coded variable for each factor. The total number of points in the design space (and also the total number of possible distinct experiments) is obtained by multiplying together the numbers of levels of each factor.

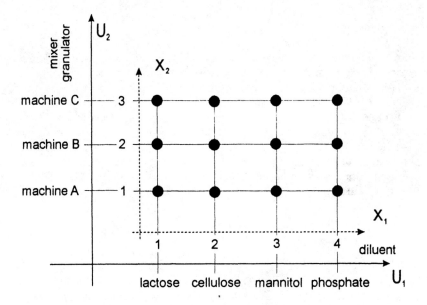

Figure 1.3. Qualitative factors and the factor space. The order, and spacing of the factors is arbitrary.

3. Experimental runs

An experimental run is a practical manipulation or series of manipulations, carried out under defined conditions (the levels of the different factors, those that are allowed to vary in the design, and those which are held constant) resulting in a (single) datum for each of the responses to be measured.

The combination of factor levels being studied in a run is represented by a point in the design space. In an experimental design, all the runs may be under different conditions (distinct points in the design space) or certain runs may be *replicates*, being carried out under the same conditions. Each experimental run is normally set up independently of every other run. Thus, even if conditions are repeated, the apparatus or machine (assuming here that we are studying a process) is set up afresh for each run. The term *experiment* in this book is normally used to mean experimental run. The term experimental *unit* is also sometimes used.

4. Response variables

Responses are measured properties of the process (for example, yield, dissolution rate, viscosity) sometimes referred to as dependent variables. The symbol for a measured response will be y. The measured response for experiment i will be written as y_i.

5. Mathematical model

Normally referred to simply as the **model**, it is an expression defining the dependence of a response variable on the independent variables – that is the variables whose influence we are studying or whose optimum values we are trying to find. For example, we have assumed above that the response could be described by a first-order model. It is generally written in terms of the *coded variables*.

The models used in this book are all **empirical**, where the system is essentially a "black box". The models are most frequently, but not invariably, polynomials of a given order. These are examples of **linear** models (not to be confused with first-order models). For example the second-order model in two factors X_1 and X_2:

$$ y = \beta_0 + \beta_1 x_1 + \beta_2 x_2 + \beta_{11} x_1^2 + \beta_{22} x_2^2 + \beta_{12} x_1 x_2 + \varepsilon $$

is a linear model, as all terms may be represented by a constant, β_i, β_{ii}, β_{ij}, multiplied by a variable, x_i, x_i^2, $x_i x_j$. ε represents the experimental error.

Theoretical or **mechanistic** models may exist, or be proposed. They might be thermodynamic or kinetic in origin. They are most often **non-linear** models (6), generally formulated as differential equations, and when an analytical solution is available this is most often exponential in form. Transformation to a linear function may occasionally be possible. It is rare to use these theoretical relationships directly in pharmaceutical experimental design and the designs are generally not adapted for determining or testing them. However, they often enter into the choice of factors chosen for study, empirical models and constraints, and in the interpretation of the results of factor-influence studies.

6. Experimental design

The *design* is the arrangement of experimental runs in design space (that is, defined in terms of the coded variables). The design we choose depends on the proposed model, the shape of the domain, and the objectives of the study (screening, factor influence, optimization ...).

7. Experimental plan

This is the design transformed back into the real or natural variables of each factor, normally with the runs in the order they are to be performed.

8. R-efficiency

We have said that experimental design is to do with efficiency, and so we need to be able to measure a cost-benefit ratio. The cost is easy to measure, and is, essentially, the number of experiments. The "benefit" is more difficult to quantify. It may be considered as the amount of information from the design. The R-efficiency is the simplest measurement of a design's efficiency. It is the ratio of the number of parameters in the model (p) to the number of experiments (N):

$$R_{eff} = \frac{p}{N} \le 1$$

This simply tells us the number of experiments required to calculate so many parameters, but it tells us nothing of the *quality* of the estimation, which must include not only the number of coefficients in the model, but also the precision to which they are calculated. There are ways of quantifying the quality of information obtained per experiment so that different experimental approaches can be compared, and these, along with other definitions of efficiency which take such considerations into account, will be discussed later in the book, especially in chapter 8.

The R-efficiency may also be expressed as a percentage.

B. Some Important Concepts

The following ideas will be developed over much of the book, but we will state them here, immediately, for emphasis.

1. Improved precision by varying all factors

It is common to determine the effects of a number of factors by varying each in turn. We can get a much better estimate (in terms of improved precision) using designs where all factors are varied at the same time.

2. Reproducibility

Before starting the experimentation of a design it is necessary to have some idea of the reproducibility of the experiments. This may be the result of a number of repeated experimental runs (repeatability), but it may also be the result of the experimenter's own experience if the method is well-known or a standard one. It is, of course, normal procedure to estimate the repeatability by replicating experiments within the design.

Repeating only part of an experimental procedure, for example testing multiple units from the same batch of tablets, is not replication and indicates only part of the repeatability. Such multiple testing is frequent and often necessary, but the response in such cases is the mean value obtained and is considered as that of a single experimental run.

3. Planning and sequencing within a project

Steps and sequences in design
The optimum strategy is rarely that of carrying out the project in a single step. It is unusual for one design to be enough. Research and development is in logical stages and one design may follow on from another. Designs may be augmented if the information they give is interesting but insufficient. The team must therefore be made aware of the ways that designs may be carried out in several stages - or that one design may re-use information from an earlier one.

Experimental work on a project may be carried out in the order:
- Screening
- Quantitative factor study
- Response surface modelling
- Optimization
- Validation

Statistical experimental design may be used at all these stages, though evidently we do not need to carry through all of these steps for every project. Screening may be omitted if the process is well enough known from analogous studies, and we can go directly to a quantitative factor study of a limited number of factors. Separate optimization stages may be required for the process and the formulation. Scaling-up studies may be needed after optimization at a laboratory scale.

Sequential and batch approaches
The sequential strategy is not always the best. The *batch* approach is particularly appropriate to problems like stability testing. Although sequential testing may well reduce the number of man-hours, much time could be lost in checking the stability of one set of experiments before doing the rest, involving this could involve unacceptable delays. Therefore a much larger design would normally be set up.

Increased speed can only be achieved here by a *concurrent batch* approach. Each batch or block of experiments is done by a different operator, probably using separate (identical or at least similar) machines. Each block will certainly have to be carried out. The batch approach is thus likely to be expensive in terms of resources, but more rapid.

Note that even in the case of stability testing it is possible to combine the advantages of batch and sequential methodology by putting the total number of experiments on store as a single batch, but by analysing them as sequential blocks, all non-analysed samples going into the deep freeze.

Choice of design
Do not adapt the problem to the method chosen. Rather, choose a design that is right for treating the problem that has been set.

4. Preliminary experimentation

Some preliminary experimentation is usually necessary for the experimenter to practice the technique, to estimate the repeatability, or ensure adequate reproducibility before beginning the main study. However, because experimental designs often comprise quite large numbers of experiments some workers are quite reasonably concerned about getting it right before beginning the design. This can give rise to the *preliminary experiments* syndrome! This can mean carrying out a few runs, without any structure, just to "get the feel of it". These can easily add up to a premature attempt at an optimization, as the experimenter does experiment run after another, knowing that the next is going to be the "good one". Now preliminary experiments are usually necessary, but there is a very good argument for structuring them, so that they themselves are part of a planned design, and their results can be interpreted. They can be done in three ways, replicated experiments, small experimental designs, and few experiments chosen from a larger plan

Repeated experiments
In this way it is possible to check that the process is reasonably reproducible before starting experimentation on the design, and may enable stabilisation of the operator's technique.

Small screening or factorial designs
A small design which varies just two or three of the factors expected to be most influential can be carried out. An example might be the 2^{3-1} factorial design described in chapter 3, section IV.E, requiring 4 runs. Replicated centre points may be included. In this way, interpretable results can be obtained without sacrificing too many resources.

Depending on our assessment of the situation, we may stop here and start a new design, perhaps increasing the number of factors. Alternatively we may decide to shift the experimental region to one centred on other values of X_i. Or, in the ideal case, this design might serve for part of a larger design for three factors by the addition of one or two further blocks: a complete 2^3 factorial (see chapter 3), a central composite design, or a small central composite design (chapter 5, section III.A), respectively.

A few experiments chosen from a larger plan
Once a probable design has been selected, preliminary experiments can be taken from this, either chosen at random or by taking especially interesting or potentially difficult runs. Thus, they may be incorporated in the plan - when and if it is completed.

5. Quality of a design

No experimental run contains any information on its own. The information it provides depends on its position (in factor space) with respect to that of other experimental runs. It is this arrangement of experiments, each with respect to all the others, that is an experimental design. Therefore, careful reflection is required before beginning experimentation.

The quality of the *design* depends on the mathematical model that is assumed for the relationship being studied and on the arrangement of the experiments in the factor space. It does *not* depend on the experimental results.

The quality of an experimental design can and should be assessed *before* the experiments are started, and not after. This does not mean that the quality of the results does not depend on the precision and accuracy of the experimental measurements. These are just as important as one would suppose, but the quality of measurement can be separated from the question of quality of design.

6. Efficiency and cost-benefit ratio

No single parameter describes a design's quality, and the choice of the "best" design will almost invariably be a compromise. We will ask how many experiments are needed per effect calculated and whether the estimates of the effects or parameters calculated are independent. Quite simply, the cost is the number of experiments and the benefit is the quality of the design. We will assess the amount of information obtained per experiment and whether the precision of a calculated response over the design space is relatively constant (as is generally preferred) or varies widely from one part of the design space to another.

We often see statements like: "Experimental design enabled the necessary information to be obtained using a small number of experiments". The reality does not always bear this out. One reason is that the statistical design method enables us to see very clearly exactly what information is available from a given design, before doing the experiments. Analysis of the design will also indicate what will remain unknown, and a perceived level of ignorance is not always acceptable.

In fact, the object of experimental design is to do the necessary experimentation efficiently. We get what we pay for. The results are, one might say, good value, but not necessarily cheap. Careful reading of this book will demonstrate ways of improving efficiency and, sometimes, of achieving rapid development.

7. Significance

We use this term in the restricted sense of *statistical significance*. In other words, if an effect is "significant", there is a high probability (95%, 99%, 99.9%) that the effect is "real" - that is, different from zero. The determination of the significance of an effect or of a mathematical model is an essential tool for the experimenter.

However, he should be careful here not to confuse (statistically) "significant" with "important" (7). A significant effect may still be quite small. It is for the researcher to decide whether statistically significant effects are of "practical" or pharmaceutical significance, or no. If the experiment is highly reproducible, the effects of certain factors which in practice are unimportant may show up as significant in the statistical analysis. He may reasonably choose not to study them further. On the other hand, there may be large effects which, if the estimated values are correct, would be highly important, but which are not found to be significant because of insufficient experimental precision. He may then come to a decision based on the estimated effect, but also on the possible inaccuracy of the estimate.

C. Recommended Books on Experimental Design

A number of excellent books have been written on statistical experimental design and references to these, and to useful articles, may be found in the bibliographies of the various chapters. This present work, which attempts a practical introduction to the range of approaches useful in pharmaceutical development, does not attempt to replace these in any way. Two works in particular are essential reading – *Statistics for Experimenters*, by Box, Hunter, and Hunter (6), and Cornell's book, *Experiments with Mixtures* (8). Other valuable texts are given here (9, 10, 11) or referenced later in the book.

D. Choosing Computer Software

The experimenter wishing to use statistical design will quickly find he needs to buy a program for multi-linear regression. Specialized software is also necessary for setting up non-standard designs in irregular shaped experimental domains and also for optimization methods such as desirability, optimum path, and possibly canonical analysis. Integrated packages for design allow the construction of standard designs as well, and enable one to switch easily from coded to natural variables.

A large number of such packages are available to assist us, both at the experimental design and the data analysis stages, and as a result some quite difficult and "advanced" methods have become possible. Many of these methods are directly applicable to the kind of problems likely to be encountered by the formulator and development scientist and, consequently, they are described here in some detail. This is so that the experimenter may best choose a design method when using such a program, and also use the method correctly, and understand the significance of the output which may refer to some quite complex statistical theory. Because we have concentrated on these particular aspects of experimental design, screening, factor studies, and optimization, our book does not cover such topics as the comparison of two or more data sets, nor linear regression with a single variable, in any detail. We have described factorial and fractional factorial designs relatively briefly.

The choice of computer software is vital, and must be made with respect to completeness of the range of designs, statistical methodology, data treatment, particularly its graphical aspects. Ease of use, flexibility, and clarity are all important, and of course the price is also to be considered. Programs are developing and changing rapidly; anything we write here about individual programs is likely to be out of date even before the time of publication. No individual program is likely to be complete as far as all users are concerned. In this book we have used three programs (12, 13, 14), but there are many others. A check-list of useful features is given in appendix IV, to help the reader in his choice.

References

1. D. E. Coleman, D. C. Montgomery, B. H. Gunter, G. J. Hahn, P. D. Haaland, M. A. O'Connell, R. V. Leon, A. C. Schoemaker and K.-L. Trui, A systematic approach to planning for a designed experiment, Discussion, Response, *Technometrics*, **35**, 1-27 (1993).
2. W. F. Stansbury, Development of experimental designs for organic synthetic reactions, *Chemom. Intell. Lab. Syst.* **36**, 199-206 (1997).
3. D. C. Montgomery, Design and Analysis of Experiments, 2nd edition, J. Wiley, N. Y., 1984.
4. J. C. Miller and J. N. Miller, Statistics for Analytical Chemistry, Ellis Horwood, Chichester, U. K., 1984.
5. J. R. Green and D. Margerison, Statistical Treatment of Experimental Data, Elsevier, Amsterdam, 1978.
6. G. E. P. Box, W. G. Hunter, and J. S. Hunter, Statistics for Experimenters, J. Wiley, N. Y., 1978.
7. D. N. McCloskey, The insignificance of statistical significance, *Scientific American*, **272**, 32 (1996).
8. J. A. Cornell, Experiments with Mixtures, 2nd edition, J. Wiley, N. Y., 1990.
9. G. E. P. Box and N. B. Draper, Empirical Model-Building and Response Surface Analysis, J. Wiley, N. Y., 1987.
10. P. D. Haaland, Experimental Design in Technology, Marcel Dekker, N. Y., 1989.
11. G. Taguchi, System of Experimental Design, Vol. 1 and 2, N. Y. Unipub., 1987.
12. RS/Discover, BBN Software Products, Bolt Beranek and Newman Inc., Cambridge, MA 02140, USA.
13. Design Expert®, Stat-Ease Inc., Minneapolis, MN 55413, USA.
14. NEMROD®, D. Mathieu and R. Phan-Tan-Luu, LPRAI SARL, Marseilles, F-13331, France.

2

SCREENING

Designs for Identifying Active Factors

I. SCREENING OF FACTORS

A. Its Place in Development

To screen is to select from the factors which may possibly influence the process or phenomenon being studied those which have a real effect, an influence that is distinguishable unequivocally from the background noise. This study is normally done very early in the life of the project in order to simplify the problem and thus enable the experimenter to concentrate his attention and resources in a more detailed examination and optimization of the principal factors. Therefore, this stage should not take up more than 20 to 30 % of the total cost of the project (1).

For example, we may have considered a total of 15 potential factors at the start, but have found, having completed our preliminary screening study, that only 4 of them were important. One might well imagine that a great deal, perhaps an excessive amount, of work would be required to study the 15 factors, only in the end to draw conclusions from the data on these last 4. This would be true were one looking for the same quality of information for all 15 of them, but in fact the screening study calls for a number of simplifying hypotheses. These may sometimes be over-simplifications, but more frequently they are entirely justified.

Although many factors can often be listed as possibly important it is not unusual that a large proportion of experimental variation can be explained by a small proportion of the factors. This is sometimes called the "sparsity" effect. Those factors on which most of the variation depends are called "active" factors, the others "inert" or "non-active" factors. We use the term "active" in preference to "significant" where we have insufficient data to test for statistical significance.

B. Recognising the Screening Situation

The object of a screening study is *not* to obtain complete and exact numerical data on the properties of a system to allow a full description, but rather to be able to answer "yes" or "no" (or "perhaps") to some simple questions about whether or not certain factors have an effect. There follows a few examples that illustrate different situations which we will develop later on.

One object of this exercise is to select factors that are probably active and to get a preliminary idea as to their effects. Another object is to eliminate as many of the factors as possible. The factor is not retained in the remainder of the study if it appears to have no effect. This usually means that from then on the factor is fixed at one of the levels used in the screening study.

Consider an extrusion-spheronization process study. Is the response (the yield of pellets of the desired particle size) significantly different when the mixing time of the wet powder mass (before extrusion) is increased from 2 minutes to 5 minutes?

This question is concerned with the factor *time*, which is **quantitative** and **continuous**. It is easy to see that the mixing time could be varied continuously (at least within the limits of the precision of the clock) and that each of the infinite number of levels might be represented by a real number.

If there is no appreciable difference (that is a difference no greater than the order of magnitude of the experimental error) between the results for 2 min. and for 5 min. mixing, one might conclude that the change in mixing time has no effect on the response. This is the *only* conclusion that we may draw, from a strict mathematical and statistical point of view. However it is probable that if we were to mix for 3 or 4 min. we would find the result to be no different from that at 2 and at 5 min. We might even, at a certain risk, extrapolate to 6 or 7 min. (but not to 15). It would, however, be unwise to attempt to draw conclusions for a mixing time of less than 2 min.

Again, considering the same extrusion-spheronization process study, is the response significantly different when the spheronization time is increased from 2 minutes to 8 minutes?

As previously stated, if there is no appreciable difference found between the results obtained at 2 min. and 8 min. we conclude that increasing the time from 2 to 8 min. has no effect on the response. We are not allowed to conclude anything, strictly speaking, concerning times other than 2 and 8 min. An intermediate time might be much more favourable to the production of pellets of the correct size and shape. A longer time might provoke the disintegration of the pellets, a short time could be insufficient.

We see that apart from the purely numerical values obtained in the experiment, our interpretation of the results can depend on our preconceived ideas of the phenomenon or process and on our existing knowledge. There is certainly a risk of error in these interpolations, still more in extrapolation, and the magnitude of these risks depends on the validity of our hypotheses.

For the initial mixing stage we have the choice of 3 different mixer-granulators (type A, B or C). Does the nature of the equipment used have an effect on the response?

Here the "nature of equipment" is a **qualitative** factor with **3 levels** (or states), each level being a different piece of equipment. The screening experiment allows us to determine if the results are different on each machine, or if 2 of the machines behave in the same way but differently from the third, or if all 3 machines may be considered identical. In this example it is evident that the conclusions drawn from the screening study cannot normally be extended to other machines (D, E...etc.).

In the above extrusion-spheronization process, can the rate setting of the extruder modify the results obtained?

The rate setting can be treated as either qualitative or quantitative, depending on the circumstances. It is this kind of variable that causes the most difficulties in formulating problems in experimental design! Let us consider four possible circumstances in turn:

  (a) Two extrusion speeds are possible, which we call *fast* and *slow*. We have here a qualitative factor at 2 levels. The screening study will allow us to decide if the results are affected by the change in the extrusion speed control.

(b) There are three possible rates (*fast*, *medium* and *slow*). The extrusion rate can be seen to be a qualitative factor at 3 levels, as in the previous question concerning the mixer-granulators, and the problem can be treated in the same way.

(c) The extrusion rate can take one of three values, each corresponding to a different rate, R_1, R_2 and R_3, in cm/s. This time the factor is *quantitative* and *discontinuous* at 3 levels. If the experimenter limits himself to these 3 values of the extrusion rate, both in the design of the experiment and his interpretation of the results, then the levels may be treated as though qualitative. But if he should try to derive a mathematical relationship between the extrusion rate and the response, then it is no longer a screening problem but one for the more detailed factor study.

(d) The extruder can work at any given rate between a minimum and a maximum value (the factor being *quantitative* and *continuous*). The experimenter can select several, sufficiently distinct, values within this range. (He might in such a case, select the maximum and minimum rates defined either by the limitations of the machine or by normal procedure in the process.)

If at the beginning he intends to eliminate extrusion rate from the list of factors, as far as further study is concerned and if the results for the different rates are not sufficiently different, then the study is one of screening. If, however, the experimenter intends to study the relationship between extrusion rate and response, then it is no longer screening, but a problem for a factor study.

These situations are summarized as follows:

Extruder controls	Type of variable
fast – slow	qualitative – 2 levels
fast – medium – slow	qualitative – 3 levels
speeds R_1 –R_2 –R_3	quantitative – discontinuous
speeds R_{min} to R_{max}	quantitative – continuous

We have a list of possible additives (excipients) for a tablet. In formulating the tablet we want to know how its properties vary when changing the relative proportions of the excipients. In particular, we want to know which of these excipients will have the greatest effect on the measured properties of the tablet and the tableting process when changing their proportions.

We would not attempt to quantify the influence of each constituent and optimize the formulation at this early stage of development. We would only try to find out if each excipient is actually useful. We could ask, for example, whether several constituents have the same role and can be replaced by a single excipient, thus simplifying the formulation. Formulation variables (these describe the percentage or fraction of the constituents in a mixture) are not independent of one another – diminishing the proportion of one excipient automatically results in an increase in the proportion of one or more others – and the methods for treating them are different from those used for other variables and are considered later in the book. In the real problems that we deal with in pharmaceutical development, there are often different kinds of factors. Unfortunately there is as yet no screening strategy that will allow us to deal with both independent factors and formulation factors at the same time.

Apart from this, for each of the various situations we have just outlined there is a suitable experimental strategy. We will now develop those appropriate for screening independent variables.

II. MATHEMATICAL MODELS FOR SCREENING INDEPENDENT FACTORS

The previous section might lead us to suppose that the mathematical treatment of a screening study is simple and that it is simply the direct comparison of two or more values. However the strategies we propose do not involve varying the factors one after the other. On the contrary, as we have indicated in the introductory chapter, we can achieve greatly improved precision and efficiency using designs where the factors are varied *simultaneously*. Interpretation by direct comparison of the data is therefore impossible and a mathematical model must be used.

A. A Single Factor at Multiple Levels

1. A factor at 2 levels

When a factor can take only 2 levels, whether these are qualitative or quantitative they are usually represented as -1 and +1. In this way, setting up the mathematical models and interpretation of the coefficients becomes much easier. Once a factor takes more than 2 quantitative levels this is no longer possible. This is why we will begin with a different approach here that might appear unduly complicated, but has the advantage that it can be extended to the study of qualitative variables at 3 or more levels.

Suppose we want to know if the factor F_1 influences the experimental response, which we will call y. If F_1 has no effect on y, all other factors being fixed, we can immediately write down the following model:

$$y = \beta_0 + \varepsilon$$

where β_0 is a constant - the true, or theoretical, response - and ε represents the (random) experimental error, with mean value of zero.

In the general case the relationship of a factor to the response is unknown. It may not even interest us, except in certain exceptional cases – for example, if a theoretical model, such as a thermodynamic equation, exists. What the experimenter will usually try to demonstrate is the change or *variation in the response* of the system resulting from the *variation of the state of the factor*. He is looking not for an **absolute** effect, but for a **differential** effect of the factor on the response.

Let us now identify the 2 levels (or states) of the factor F_1 as A and B. The factor or variable might be the choice of excipient, as in a previous example. Thus tablets may be formulated using 1% magnesium stearate ($F_1 = A$) or another lubricant such as 2.5% hydrogenated castor oil ($F_1 = B$). The disintegration time or other properties of the resulting tablets are then measured.

The effect of the factor may be written either in terms of a real or a hypothetical reference state: the reference state model and the presence-absence model, respectively.

2. The reference state model

One of the states is taken as reference and the variation of the response is described with respect to that state or level. Taking for example A as reference the model may be written as:

$$y = \beta'_0 + \varepsilon \qquad\qquad F_1 \text{ in state A}$$
(tablets with magnesium stearate)
$$y = \beta'_0 + \beta'_{1,B/A} + \varepsilon \qquad F_1 \text{ in state B}$$
(tablets with hydrogenated castor oil)

Thus, $\beta'_{1,B/A}$ describes the effect on the response of changing F_1 from state A to B. In our example, it is the increase of the tablet disintegration time resulting from the replacement of 1% magnesium stearate in the formulation by 2.5% hydrogenated castor oil. This method introduces a certain asymmetry between the 2 levels, which is perhaps not justified. Why choose one level, or one lubricant, as the reference state rather than another?

3. The presence-absence model

Alternatively, we might consider that each level of factor F_1 introduces an equal and opposite variation from a hypothetical mean value of the response – hypothetical because it is a value that does not correspond to any accessible level of the factor (neither A nor B), but which would be the value that would be obtained (for all states of F_1) if the change in the state of the factor F_1 had no effect on the response:

$$y = \beta_0 + \beta_{1,A} + \varepsilon \qquad \text{when the factor } F_1 \text{ is in state A}$$

$$y = \beta_0 + \beta_{1,B} + \varepsilon \qquad \text{when the factor } F_1 \text{ is in state B}$$

One could say that $\beta_{1,A}$ represents the differential effect of the factor F_1 in state A, and $\beta_{1,B}$ is the differential effect of factor F_1 in state B, both with respect to the mean value. Symmetry implies that:

$$\beta_{1,A} + \beta_{1,B} = 0 \quad \text{or} \quad \beta_{1,A} = -\beta_{1,B}$$

In practice it would be difficult to use this notation, as the equation would be different according to the level taken by the factor. We therefore define a variable $x_{1,A}$ which takes the value 1 when F_1 is at level A, and otherwise equals zero. Similarly $x_{1,B}$ equals 1 if F_1 is at level B, and otherwise equals zero.

If $F_1 = $ "A" then $x_{1,A} = 1$ and $x_{1,B} = 0$
If $F_1 = $ "B" then $x_{1,A} = 0$ and $x_{1,B} = 1$
and
$$x_{1,A} + x_{1,B} = 1$$

We may thus write the model known as the ***presence-absence model*** as:

$$y = \beta_0 + \beta_{1,A}x_{1,A} + \beta_{1,B}x_{1,B} + \varepsilon \tag{2.1}$$

This is otherwise known as the ***Free-Wilson model***, especially in structure-activity studies. The models are shown graphically in figure 2.1.

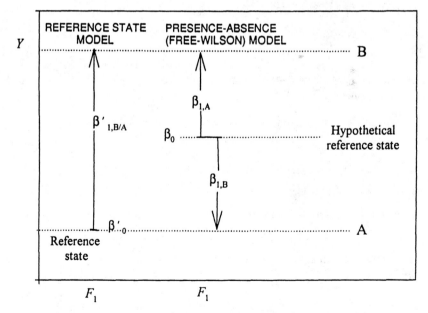

Figure 2.1   Reference state and "presence-absence" models.

As an example, consider the percentage yield of a chemical reaction, the response y, measured when the factor "reaction temperature", F_1 can take one of the two levels 25°C and 40°C. The two experiments are represented in the plan below, and a further column has been added for the experimental results. This is a non-pharmaceutical example, but the reasoning would be identical for the case of tablet disintegration times resulting from the use of different lubricants.

	Temperature (°C)	y: % yield
1	25	32
2	40	60

The experimental design (presence-absence variables) can be written as:

	$X_{1,A}$	$X_{1,B}$	y: % yield
1	1	0	32
2	0	1	60

and the three coefficients of the model are estimated simply by:

$$b_0 = \frac{1}{2}(32 + 60) = 46$$

$$b_{1,A} = \frac{1}{2}(32 - 60) = -14$$

$$b_{1,B} = \frac{1}{2}(60 - 32) = +14$$

b_0, $b_{1,A}$ and $b_{1,B}$ are *estimates* of the true values of the coefficients β_0, $\beta_{1,A}$ and $\beta_{1,B}$. They include experimental error. If the unknown error of experiment 1 is ε_1 and that of experiment 2 is ε_2 then:

$$b_0 = \beta_0 + \frac{1}{2}(\varepsilon_1 + \varepsilon_2) \qquad \text{and} \qquad E(b_0) = \beta_0$$

$$b_{1,A} = \beta_{1,A} + \frac{1}{2}(\varepsilon_1 - \varepsilon_2) \qquad \text{and} \qquad E(b_{1,A}) = \beta_{1,A}$$

$$b_{1,B} = \beta_{1,B} + \frac{1}{2}(-\varepsilon_1 + \varepsilon_2) \qquad \text{and} \qquad E(b_{1,B}) = \beta_{1,B}$$

The experimental errors ε are assumed to be normally distributed. Their mean value is zero. This mean value is the *expectation* of the random error. Thus the more times we carry out the experiment the more likely it is that our mean estimate b of each coefficient will be close to the true value β. We say therefore that β is the *expectation* of b, written as $E(b)$. The result can be shown graphically as in figure 2.2.

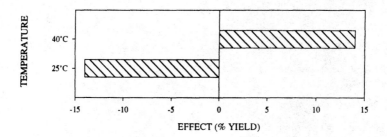

Figure 2.2   Effect of temperature on yield (presence-absence model).

The two models are mathematically equivalent, and the above data may also be analysed in terms of the reference state model.

$b'_0 = 32\%$ (value of the response in reference state, A)

$b'_{1,B/A} = 28\%$ (differential effect of raising the temperature 25°C to 40°C)

b'_0 and $b'_{1,B/A}$ are, as before, estimates of the "true" values β'_0 and $\beta'_{1,B/A}$.

4. More than 2 levels

To illustrate what happens when the factor takes more than 2 levels, we look at an extension of the previous example. Consider the response y, the percentage yield of the reaction measured when a second factor, the nature of the catalyst F_2 is investigated. Four different catalysts C1, C2, C3, C4 are studied, corresponding to levels A, B, C, D as before. The experimental plan (with results) is:

No.	Nature of catalyst	F_2	y: % yield
1	C1	A	32
2	C2	B	60
3	C3	C	42
4	C4	D	34

and the experimental design, expressed in terms of the presence-absence variables is given below (enclosed in double lines):

No.	$x_{2,A}$	$x_{2,B}$	$x_{2,C}$	$x_{2,D}$	y: % yield
1	1	0	0	0	32
2	0	1	0	0	60
3	0	0	1	0	42
4	0	0	0	1	34

We postulate the mathematical model:

$$y = \beta_0 + \beta_{2,A}x_{2,A} + \beta_{2,B}x_{2,B} + \beta_{2,C}x_{2,C} + \beta_{2,D}x_{2,D} + \varepsilon$$

The estimates of the 5 coefficients in the model (plotted in figure 2.3) are:

$$
\begin{aligned}
b_0 &= \tfrac{1}{4}(32 + 60 + 42 + 34) = 42 \\
b_{2,A} &= 32 - 42 \quad = -10 \\
b_{2,B} &= 60 - 42 \quad = +18 \\
b_{2,C} &= 42 - 42 \quad = 0 \\
b_{2,D} &= 34 - 42 \quad = -8
\end{aligned}
$$

and plotted in figure 2.3.

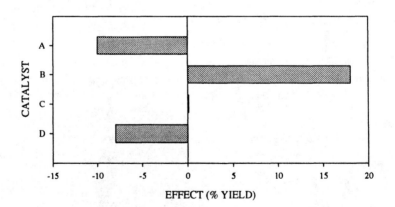

Figure 2.3   Effect of catalyst on reaction yield: presence-absence model.

We see here that the differential effect of the catalyst C3 (level C) is estimated as zero. This does not mean that C3 has no catalytic effect. It is the *differential* effect that is zero with respect to a mean value. We could say that the activity of catalyst C3 is "average". Similarly we conclude that catalysts C1 and C4 have similar (low) effects on the reaction yield and the effect of C2 is high The order of the catalysts on the graph is completely arbitrary, as is the spacing between them on the graph. The representation in figure 2.4 is equivalent to the previous diagram.

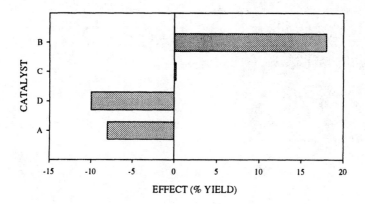

Figure 2.4 Effect of catalyst on the yield: modified order.

B. Several Factors with Various Numbers of States

Our experimental strategy involves studying the effects of *concurrent* variation of *all* of the factors. Thus the previous examples for the effect of a temperature (factor 1, levels A, B) and catalyst (factor 2, levels A, B, C, D) on a reaction may be combined into an overall first order equation.

$$y = \beta_0 + \beta_{1,A}x_{1,A} + \beta_{1,B}x_{1,B} + \beta_{2,A}x_{2,A} + \beta_{2,B}x_{2,B} + \beta_{2,C}x_{2,C} + \beta_{2,D}x_{2,D} + \varepsilon$$

where the $x_{1,A}$ etc. take values 0 or 1 and the coefficients are not independent but are related to one another by:

$$\beta_{1,A} + \beta_{1,B} = 0 \qquad\qquad x_{1,A} + x_{1,B} = 0$$

$$\beta_{2,A} + \beta_{2,B} + \beta_{2,C} + \beta_{2,D} = 0 \quad x_{2,A} + x_{2,B} + x_{2,C} + x_{2,D}$$

The total number of coefficients of this model (and thus the number of effects of the different factors) is equal to the total number of levels, plus one for the constant term, thus $2 + 4 + 1 = 7$. However, since there is a single constraint associated with each factor and the coefficients are not independent we must subtract 1 for each factors. Thus the total number of independent effects in this case is $7 - 2 = 5$.

Therefore, if when carrying out a preformulation study we wished to compare the effects of 4 diluents (e.g. lactose, calcium phosphate, microcrystalline cellulose, and mannitol), 3 disintegrants, 2 binders, 2 lubricants, and a glidant (e.g. presence or absence of colloidal silica), it is easy to verify that the model contains 14 coefficients (including the constant term), but only 9 of these are independent.

See appendix II for the generalisation of this model and its conversion to the reference state model.

C. Alternatives to the General (Presence-Absence) Model

1. All factors at 2 levels

This is a very frequent case. If all factors are studied at 2 levels only, we can use a mathematical model rather different from the one established in the last paragraph. We re-examine first of all the example of a single factor at 2 levels (as described in section A: equation 2.1):

$$y = \beta_0 + \beta_{1,A} x_{1,A} + \beta_{1,B} x_{1,B} + \varepsilon$$

Of the 3 coefficients in the model, 2 are connected by the relationship:

$$\beta_{1,A} = - \beta_{1,B}.$$

It is much simpler for the subsequent mathematical treatment to construct a model with only two independent coefficients.

$$y = \beta_0 + \beta_1 x_1 + \varepsilon \tag{2.2}$$

In this model the single variable x_1 replaces the two presence-absence variables $x_{1,A}$, $x_{1,B}$. Unlike the presence-absence variables that can take values 1 and 0, the new variable x_1 takes one of two equal and opposite values, usually 1 and -1. Thus:

when $F_1 = $ "A" $x_1 = -1$ $[x_{1,A} = 1, x_{1,B} = 0]$

when $F_1 = $ "B" $x_1 = +1$ $[x_{1,A} = 0, x_{1,B} = 1]$

The coefficient β_1 is called the ***main effect*** of the factor F_1. We see that

$$\beta_1 = -\beta_{1,A} = \beta_{1,B}$$

The design may be reformulated as:

N°	X_1	y: % yield
1	-1	$y_1 = 32$
2	+1	$y_2 = 60$

and the two coefficients in the model estimated simply as:

$$b_0 = \frac{1}{2}(y_1 + y_2) = 46\%$$

$$b_1 = \frac{1}{2}(y_2 - y_1) = 14\%$$

This method of representing the levels will be familiar to many readers who have used factorial design. It might seem that we have gone rather a long way around to arrive at this apparently trivial result.

We have taken this "round trip" in order to insist on the fact that there is a fundamental difference between the meaning of the expressions "$x_1 = +1$" or "$x_1 = -1$" referring to the level of a qualitative variable (with only 2 possible states) and that of the same expression when x_1 is the coded value of a continuous quantitative variable that has undergone a linear transformation and can therefore take all values between -1 and +1.

For this reason we have shown that two qualitative levels may be treated as a special simple case of the general case of multiple qualitative levels. Examples of screening experiments of a number of variables each at two levels are given in section IV of this chapter.

2. Conversion to the reference state model

This simplifying treatment cannot be used when one of the variables takes more than two levels. However if each of the k factors can take any number of levels, the general model may be simplified by taking into account the interdependence of the (presence-absence) variables.

Let us consider for example two factors, the factor F_1 being studied at 3 levels (called A, B, C) and the second factor F_2 at 4 levels (which we refer to as A, B, C, D). Of course, levels A, B, ... of F_1 and levels A, B, ... of F_2 are not the same! The model (from section II.B) is the following:

$$\begin{aligned}
y = \beta_0 &+ \beta_{1,A}x_{1,A} + \beta_{1,B}x_{1,B} + \beta_{1,C}x_{1,C} \\
&+ \beta_{2,A}x_{2,A} + \beta_{2,B}x_{2,B} + \beta_{2,C}x_{2,C} + \beta_{2,D}x_{2,D} \\
&+ \varepsilon
\end{aligned}$$

but of its 8 coefficients, only 6 are independent. The vast majority of computer programs for multi-linear regression do not allow coefficients to be calculated where there are constraints. The coefficients must be calculated using the reduced reference state model:

$$y = \beta'_0 + \beta'_{1,A}x_{1,A} + \beta'_{1,B}x_{1,B} + \beta'_{2,A}x_{2,A} + \beta'_{2,B}x_{2,B} + \beta'_{2,C}x_{2,C} + \varepsilon \qquad (2.3)$$

where we have replaced combinations of coefficients by new coefficients, distinguished as β'. The variables refering to the reference states have been eliminated from the equation, and there are now 6 independent variables.

It is this reduced model that we use to calculate the values of the coefficients by least squares regression. It can be seen that we have gone back to a representation in terms of standard states for the variables, in this level case "C" for F_1 and level "D" for F_2.

But we have already objected that there is no reason to choose one state for each variable as reference rather than another. The physical significance of the coefficients is rather easier to interpret if we use the presence-absence model, so it is more usual after calculating the alternative (arbitrary reference state) coefficients to return to the original coefficients by the inverse transformation.

The equations for converting between the presence-absence and reference-state models, both for the above example and the general case, are given in appendix II.

3. Multilinear regression and analysis of screening designs

Estimation of the model coefficients in the above equations from the results of screening designs is normally by a computer program, using multilinear regression. It will not be described in this chapter. The results of multilinear regression will be *assumed* where necessary, both here and in chapter 3. The method is covered in chapter 4 along with a number of other important statistical topics. For designs at 2 levels only there is an alternative method, that of linear combinations, and that is described below.

Since the calculations are done by computer the experimenter does not normally need to use the equations given above and in appendix II. However he should make the effort to understand the models, and in particular to identify which particular model of the two corresponds to the output of his own software package.

D. General Properties of Experimental Designs for Screening

The number of experiments required must be at least equal to the number of independent coefficients p that we need to calculate. This is given by the formula:

$$p = 1 + \sum_{i=1}^{k} (s_i - 1)$$

where s_i is the number of levels of the ith factor and there are k factors in all.

Experimental designs for screening have three essential properties:

- the number of experiments is kept as small as possible so as to limit the amount of work carried out in this initial stage of the project to no more than 20%–30% of the total investment necessary. Such designs are termed *saturated* if they consist of exactly p experiments, or *nearly saturated* where the number of experiments is very close to p,

- the estimators of the coefficients are of the best possible "quality" – that is they are of maximum precision,

- these estimators are also as nearly *orthogonal* as possible, that is they are not correlated with one another.

The main experimental designs for screening independent factors are summarised in table 2.1.

Table 2.1 Summary of Experimental Designs for Screening

Type of design	Number of experiments*	Levels	Methods of construction
Hadamard (or Plackett & Burman)	$4n$: 4, 8, 12, 16, 20, 24, 28, 32 ...	2 levels of each factor	See ref. (2) and appendix II.
Symmetrical fractional-factorial	$s_n^{k'}$: 4, 8,16, 32, 64, 9, 27, 16, 64, 25, 49	Each (qualitative) factor has s_n levels	Special methods . See also ref. (2)
Asymmetrical fractional-factorial	Various numbers	Factors have different numbers of levels	Special methods
D-optimum	Any number equal to or greater than the number of independent coefficients in the model.	Factors have different numbers of levels	Experiments selected from the full factorial design using an exchange algorithm.

* Limited to designs with ≤ 64 runs

III. EXAMPLES OF SCREENING FACTORS AT 2 LEVELS

In general, if we do N experiments with each factor at 2 levels, the best possible precision obtainable for estimating a coefficient in the model is given by:

$$\text{var}(b_i) \geq \frac{\sigma^2}{N}$$

where σ is the standard deviation for the result of an individual experiment.

So for 4 experiments, which enable us to determine the effects of up to 3 variables we cannot obtain a variance less than $\sigma^2/4$. The best (optimum) matrices have:

$$\text{var}(b_i) = \frac{\sigma^2}{N} \qquad \text{or} \qquad \sigma_{b_i} = \frac{\sigma}{\sqrt{N}}$$

where N is 2 or a multiple of 4. For the example given in the next section we will list 7 possible factors. The additive (first order) mathematical model will contain 8 terms, including the constant. Thus if we want an experimental design to estimate the effects of these factors we must carry out at least 8 experiments. The smallest optimum design consists of that number of experiments. These optimum designs are **Hadamard** (often called Plackett-Burman) designs (2).

A. Screening in a Pharmaceutical Process – Extrusion-Spheronization

1. Defining the problem

We continue with the example of extrusion-spheronization. A formulation has been developed containing the drug substance, a plastic diluent and a binder. We want an optimized process and the response that most interests us is the percentage mass yield of pellets having a particle size between 900 µm and 1100 µm. We are not concerned for the present with other characteristics such as the shape, surface quality or friability of the pellets (potentially interesting as these may be). Factors that may affect the yield (and quality) of the resulting pellets are:

- the granulation conditions: amount of water
                                     granulation time
                                     mixer speed
- the extrusion conditions: extruder grill size
                                     extruder speed
- the spheronization conditions: spheronization time
                                     spheronization speed

Our objective is to determine the effects of each of these factors and thus establish optimum conditions for manufacture. An experimental design to optimize all of them would require at least 36 experiments (the number of coefficients in the second order model)! A standard design for optimizing on 7 parameters would require over 50 experiments. This is rather too many. Instead it was decided to try to find out initially which factors affect the yield and require further study, in order to describe and later optimize the manufacturing process. We want to do this with a minimum of experiments. It requires a screening design.

2. Experimental domain

For simplicity we use coded variables: all of the variables are quantitative and take numerical values, but in each case we examine only the variation in the effect of changing from the lower to the upper value. The higher value of each variable is set equal to +1 and the lower limit to -1.

Table 2.2   Variables for Extrusion-Spheronization Study

Factor (Process variable)		Associated variable	Lower limit (coded -1)	Centre (coded 0)	Upper limit (coded +1)
amount of binder	%	X_1	0.5	0.75	1
amount of water	%	X_2	40	45	50
granulation time	min	X_3	1	1.5	2
spheronization load	kg	X_4	1	2.5	4
spheronization speed	rpm	X_5	700	900	1100
extruder rate	rpm	X_6	15	37.5	60
spheronization time	min	X_7	2	3.5	5

3. Design of experiment in coded and original (natural) variables

The Plackett and Burman design of 8 experiments is given below (in reality the values -1 and +1 have been inverted with respect to those of the original design given in appendix II).

Table 2.3   Plackett and Burman Screening Design for Extrusion-Spheronization Study: Coded Variables

No.	X_1	X_2	X_3	X_4	X_5	X_6	X_7
1	-1	-1	-1	+1	-1	+1	+1
2	+1	-1	-1	-1	+1	-1	+1
3	+1	+1	-1	-1	-1	+1	-1
4	-1	+1	+1	-1	-1	-1	+1
5	+1	-1	+1	+1	-1	-1	-1
6	-1	+1	-1	+1	+1	-1	-1
7	-1	-1	+1	-1	+1	+1	-1
8	+1	+1	+1	+1	+1	+1	+1

If we replace the coded values by the corresponding values of the natural variables, that is the real states of each factor, then we obtain the experimental plan (table 2.4) given below with the yield of pellets for each experiment.

Table 2.4   Experimental Plan for Extrusion-Spheronization Study (Natural Variables) and Response (Yield)

No.	Binder (%)	Water (%)	Granul. time (min)	Spher. load (kg)	Spher speed (rpm)	Extrud. rate (rpm)	Spher. time (rpm)	Yield y_j (%)
1	0.5	40	1	4	700	60	5	55.9
2	1.0	40	1	1	1100	15	5	51.7
3	1.0	50	1	1	700	60	2	78.1
4	0.5	50	2	1	700	15	5	61.9
5	1.0	40	2	4	700	15	2	76.1
6	0.5	50	1	4	1100	15	2	59.1
7	0.5	40	2	1	1100	60	2	50.8
8	1.0	50	2	4	1100	60	5	62.1

Each row describes an experimental run, and each column describes one of 7 variables tested at two levels (+1 and -1). The sum of the coded variables in each column is zero. The yield obtained, the percentage mass of material consisting of pellets of the desired size, is given in the final column, y.

4. Mathematical model

We have postulated that the result of each experiment is a linear combination of the effect of each of the variables, X_1, X_2,.. X_k. The response measured for each experiment j is y_j. A first order or additive model for 7 variables is proposed:

$$y = \beta_0 + \beta_1 x_1 + \beta_2 x_2 + \beta_3 x_3 + \beta_4 x_4 + \beta_5 x_5 + \beta_6 x_6 + \beta_7 x_7 + \varepsilon \qquad (2.4)$$

5. Calculation of the effects

If y_{ij} is the value for the response of the jth experiment, we can successively substitute the values of x_i in each row of table 2.3 into the model, equation 2.4, obtaining:

$$y_1 = \beta_0 - \beta_1 - \beta_2 - \beta_3 + \beta_4 - \beta_5 + \beta_6 + \beta_7 + \varepsilon_1$$
$$y_2 = \beta_0 + \beta_1 - \beta_2 - \beta_3 - \beta_4 + \beta_5 - \beta_6 + \beta_7 + \varepsilon_2$$
$$y_3 = \beta_0 + \beta_1 + \beta_2 - \beta_3 - \beta_4 - \beta_5 + \beta_6 - \beta_7 + \varepsilon_3$$

$$y_4 = \beta_0 - \beta_1 + \beta_2 + \beta_3 - \beta_4 - \beta_5 - \beta_6 + \beta_7 + \varepsilon_4$$
$$y_5 = \beta_0 + \beta_1 - \beta_2 + \beta_3 + \beta_4 - \beta_5 - \beta_6 - \beta_7 + \varepsilon_5$$
$$y_6 = \beta_0 - \beta_1 + \beta_2 - \beta_3 + \beta_4 + \beta_5 - \beta_6 - \beta_7 + \varepsilon_6$$
$$y_7 = \beta_0 - \beta_1 - \beta_2 + \beta_3 - \beta_4 + \beta_5 + \beta_6 - \beta_7 + \varepsilon_7$$
$$y_8 = \beta_0 + \beta_1 + \beta_2 + \beta_3 + \beta_4 + \beta_5 + \beta_6 + \beta_7 + \varepsilon_8 \qquad (2.5)$$

To estimate β_1, take a linear combination of the values of y_j (listed in the right hand column of table 2.4) with the same signs as β_1 in equations 2.5:

$$- y_1 + y_2 + y_3 - y_4 + y_5 - y_6 - y_7 + y_8$$
$$= -55.9 + 51.7 + 78.1 - 61.9 + 76.1 - 59.1 - 50.8 + 62.1 = 40.3$$
$$= 8\beta_1 - \varepsilon_1 + \varepsilon_2 + \varepsilon_3 - \varepsilon_4 + \varepsilon_5 - \varepsilon_6 - \varepsilon_7 + \varepsilon_8$$

or:

$$\beta_1 + \tfrac{1}{8}(-\varepsilon_1 + \varepsilon_2 + \varepsilon_3 - \varepsilon_4 + \varepsilon_5 - \varepsilon_6 - \varepsilon_7 + \varepsilon_8) = \tfrac{1}{8}(- y_1 + y_2 + y_3 - y_4 + y_5 - y_6 - y_7 + y_8)$$

Let the combination of response values on the right hand side of the last equation be represented by b_1, which is an unbiased estimated value of β_1. This is because the random errors ε_i are normally distributed about zero. More formally:

$$E[b_1] = E[\beta_1] + \tfrac{1}{8} E[-\varepsilon_1 + \varepsilon_2 + \varepsilon_3 - \varepsilon_4 + \varepsilon_5 - \varepsilon_6 - \varepsilon_7 + \varepsilon_8]$$

where $E[\ ]$ denotes the *mathematical expectation* of the function within the brackets. The linear combination of the random errors, ε_i, may be positive or negative, but its expectation is obviously zero. Therefore:

$$E[b_1] = E[\beta_1] + 0 = E[\beta_1] = \beta_1$$

and $b_1 = 5.0$ is an unbiased estimator of β_1.

A similar reasoning gives us:

$$b_2 = \tfrac{1}{8}(- y_1 - y_2 + y_3 + y_4 - y_5 + y_6 - y_7 + y_8) = 3.3$$

Analogous calculations may be carried out for the remaining coefficients b_3 to b_7. It should be noted that for each coefficient b_1 to b_7, the signs in y_j in the expression enabling its calculation are the same signs as those in the corresponding column X_j.

The constant term b_0 is the mean response for all the experimental runs:

$$b_0 = \tfrac{1}{8} \sum_{i=1}^{8} y_i$$

The estimates of the coefficients are therefore:

constant term	$b_0 = 62.0\%$
% binder	$b_1 = 5.0\%$
water content	$b_2 = 3.3\%$
granulation time	$b_3 = 0.8\%$
spheronization load	$b_4 = 1.3\%$
spheronization speed	$b_5 = -4.1\%$
extruder rate	$b_6 = -0.2\%$
spheronization time	$b_7 = -6.0\%$

These are shown in figure 2.5, using two different representations. The first, an effects plot, shows the magnitude and sign of each effect in the original order of the variables. The second puts the variables in the order of decreasing absolute magnitude of the effects (a Pareto chart, see chapter 3) enabling the most important variables to be identified immediately.

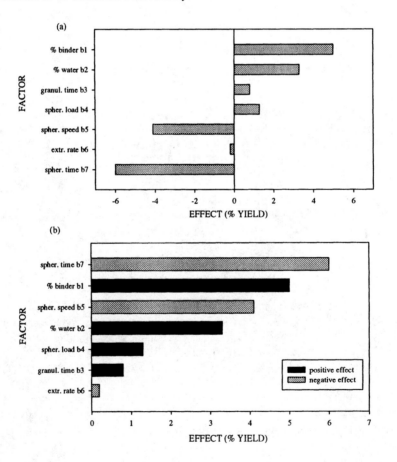

Figure 2.5. Extrusion-spheronization: (a) Effects plot, (b) Pareto chart.

Figure 2.5b indicates that the coefficients most likely to be active are b_7 and b_1 (spheronization time and % binder), followed by b_5 and b_2.

6. Precision of estimation of the coefficients

Eight experiments were carried out according to this design to estimate 7 coefficients (effects) plus a constant term. There is a unique solution to the problem and the results themselves do not give any clear evidence of the experimental precision. Nor can we tell if these values are statistically significant or whether the differences between them can be attributed to random error. We would suppose that the factors whose coefficients have high absolute values - % binder, spheronization time and spheronization speed - are probably significant, whereas the effects of granulation time, extrusion rate and spheronizer load are small, even though they might possibly be found on replicating experiments to be statistically significant.

The design is therefore described as **saturated**. The experimental data are not exact (because of random variation in the conditions, uncontrolled factors, measurement imprecision) so the calculated coefficients, as always, are not exact either, but are *estimates* of the true values β_i.

We may define the imprecision of each experimental datum by a variance σ^2 and a standard deviation σ. From this σ, the corresponding variance $var(b_i)$ of the estimate of β_i, can be calculated:

$$
\begin{aligned}
var(b_i) &= var\{(-y_1 + y_2 + y_3 - y_4 + y_5 - y_6 - y_7 + y_8)/8\} \\
&= var\{(-y_1 + y_2 + y_3 - y_4 + y_5 - y_6 - y_7 + y_8)\}/64 \\
&= [var(y_1) + var(y_2) + var(y_3) + var(y_4) + var(y_5) + var(y_6) + var(y_7) + var(y_8)]/64 \\
&= 8\sigma^2/64 \\
&= \sigma^2/8
\end{aligned}
$$

The above argument is only valid provided all the measurements are independent, and not correlated in any way; that is the covariance of y_i and y_j equals zero for all $i \neq j$. The relationship is then valid for all the coefficients:

$$var(b_i) = \sigma^2/8 \quad \text{or} \quad \sigma_{b_i} = \sigma/\sqrt{8}$$

This is the value given previously for the standard deviation of an optimal design with each factor at two levels. To determine the precision of each estimate b_i we need to know the experimental precision, in this case of the yield.

Instead of using the above experimental design, we could have tried the "one-factor-at-a-time approach". The effect of changing the percentage binder, associated variable X_1 from 0.5% (level -1) to 1.0% (level +1) may be determined by carrying out two experiments each with all other factors held constant, the responses being y'_1 and y'_2. We calculate the effect of X_1 by:

$$b_1 = \frac{1}{2}(y'_2 - y'_1)$$

The precision (variance) of the estimate is given by:

$$\text{var}(b_1) = \text{var}\{(y'_2 - y'_1)/2\} = \tfrac{1}{4}[\text{var}(y'_2) + \text{var}(y'_1)] = \sigma^2/2$$

and the standard deviation (or standard error) of the estimate is $\sigma/\sqrt{2}$.

The 8-run design gives us a twofold decrease in the standard deviation, with respect to that of a direct comparison. The calculation demonstrates clearly that small effects may be determined with a considerably better precision than by changing one factor at a time. Any error in one of the experiments is "shared" equally between the estimates of the different parameters. The design is robust.

We recall that experimental designs for screening have these 3 essential properties:

- *They consist of a minimum number of experiments.* The above design consists of 8 experiments, equal to the number of coefficients. It is therefore saturated.
- *The coefficients are estimated with the minimum possible standard error* (or standard deviation). This is the case; the standard deviation is $\sigma/\sqrt{8}$.
- *These estimators should be as nearly orthogonal as possible.* In Plackett and Burman designs the variables studied are not correlated in any way and the estimators of the coefficients are independent of one another.

We have no way of knowing whether the model itself is adequate in describing the real system accurately. Screening designs are not intended for that, but are used to show which factors have a large effect on the system (and require further more detailed study) and the factors that can probably be ignored.

7. Determining active effects

The precision of measurement is defined as the standard deviation of measurement (σ). If σ (for each experiment) is known or can be estimated then the standard deviation of the estimate of each coefficient b_i in the model can also be calculated or estimated. It can be determined whether the coefficient is statistically significant, that is whether there is a 95% probability that the absolute value of the effect is greater than zero. We will call the standard deviation estimation of the coefficient parameter b_i its **standard error**.

In fact, we can sometimes obtain useful information even with a saturated design that does not allow any estimate of the precision. If the true values of all of the parameters were zero, it would be most unlikely that their *calculated* values would also be zero. They would be distributed about zero according to an approximately normal distribution, with a standard deviation of $\sigma/\sqrt{8}$. The data for the coefficients can thus be converted to cumulative probabilities and may be plotted on "probability paper" as in probit analysis. Many computer programs for experimental design and analysis have such a facility. A normal distribution gives a linear plot, and active effects would appear as outliers. A description of this and other methods for identifying significant effects in saturated designs is given in the following chapter, section III.D. However 8 experiments is usually too few for these methods to be reliable.

The plots of the coefficients in figure 2.5 show that it is b_1 and b_7 which are most likely to be active.

8. Estimating the precision

To understand more about the system and which factors are statistically significant it would be necessary to determine the precision of the experiment. An estimation s of the standard deviation σ of the experimental response y is best determined by replication of the experiment.

$$s^2 = \frac{1}{n-1} \sum_{i=1}^{n} (y_i - \bar{y})^2$$

where $\bar{y}$ is the mean value of y_i for n replications.

This might not be considered necessary at the screening stage. It might be thought that the effects of spheronizer load, granulation time, and extrusion rate were so small that they could be neglected, whether statistically significant or not, and that one could go on to the next stage of optimization of the remaining 4 variables. However an experiment at the centre of the domain was repeated 4 times. The conditions were given in table 2.2 (coded level = 0). It is possible to test at the centre because the factors are quantitative.

No.	X_1	X_2	X_3	X_4	X_5	X_6	X_7	Yield
9	0	0	0	0	0	0	0	64.3%
10	0	0	0	0	0	0	0	67.9%
11	0	0	0	0	0	0	0	66.0%
12	0	0	0	0	0	0	0	63.8%

The mean value is 65.5% and the standard deviation is estimated as 1.86%. This is only an estimate, obtained with 3 degrees of freedom. The estimate of the standard error for each coefficient is $1.86/\sqrt{8} = 0.66\%$.

The value of Student's t for a probability of 95% and 3 degrees of freedom is 3.18. Thus parameters that exceed a limiting value of $0.66 \times 3.182 = 2.09\%$ are significant. As we concluded was probably the case, 4 factors are statistically significant at a confidence level better than 95%.

9. Ordering the experiments

In the above design, the experiments are ordered in a regular manner so that the structure of the design may be seen. It would be normal to do the experimental runs in a random order, because of the possibility of time effects (see also the section on *blocking* in the following chapter). The warming up of a piece of equipment from one run to another might well affect its functioning. The first run of the day using the spheronizer could well give slightly different results from the last of the day. Processes may be dependent on external, uncontrolled factors, such as the relative humidity on the day of the experiment.

The 8 experiments in the Plackett-Burman design and the 4 repeated experiments used for calculating the precision and testing the model were therefore done in a random order.

10. Concluding remarks

The results of this study may be used to set *improved* (but not optimum) conditions for the process. The three inactive factors may be set according to our convenience – for example, maximum load, maximum extrusion rate and minimum granulation time in order to shorten the time of the process. The percentage binder is fixed at its maximum value of 1%. The spheronization speed and spheronization time may be set to their optimum (minimum) values, but we might well consider that these parameters, along with the water content would repay further study, perhaps changing the limits of the experimental domain and investigating possible interactions between variables.

In this study we have been trying to identify which effects are active rather than determine the yield of pellets within the experimental domain. In this there is no practical difference in our treatment of qualitative and quantitative variables. But if we also postulate that the variation between the upper and lower limits of one of the variables (all other variables being held constant) is linear then we can predict the yield at the centre of the domain and compare it with the measured value.

The mean value of the experiments at the centre, 65.5%, is an estimate of β_0. The previous estimate, obtained from the 8 experiments of the Plackett-Burman design, is 62.0%. The difference between the two figures is 3.5%. The yield at the centre of the domain is thus rather greater than that predicted, using the first order model, from the results at the edge of the domain. We conclude that the linear model does not hold exactly and that further experiments may be necessary to describe the system adequately.

We have already seen that had we wished to screen an 8th variable (the extruder grill diameter for example) we would have needed a Plackett-Burman design of 12 experiments.

B. Validation of an Analytical Method – Dissolution Test

1. Ruggedness testing of analytical methods

In validating an analytical method we need to know if it performs correctly under normal conditions. Many factors can vary from one experimental run to another – different apparatus, a different chromatography column, changes in temperature, another batch of reagent, a different operator. The sensitivity of the method to these changes should be assessed as part of the method validation. This is called *ruggedness* testing.

These are not strictly speaking screening experiments, in the sense that these were defined at the beginning of the chapter as the variation imposed on the factors is generally rather small. However they are discussed here because suitable designs for treating this kind of problem (3) are identical to those used for screening (4-6).

Dissolution testing is a key technique in pharmaceutical development laboratories. Methods are described in various pharmacopoeias, along with the calibration of the testing apparatus. Specifications of the precision of the testing conditions are given – for example the temperature of the dissolution medium, nominally 37°C, must be between 36.5°C and 37.5°C. The stirring speed must be accurate within 4%.

2. Experimental domain – measured response

In this section we will consider a dissolution method that has been developed for tablets. It is carried out in 0.01 M hydrochloric acid with a small quantity of surfactant added to improve wetting. The percentage dissolution from the tablet is measured by ultra-violet spectrophotometry as a function of time. As part of the method validation we wanted to test the effect of *small* changes in the concentration of these constituents as well as other random and systematic modifications. These variations and modifications should be in general of the same order as those likely to be encountered in day to day use of the method. These are summarised in table 2.5. The measured response for each experimental run was the time in minutes for 75% liberation of the active substance in the tablets tested.

3. Experimental design – a Hadamard design of 12 experiments

As part of the validation procedure, as well as examining such properties as accuracy, precision, linearity, selectivity, we looked for a design that would enable us to verify the ruggedness of the analytical method to minor changes in the operating conditions, and where necessary to identify factors requiring tighter control.

The quantitative variables, X_1, X_2, X_3 and X_4 are set at their extremes, ± 1 in the coded variables. We might wish to validate the method using several operators, and perhaps 3 different sets of apparatus. However we saw in the introduction to this chapter that the problem is considerably simpler if we limit ourselves to 2 levels of each qualitative variable, which are then set at levels -1, and +1. The problem of more than 2 levels of a qualitative variable will be treated in the following section.

The upper and lower limits of the quantitative, continuous variables, X_1, X_2, X_3 and X_4 were close to one another so it was considered probable that a linear model can be used to realistically describe the system. For the same reason we expected interactions between these variables to be negligible, that is, the effect of changing any one variable (such as temperature) will not be dependent on the current value of another variable (such as the stirrer speed). Situations where this cannot be assumed are described in the next chapter.

With 8 coefficients to be determined, we needed to do more than 8 experimental runs. The smallest screening design with a multiple of 4 experiments is a 12 experiment Hadamard (Plackett-Burman) design (table 2.6). Its derivation and structure is described in appendix II, along with the other 2-level Plackett-Burman designs.

Table 2.5 Experimental Domain for a Dissolution Method Validation Experiment

Factor	Associated variable	Normal value	Levels -1	Levels +1
Concentration of acid	X_1	0.01M	0.009M	0.011M
Concentration of polysorbate 80	X_2	0.05%	0.04%	0.06%
Stirring speed	X_3	50 rpm	45 rpm	55 rpm
Temperature	X_4	37.0°C	35°C	39°C
Degassing of dissolution medium	X_5	Yes	No	Yes
Position of filters for sampling	X_6	Middle	Low	High
Operator	X_7		A	B
Apparatus	X_8		V	W

Each line in the design of table 2.6 describes an experimental run, and each column describes one of 8 variables, corresponding to a factor tested at two levels (+1 and -1). The sum of each column is zero. The 3 last columns X_9, X_{10} and X_{11}, do not correspond to any real factor and are therefore omitted. They have been included in the table to indicate how the design was constructed. The experimental plan, the design in terms of natural variables, and the measured response for each experimental run are given in table 2.7.

Table 2.6   Plackett-Burman (Hadamard) Design of 12 Experiments and 11 Factors for Dissolution Method Validation

No.	X_1	X_2	X_3	X_4	X_5	X_6	X_7	X_8	X_9	X_{10}	X_{11}
1	+1	+1	-1	+1	+1	+1	-1	-1	-1	+1	-1
2	-1	+1	+1	-1	+1	+1	+1	-1	-1	-1	+1
3	+1	-1	+1	+1	-1	+1	+1	+1	-1	-1	-1
4	-1	+1	-1	+1	+1	-1	+1	+1	+1	-1	-1
5	-1	-1	+1	-1	+1	+1	-1	+1	+1	+1	-1
6	-1	-1	-1	+1	-1	+1	+1	-1	+1	+1	+1
7	+1	-1	-1	-1	+1	-1	+1	+1	-1	+1	+1
8	+1	+1	-1	-1	-1	+1	-1	+1	+1	-1	+1
9	+1	+1	+1	-1	-1	-1	+1	-1	+1	+1	-1
10	-1	+1	+1	+1	-1	-1	-1	+1	-1	+1	+1
11	+1	-1	+1	+1	+1	-1	-1	-1	+1	-1	+1
12	-1	-1	-1	-1	-1	-1	-1	-1	-1	-1	-1

Table 2.7   Experimental Plan for Dissolution Method Validation – Results

No	Conc. acid (M)	Conc. Polysorb. (%)	Agit. (rpm)	Temp. (°C)	Degas.	Filter position	Operat	App.	Time for 75% dissol.
1	0.011	0.06	45	39	Yes	High	A	V	20.5
2	0.009	0.06	55	35	Yes	High	B	V	16.5
3	0.011	0.04	55	39	No	High	B	W	16.7
4	0.009	0.06	45	39	Yes	Low	B	W	19.3
5	0.009	0.04	55	35	Yes	High	A	W	14.7
6	0.009	0.04	45	39	No	High	B	V	18.1
7	0.011	0.04	45	35	Yes	Low	B	W	15.5
8	0.011	0.06	45	35	No	High	A	W	17.9
9	0.011	0.06	55	35	No	Low	B	V	16.5
10	0.009	0.06	55	39	No	Low	A	W	19.1
11	0.011	0.04	55	39	Yes	Low	A	V	16.7
12	0.009	0.04	45	35	No	Low	A	V	20.5

Using a similar argument to that of section III.A.5 it can be shown that:

$$b_1 = (y_1 - y_2 + y_3 - y_4 - y_5 - y_6 + y_7 + y_8 + y_9 - y_{10} + y_{11} - y_{12})/12$$

Similar calculations are carried out for the 7 remaining coefficients. The mean value of the responses b_0 is an unbiased estimate of the constant term of the model. The resulting estimations of the coefficients b_i which as before are unbiased estimates of the β_i are:

$b_0 = 17.9$ min $b_1 = -0.4$ min $b_5 = -0.6$ min

$\qquad\qquad\qquad\qquad\quad b_2 = +1.6$ min $b_6 = -0.2$ min

$\qquad\qquad\qquad\qquad\quad b_3 = -1.6$ min $b_7 = -0.8$ min

$\qquad\qquad\qquad\qquad\quad b_4 = +1.8$ min $b_8 = -0.6$ min

Thus, for example, changing the concentration of polysorbate 80 from 0.04 to 0.06% leads to an increase of $1.6 \times 2 = 3.2$ min in the time for 75% dissolution. Increasing the temperature from 35°C to 39°C also (unusually) gave an increase in the dissolution time, of $1.8 \times 2 = 3.6$ min.

5. Interpretation of results

We cannot judge the statistical significance of the results from these figures alone, as some measure of the reproducibility of the method is required. If the standard deviation of repeated measurements is σ then the standard deviation for the estimate of each effect is $\sigma/\sqrt{12}$. What we need is an estimate of σ.

The ideal is to determine the reproducibility from repeated dissolution experiments under identical conditions, on the same batch. This will add to the cost of the experiment but it will probably be necessary. Nevertheless before considering doing this we should use at least what information we have to the full, and there are two important sources of information on the reproducibility of the experiment already available.

Firstly the unused columns X_9, X_{10} and X_{11} of table 2.6 are "dummy variables" that do not correspond to any real factor. However, we may estimate the corresponding coefficients b_9, b_{10} and b_{11} in exactly the same way as we did for the effects of the real factors. The coefficients are independent of all the others, and, except for experimental error, they should be equal to zero. Their values should be thus indicate random fluctuations. It can be shown that the square of each of these coefficients is an independent estimate of the variance of the coefficients σ_b^2 provided the additive model is correct. The mean of the 3 squares is an estimate of σ_b^2 with 3 degrees of freedom.

The assumption that the linear model is correct and these columns may be used to derive estimates of the random variation is only reasonable because in this ruggedness study the changes in the variables have been kept small. It is therefore assumed that interactions between the variables may be neglected. The values found are respectively 0.1, 0.2, and 0.4. Therefore:

$$s_b^2 = (0.1^2 + 0.2^2 + 0.4^2)/3 = 0.07 \quad \text{and} \quad s_b = 0.26$$

This is an estimate of the standard error of b_i (always assuming the first-order model to be valid). It allows us to decide which coefficients are statistically significant. Referring to a table of values for Student's t gives a value of 3.182 for 95% probability and 3 degrees of freedom. The critical value is therefore $0.26 \times 3.182 = 0.827$. All coefficients whose absolute values are equal or greater than this value can be considered as statistically significant. Important variables are the concentration of surfactant, the stirring speed and the temperature. However the absolute effects are not large, even if statistically significant, and it can be concluded that the method is robust.

Multiplying s_b by $\sqrt{12}$ gives an estimate s of the run-to-run repeatability of the method (σ). In addition the experiment also shows that there is little difference between results obtained by the different operators (effect b_7) or using different apparatus (effect b_8).

IV. SYMMETRICAL DESIGNS FOR FACTORS AT 2 OR MORE LEVELS

The mathematical models introduced in section II were general ones, in the sense that each factor could take any number of levels. The Plackett-Burman designs discussed previously, where all factors take 2 levels, are not always adequate - in particular for the screening of *qualitative* variables. Designs are needed which are (a) suitable for testing any number of levels and (b) where the number of levels is different from one variable to another. Such designs are, for example, useful in drug-excipient compatibility testing, introduced below.

A. Excipient Compatibility Testing

Experimental designs to establish the effect of the composition of a mixture on its properties differ from other designs in that the initial parameters are not usually independent of one another. The properties of a mixture of three diluents, lactose, calcium phosphate and microcrystalline cellulose, for example, may be defined in terms of the percentage of each component. If we know the concentration of the first two components we can automatically find the concentration of the third. This non-independence of the variables means that designs of the type outlined above are unsuitable for many problems involving mixtures. However the classical designs may be used in two circumstances – for the *choice* of excipients, where the factors are purely qualitative, and also in problems where the proportions of all but one of the excipients are constrained to be relatively small.

We consider here the former case, where the factors to be investigated are the nature of the excipients. The factors are qualitative and independent.

The compatibility of a drug substance with a range of excipients has often been investigated by preparing binary mixtures, storing the mixtures at various conditions for several weeks or months and then analysing the mixture by a stability-indicating method (usually chromatographic). A disadvantage of this approach is that real formulations are not usually binary mixtures and the effect of an excipient under these circumstances may be very different from its effect in a quite complex mixture. How are we to test for the effect of magnesium stearate using a binary mixture, when its concentration in a real formulation is likely to be about 1% and that of the drug substance may also be quite low?

It has become standard practice, instead of testing binary mixtures (though there is still a place for such studies), to prepare various mixtures, containing drug substance, a binder, a disintegrant, a lubricant and a diluent to investigate their stabilities (7). They are sometimes called "mini-formulations" though "proto-formulations" might be a better term. A possible list of excipients for screening for conventional tablet and capsule formulations is given in table 2.8. Other variables could be studied previous to investigating possible sustained-release or other types of formulations.

Table 2.8   Excipients for Testing in Proto-Formulations

Function	Excipients (levels)			
	0	1	2	3
diluent	lactose	mannitol	calcium phosphate	microcrystalline cellulose
disintegrant	starch	sodium starch glycolate	none	
binder	PVP	HPMC	none	
lubricant	magnesium stearate	glyceryl behenate	stearic acid	hydrogenated castor oil
glidant	colloidal silica	none		
capsule shell	yes	none		

All the mixtures contain the drug substance and a diluent. They do not necessarily include one of each class of excipient, as we may well wish to discover the effect of missing out a disintegrant or a glidant. The concentration of drug substance is usually constant. Each excipient is included at a constant realistic functional level. For example, if magnesium stearate is fixed at 2% (a rather higher value than in most formulations, here chosen for convenience) there is no need for all other lubricants investigated to be also at 2%. The concentration of diluent will vary between formulations, but not by very much.

We will assume (as a first approximation) that the effect of each excipient on the stability is a constant depending only on the nature of the excipient and independent of all other excipients present. That is, the additive (first order) model introduced in the previous sections applies. If all variables are studied, including absence of binder, disintegrant and glidant, there will be 13 independent coefficients in the model.

It is evident that for treating this problem we require designs where the variables can take more than 2 levels. We begin here with the *symmetrical* problem where all factors have the same number of levels. This case is quite rare, other than for variables with two levels, but a wide variety of asymmetric designs can be derived from the symmetrical ones.

B. Fractional Factorial Designs

In the previous section we saw how factors at two levels may be screened using a Plackett-Burman design. In their paper (2), Plackett and Burman constructed experimental designs for factors at 2, 3, 5, and 7 levels. Such designs are termed *symmetric*, because all factors have the same number of levels. The designs may be set out in another way, the so-called *factorial arrangement* (8), but they are still totally equivalent to the Plackett-Burman designs. We will demonstrate their use in the case of the excipient compatibility problem introduced above.

The factorial arrangement of k' factors at s levels is that of all $N = s^{k'}$ combinations of all the levels of all the factors. For now on we will identify these levels as 0, 1, 2, 3, ..., s-1. (Note that the levels are qualitative, so the numbers describing the levels are not to be interpreted numerically.) Let us take for example the factorial arrangement of 3 factors at 2 levels. It comprises $2^3 = 8$ combinations of factors and levels. Four factors, each at 3 levels gives $3^4 = 81$ combinations. A factorial design therefore consists of $s^{k'}$ experiments, but it is frequently possible to use such a design to investigate more than k' factors.

In section II we showed that a variable at s levels could be decomposed into s presence-absence variables (whose sum equals unity) and that s-1 independent effects could be calculated. Thus for k factors the total number of independent effects p, which includes the constant term in the model, is given by:

$$p = 1 + k(s - 1).$$

A design enabling us to estimate these p coefficients must consist of at least p experiments. In the case of a factorial design of $s^{k'}$ combinations this implies that:

$$s^{k'} \geq 1 + k(s - 1)$$

Table 2.9 List of Symmetrical Designs for Screening

Design	k'	$N = s^{k'}$ (n° of expts)	k (n° of factors)
$2^7 // 2^3$	3	$2^3 = 8$	4 - 7
$2^{15} // 2^4$	4	$2^4 = 16$	8 - 15
$2^{31} // 2^5$	5	$2^5 = 32$	16 - 31
$3^4 // 3^2$	2	$3^2 = 9$	2 - 4
$3^{13} // 3^3$	3	$3^3 = 27$	5 - 13
$3^{40} // 3^4$	4	$3^4 = 81$	14 - 40
$4^5 // 4^2$	2	$4^2 = 16$	2 - 5
$4^{21} // 4^3$	3	$4^3 = 64$	6 - 21
$5^6 // 5^2$	2	$5^2 = 25$	2 - 6
$7^8 // 7^2$	2	$7^2 = 49$	2 - 8

Thus, from the $s^{k'}$ experiments of a factorial design of k' factors at s levels we can screen up to k factors, also at s levels where: $k \leq (s^{k'}-1)/(s-1)$. For example, beginning with the 8 experiments of a factorial design of 3 factors at 2 levels, we can screen up to 7 factors at 2 levels (as we have already seen with the Plackett-Burman design of section III.A.3). We use the following notation for such a design: $2^7//2^3$. Similarly with the 27 experiments of a 3^3 design we can screen up to 13 factors at 3 levels. The design is written as $3^{13}//3^3$. The designs (those for up to 7 levels and with less than 100 experiments) are listed in table 2.9, along with the maximum number k of factors that can be treated in each case.

Similarly, we can calculate the minimum size of factorial design for treating a given problem. k' is the smallest whole number so that $N = s^{k'} > k(s-1)$. If, for example, we want to study $k=8$ factors, each at $s=3$ levels, then the 3^3 design of $N = 27$ experiments is the smallest factorial design $3^{k'}$ that can be used: $3^8//3^3$. Details of these designs are given in this section or in appendix II.

We will now look at how the excipient compatibility problem introduced in section A may be treated, for factors taking increasing numbers of levels.

C. A $2^5//2^3$ Symmetrical Factorial Design

1. Defining the problem and the experimental domain

The problem of preformulation studies for screening of excipients to know their compatibility with the drug substance, given in table 2.8, is here simplified by reducing the number of possibilities in each class of excipient (factor) to 2 (9). For this illustration we will also eliminate the final variable, the presence or not of a gelatine capsule shell. The excipients tested and their concentrations are therefore those of table 2.10.

Table 2.10 Compatibility Testing of 5 Excipients at 2 Levels – Experimental Domain

Factor	Associated variable	Level coded 0	Level coded 1
diluent	X_1	lactose	micro-crystalline cellulose
lubricant	X_2	2% magnesium stearate	2% stearic acid
binder	X_3	5% PVP	5% HPMC
disintegrant	X_4	7% maize starch	3% sodium starch glycolate
glidant	X_5	1% colloidal silica	none

We are interested in knowing the average stability of the drug substance in these proto-formulations and also in finding out if the stability is improved or compromised by including one excipient rather than another in the same class.

2. Experimental design

A $2^5//2^3$ symmetrical reduced (or fractional) factorial design can be used for this problem. We derive this from the $2^7//2^3$ design given in appendix II. There are only 5 variables, so only 5 out of the 7 columns are needed. We take the first 5 columns. (We will see in the next chapter that it is possible to obtain further information and on assuming different models we would select different designs, in eliminating different columns. But for the time being in assuming the additive model, the choice of columns is entirely arbitrary.) The resulting design is shown in table 2.11.

The experimental plan (that is, the design in terms of the original, or natural variables) of the 8 mixtures is shown in table 2.12, with the results after storage at two different conditions of temperature and humidity.

Table 2.11 $2^5//2^8$ Design

No.	X_1	X_2	X_3	X_4	X_5
1	0	0	0	1	1
2	1	0	0	0	0
3	0	1	0	0	1
4	1	1	0	1	0
5	0	0	1	1	0
6	1	0	1	0	1
7	0	1	1	0	0
8	1	1	1	1	1

3. Mathematical model and calculation of the coefficients

The model for the degradation can be defined in terms of an arbitrary reference state (all factors at level coded zero: lactose, magnesium stearate, povidone, maize starch, and colloidal silica):

$$y = \beta'_0 + \beta'_{1,1}x_{1,1} + \beta'_{2,1}x_{2,1} + \beta'_{3,1}x_{3,1} + \beta'_{4,1}x_{4,1} + \beta'_{5,1}x_{5,1} + \varepsilon$$

$x_{i,j}$ are presence-absence variables. For example, where the disintegrant is sodium starch glycolate, $x_{3,1} = 1$, otherwise $x_{3,1} = 0$. The coefficient $\beta'_{3,1}$ is the differential effect of replacing starch in the formulation by starch sodium glycolate on the stability of the drug. Estimations of the coefficients obtained by multi-linear regression, or by direct calculation, are as follows:

$$b'_0 = 1.90 \qquad\qquad b'_{2,1} = -0.18 \qquad\qquad b'_{4,1} = 0.92$$
$$b'_{1,1} = 1.18 \qquad\qquad b'_{3,1} = 0.22 \qquad\qquad b'_{5,1} = 0.52$$

They can be transformed to give the coefficients in the presence-absence model. So in the case of the disintegrant we have:

$$b_{4,0} = -0.462 \text{ (for maize starch)}$$
$$\text{and} \quad b_{4,1} = 0.462 \text{ (for starch sodium glycolate)}$$

describing the action of these disintegrants with respect to a hypothetical mean value. The constant term b_0 becomes 3.238%.

It is very important to note that the magnitude of the coefficients depends on how we have written the design. Here the levels are 0 and 1 and the reference state coefficients describe the effect of changing from level 0 to 1. In the case of the otherwise equivalent Plackett-Burman designs where the levels were designated as -1 and +1 the coefficients were the half the effect on the response of changing from -1 to level +1.

Table 2.12 Experimental Plan (Original Variables) and Responses

No.	Diluent	Lubricant	Binder	Disintegrant	Glidant	degradation	
						D1	D2
1	lactose	Mg stearate	PVP	SSG	none	2.8%	2.2%
2	cellulose	Mg stearate	PVP	starch	silica	3.8%	2.6%
3	lactose	stearic acid	PVP	starch	none	2.2%	1.5%
4	cellulose	stearic acid	PVP	SSG	silica	4.6%	2.8%
5	lactose	Mg stearate	HPMC	SSG	silica	3.6%	2.0%
6	cellulose	Mg stearate	HPMC	starch	none	3.1%	2.5%
7	lactose	stearic acid	HPMC	starch	silica	2.0%	1.5%
8	cellulose	stearic acid	HPMC	SSG	none	3.8%	2.5%

D1: samples stored for 4 weeks at 50°C/50% relative humidity
D2: samples stored for 12 weeks at 35°C/ 80% relative humidity

4. Interpretation of the results

Analysis of the coefficients indicates that the drug is relatively unstable when formulated. Formulations containing cellulose are less stable than those with lactose and sodium starch glycolate has a deleterious effect on stability compared with starch. There appears to be no difference between the lubricants and the binders tested and the presence of colloidal silica has no effect on the stability.

This kind of design is useful for screening a large number of variables but the limitation to 2 excipients of each type may be a particular disadvantage in compatibility testing. Such an experiment will give information on whether there are likely to be stability problems in conventional solid dosage forms. But if the formulator does find a certain instability he will most likely wish to screen a much larger range of excipients and combinations of excipients, expanding the variety of excipients in certain categories.

Mixtures may be prepared containing several diluents, disintegrants, and lubricants. The presence or not of each of these excipients may be treated as a separate variable. Durig and Fassihi (10) investigated the effect of 11 excipients as well as temperature and relative humidity on the stability of pyridoxal hydrochloride. They used a Hadamard (Plackett-Burman) design of 24 experiments to determine main effects for the 13 variables. This large number of experiments, 8 more than the minimum required, allowed effects of "dummy" or "pseudo-variables" to be calculated so that some idea of the validity of the data could be obtained. Assuming all pseudo-variable effects to be a result of random experimental variation the standard deviation was estimated, and significant results could be identified (compare section III.B.5).

A disadvantage of this approach is the large variation in the concentrations of excipients depending on the number of diluents present in the mixture. This problem may be treated by using the mixture screening designs described in chapters 9 and 10. Otherwise, the excipients are grouped into classes and each protoformulation run will contain not more than one excipient in each class. Then, as we have already seen, the concentration of the diluent may vary between runs, but not by a significant amount. The following sections will describe how this is done (11).

It may also be useful to introduce other formulation or process variables. An important factor, to be studied very early in formulation, is the effect of wet granulation on the stability. Another possible factor is the effect of concentration of the drug substance. Mixtures at a low concentration are frequently less stable than more concentrated mixtures. However, designs as simple as the one we have just looked at must be used with caution in such a situation. It has frequently been our experience that a drug substance that is stable in the pure solid state becomes unstable when formulated, and in particular following a wet granulation. Thus we might find all of the dry mixtures are stable, but the granulated mixtures could be unstable with considerable differences between them according to the excipients used. There is an interaction between the "granulation" factor and the excipient factors.

Different storage conditions, temperature, and relative humidity are often tested. It is advisable to store each sample at all of the conditions tested and not to try to reduce the number of experiments by including temperature and humidity as variables in the design. Temperature and humidity almost always affect the stability and their effects are often greater than those of the excipients. Frequently, their effects are so large that they would "drown" many of the effects of excipients. For this reason we would recommend testing all samples at the same storage conditions. The study of their effects is useful for selecting suitable conditions for screening.

D. A $3^4//3^2$ Symmetrical Factorial Design

1. Defining the problem and the experimental domain

Consider a problem where we want to study the compatibility of a drug substance with 4 classes of excipient, each at 3 levels, as in table 2.13. The measured response here is the degradation observed after 1 month at 50°C (50% relative humidity).

Table 2.13 Domain for Excipient Compatibility screening: 4 Factors at 3 Levels

Factor		Excipients (levels of each factor)		
		level 0	level 1	level 2
diluent	X_1	lactose	calcium phosphate	microcrystalline cellulose
disintegrant	X_2	starch	starch sodium glycolate	crospovidone
binder	X_3	povidone (PVP)	hydroxypropylmethyl-cellulose (HPMC)	none
lubricant	X_4	magnesium stearate	stearic acid	glyceryl behenate

2. Mathematical model and experimental design

As we have seen the additive model will contain 8 terms plus the constant, making 9 coefficients in all. The number of experiments required is calculated from the inequality: $N = s^{k'} > k(s-1)$.

Reference to table 2.9 shows that a suitable design is the $3^4//3^2$ design of 9 experiments (table 2.14). The experimental plan and the results are listed in table 2.15. The reference state model is:

$$y = \beta'_0 + \beta'_{1,1}x_{1,1} + \beta'_{1,2}x_{1,2} + \beta'_{2,1}x_{2,1} + \beta'_{2,2}x_{2,2} + \beta'_{3,1}x_{3,1} + \beta'_{3,2}x_{3,2}$$
$$+ \beta'_{4,1}x_{4,1} + \beta'_{4,2}x_{4,2} + \varepsilon$$

which describes the effects on the stability of replacing excipients with respect to a formulation containing those excipients coded 0 in the design: lactose, starch, PVP, and magnesium stearate. $x_{i,j}$ are presence-absence variables. For example, in all cases when the diluent is calcium phosphate, $x_{1,1} = 1$. Otherwise $x_{1,1} = 0$. The coefficient $\beta'_{1,1}$ defines the differential effect on the stability of the drug of replacing lactose in the formulation by calcium phosphate. Similarly, $\beta'_{1,2}$ describes the effect of replacing lactose by microcrystalline cellulose.

Table 2.14 $3^4//3^2$ Fractional Factorial Design in Coded Variables

No.	X_1	X_2	X_3	X_4
1	0	0	0	0
2	0	1	1	2
3	0	2	2	1
4	1	0	1	1
5	1	1	2	0
6	1	2	0	2
7	2	0	2	2
8	2	1	0	1
9	2	2	1	0

Table 2.15 Results of Excipient Screening with a Fractional Factorial Design at 3 Levels

No.	Diluent	Disintegrant	Binder	Lubricant	% degradation
1	lactose	starch	PVP	Mg stearate	4.0
2	lactose	Na starch glycolate	HPMC	gly. behenate	3.9
3	lactose	crospovidone	none	stearic acid	3.3
4	Ca phosphate	starch	HPMC	stearic acid	4.0
5	Ca phosphate	Na starch glycolate	none	Mg stearate	4.2
6	Ca phosphate	crospovidone	PVP	gly. behenate	3.8
7	cellulose	starch	none	gly. behenate	0.4
8	cellulose	Na starch glycolate	PVP	stearic acid	3.1
9	cellulose	crospovidone	HPMC	Mg stearate	1.6

4. Calculation of the coefficients

Inspection of the experimental design and the model shows that we may estimate the coefficients by linear combinations of the experimental results.

$$b'_{1,1} = \tfrac{1}{3}(y_4 + y_5 + y_6 - y_1 - y_2 - y_3)$$

is an estimate of $\beta'_{1,1}$. To obtain the effect of replacing lactose by calcium phosphate in the formulation we sum the degradation for all formulations containing phosphate and subtract from it the sum of the degradation for the formulations containing lactose. As there are 3 experiments in each class, we divide by 3. Main effects of all other excipient classes cancel each other out.

Similarly, to find the effect of removing povidone (PVP) from the formulation (replacing povidone by the absence of a binder) we subtract the results of all experiments with no binder from those with povidone:

$$b'_{3,2} = \tfrac{1}{3}(y_3 + y_5 + y_7 - y_1 - y_6 - y_8)$$

The results of these calculations are summarised in table 2.16.

Table 2.16   Excipient Compatibility: Coefficient Estimates from $3^4 /\!/ 3^2$ Design

Coefficient	Estimated value	Physical interpretation	
β'_0	+4.00%	reference state value	
$\beta'_{1,1}$	+0.27%	replacement of:	lactose by phosphate
$\beta'_{1,2}$	-2.03%	:	lactose by cellulose
$\beta'_{2,1}$	+0.93%	replacement of:	sodium starch glycolate by starch
$\beta'_{2,2}$	+0.10%	:	starch by crospovidone
$\beta'_{3,1}$	-0.46%	replacement of:	povidone by HPMC
$\beta'_{3,2}$	-1.00%	:	povidone by no binder
$\beta'_{4,1}$	+0.20%	replacement of:	magnesium stearate by stearic acid
$\beta'_{5,1}$	-0.52%	:	magnesium stearate by glyceryl behenate

As we have already noted, it is most convenient to calculate independent coefficients. This would not normally be done by this tedious method of calculating differences or contrasts, but by least squares regression on a computer. To interpret the coefficients, it is normal to go from the reference state, which is an arbitrary combination of levels for each variable, to a hypothetical reference state, as we saw earlier in the chapter. This hypothetical reference state corresponds to the mean value of the levels of each factor. These non-independent parameters, mathematically equivalent to the independent parameters in the table above, are calculated using equations A2.7 in appendix II and are shown graphically in figure 2.6.

The coefficients of the non-independent variables may also be calculated directly for these orthogonal experimental designs by taking, for example, the mean degradation of all experiments with lactose and subtracting the mean of all 9 experiments.

We have no direct way of knowing the precision of the data, so no statement may be made as to their reliability. The results indicate that lactose and calcium phosphate should be avoided as diluents, and that microcrystalline cellulose gives rather better stability than either of these. Further screening of a larger number of diluents could well be useful. The effects of other variables are much less important. There is little difference between formulations containing magnesium stearate as lubricant and those with stearic acid but both are less stable than those containing glyceryl behenate.

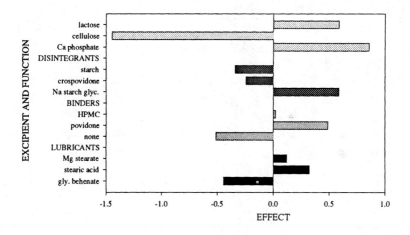

Figure 2.6 Coefficients in the screening model: hypothetical reference state.

E. Other Symmetrical Designs for Screening

All the useful symmetrical fractional factorial designs for screening are listed in table 2.9. For example a similar design to the one described above exists for 27 experiments, and for up to 13 variables, each at 3 levels.

One of the most useful symmetrical designs is based on the 4^2 factorial design (table 2.17), described (using the previous notation) as $4^5//4^2$. 5 factors at 4 levels are screened in $4^2 = 16$ experiments. We again emphasize that the numbers 0, 1, 2, and 3 identify qualitative levels of each variable and have no quantitative significance whatsoever.

It is to be noted that if the (qualitative) variable X_i is the nature of a certain class of excipient (for example, disintegrants), one of its levels may be the *absence* of a disintegrant.

Any one of the columns may be left out. Therefore the 9-experiment design can be used to screen just 3 instead of 4 variables, each at 3 levels, and the 16-experiment design to screen 3 or 4 variables each at 4 levels. But if we wanted to study 6 factors at 4 levels, the smallest *symmetrical* factorial design possible would be the $4^{21}//4^3$ design with 64 experiments. These designs are listed in appendix II.

In favourable cases therefore, these designs can be efficient, saturated, or nearly saturated. The precision of the resulting estimates is good, with minimized standard error. The designs are also orthogonal, the estimates of the coefficients of the different variables being uncorrelated with one another.

Table 2.17 The 4-level Symmetrical Design

No.	X_1	X_2	X_3	X_4	X_5
1	0	0	0	0	0
2	0	1	1	1	1
3	0	2	2	2	2
4	0	3	3	3	3
5	1	0	1	2	3
6	1	1	0	3	2
7	1	2	3	0	1
8	1	3	2	1	0
9	2	0	2	3	1
10	2	1	3	2	0
11	2	2	0	1	3
12	2	3	1	0	2
13	3	0	3	1	2
14	3	1	2	0	3
15	3	2	1	3	1
16	3	3	0	2	0

These favourable cases are somewhat few and far between. However, the designs are also useful in that they can be modified to give a variety of asymmetrical factorial designs (12).

V. ASYMMETRICAL SCREENING DESIGNS

A. Asymmetrical Factorial Designs Derived by Collapsing

1. An example of "collapsing A" derived from the $3^4//3^2$ design

We consider the previous example of compatibility testing with the modification that only 2 lubricants, magnesium stearate and glyceryl behenate are to be tested. One of the levels of this variable X_4, in table 2.14, must be eliminated.

Any variable at s levels can be transformed to a variable of $(s-1)$ levels. Assuming the levels are indexed from 0 to $s-1$, all that is needed is to replace the level $s-1$ by the level 0. Depending on the value of s, this transformation may be repeated, replacing level $s-2$ by level 1 and so on. In the case of the design of table 2.14, we replace level 2 of X_4 by 0, giving the design shown in table 2.18. This is described, using the same notation as before, as $3^3 2^1//3^2$. We may simplify the notation to $3^3 2^1$.

Table 2.18 $3^3 2^1//3^2$ Design
Derived by Collapsing A

No.	X_1	X_2	X_3	X_4
1	0	0	0	0
2	0	1	1	0
3	0	2	2	1
4	1	0	1	1
5	1	1	2	0
6	1	2	0	0
7	2	0	2	0
8	2	1	0	1
9	2	2	1	0

We may then choose to set level 0 to "magnesium stearate" and level 1 to "glyceryl behenate" in table 2.18. So the design is no longer totally "balanced" (the same number of experiments for each level), because the level "magnesium stearate" occurs twice as frequently as does the level "glyceryl behenate".

It is efficient (nearly saturated), and gives precise estimates. It remains orthogonal, in that estimates of the effects of the variables are independent of one another. However the lack of balance of the design results in a correlation between errors in the estimates of the effect of the lubricant and of the constant term.

The designs that may be derived from the $3^4 /\!/ 3^2$ design are summarised in table 2.19 below. These are not always the best possible designs. The balanced and more efficient $2^4 /\!/ 2^3$ design, requiring only 8 experiments is to be preferred to the 9 experiment $2^4 /\!/ 3^2$ design. It will be seen later how a $3^1 2^3$ design of 8 experiments can be derived from the $2^7 /\!/ 2^3$ design by another method of collapsing (C).

Table 2.19 9 Run Designs Derived from the $3^4 /\!/ 3^2$ Design by Collapsing A

$$3^4 /\!/ 3^2 \quad \rightarrow \quad 3^3 2^1 /\!/ 3^2 \quad \rightarrow \quad 3^2 2^2 /\!/ 3^2 \quad \rightarrow \quad 3^1 2^3 /\!/ 3^2 \quad \rightarrow \quad 2^4 /\!/ 3^2$$

2. An example derived from the $4^5 /\!/ 4^2$ design: Collapsing B

Consider the screening problem of table 2.20. Here we want to know the effect of various diluents, disintegrants, lubricants etc. on the stability of the drug substance. We have not imposed any restraint whatsoever on the numbers of levels for each variable. We could of course test all possible combinations with the full factorial design, which consists of 384 experiments ($4 \times 2 \times 3 \times 4 \times 2 \times 2$)! Reference to the general additive screening model for different numbers of levels will show that the model contains 12 independent terms. Therefore the minimum number of experiments needed is also 12.

The fact that the maximum number of levels is 4 (for 2 of the variables) suggests that we derive a design from the 4-level symmetrical factorial design of 16 experiments, the $4^5 /\!/ 4^2$ design. This was shown in table 2.17 for variables X_1 to X_5, the levels being identified as 0-4. We shall see that the column of the 5th variable in the design of table 2.17 is not required and is therefore omitted from the design. Let the diluent be X_1 and the lubricant be X_2. The third variable, the binder, takes 3 levels (no binder, HPMC, PVP). Here as in the previous section we simply replace one level with another so that it will appear 8 times instead of 4, by collapsing A. Here we set both level 0 and level 3 to "no binder". The result is shown in table 2.21.

We examine the fourth column of table 2.21 in detail. The 4 levels of X_4 can be used to identify combinations of levels of the 3 remaining variables, the nature of the disintegrant, presence of glidant and capsule shell, as shown in table 2.22a. Replacing the 4 levels of X_4 in the design of table 2.21 with these 3 variables each at 2 levels results in the design of table 2.22b.

The characteristics of this design are excellent in terms of precision and orthogonality. The efficiency is 75%. It is interesting to note that the final column of the symmetrical $4^5 /\!/ 4^2$ design was not needed. It could have been used to study a 7th variable with up to 4 levels.

Table 2.20 Drug-Excipient Compatibility Problem Illustrating Collapsing B

Factor	(n° levels)	Excipients (levels)			
		0	1	2	3
diluent	(4)	lactose	mannitol	calcium phosphate	microcrystalline cellulose
disintegrant	(2)	starch	starch sodium glycolate SSG		
binder	(3)	povidone (PVP)	hydroxypropyl-methylcellulose HPMC	(none)	
lubricant	(4)	magnesium stearate	glyceryl behenate	stearic acid	hydrogenated castor oil HCO
glidant	(2)	colloidal silica	(none)		
capsule shell	(2)	yes	no		

Table 2.21 Derivation of a $4^2 3^1 2^3$ Design from a $4^5 /\!/ 4^2$ Symmetric Design by Collapsing A from 4 to 3 Levels

N°	X_1 Diluent	X_2 Lubricant	X_3 Binder	X_4
1	0 = lactose	0 = stearate	0 = none	0
2	0	1 = gly. behenate	1 = PVP	1
3	0	2 = stearic acid	2 = HPMC	2
4	0	3 = HCO	3 = 0	3
5	1 = mannitol	0	0	2
6	1	1	1	3
7	1	2	3 = 0	0
8	1	3	2	1
9	2 = phosphate	0	2	3
10	2	1	3 = 0	2
11	2	2	1	1
12	2	3	0	0
13	3 = cellulose	0	3 = 0	1
14	3	1	2	0
15	3	2	1	3
16	3	3	0	2

Table 2.22 Derivation of a $4^23^12^3$ Design from a $4^5/\!/4^2$ Symmetric Design

(a) Collapsing B of 1 variable at 4 levels to 3 variables each at 2 levels

Levels in X_4 column of the $4^5/\!/4^2$ design	Corresponding levels in new experimental design		
	Disintegrant (X'_4)	Glidant (X'_5)	Gelatine capsule (X'_6)
0	0 = starch	0 = silica	1 = capsule
1	1 = SSG	0 = silica	0 = none
2	0 = starch	1 = none	0 = none
3	1 = SSG	1 = none	1 = capsule

(b) The final $4^23^12^3$ design

No.	X_1	X_2	X_3	X'_4	X'_5	X'_6
1	0	0	0	0	0	1
2	0	1	1	1	0	0
3	0	2	2	0	1	0
4	0	3	0	1	1	1
5	1	0	1	0	1	0
6	1	1	0	1	1	1
7	1	2	0	0	0	1
8	1	3	2	1	0	0
9	2	0	2	1	1	1
10	2	1	0	0	1	0
11	2	2	0	1	0	0
12	2	3	1	0	0	1
13	3	0	0	1	0	0
14	3	1	2	0	0	1
15	3	2	1	1	1	1
16	3	3	0	0	1	0

3. Combination of collapsing A and collapsing B

Table 2.23 illustrates the large number of possible experimental designs that may be derived from the $4^5/\!/4^2$ design by these methods. A simplified notation is used, where for example the $4^5/\!/4^2$ design is represented simply by 4^5, or the $4^13^22^6/\!/4^2$ design by $4^13^22^6$. All experimental designs matrices in the table consist of 16 experiments. If one of the designs is suitable for the problem to be treated then it may be derived from the 4^5 design by a combination of the two main methods of collapsing.

- We move down the table by replacing 1 variable at 4 levels by 3 variables each at 2 levels (collapsing B). The R-efficiency of the design remains unchanged.

- We move across the table by omitting one of the 4 levels of a variable (collapsing A). The design becomes unbalanced and its R-efficiency is also reduced.

Certain of the designs have large numbers of factors at 2 levels. If there are too many variables in the design (whether at 2, 3, or 4 levels), one or more of them may be omitted without any prejudice to the design's quality.

Table 2.23 Designs of 16 Experiments from the $4^5/\!/4^2$ Design by Collapsing A/B

4^5	$\to$	4^43^1	$\to$	4^33^2	$\to$	4^23^3	$\to$	4^13^4	$\to$ 3^5
$\downarrow$		$\downarrow$		$\downarrow$		$\downarrow$		$\downarrow$	
4^42^3	$\to$	$4^33^12^3$	$\to$	$4^23^22^3$	$\to$	$4^13^32^3$	$\to$	3^42^3	
$\downarrow$		$\downarrow$		$\downarrow$		$\downarrow$			
4^32^6	$\to$	$4^23^12^6$	$\to$	$4^13^22^6$	$\to$	3^32^6			
$\downarrow$		$\downarrow$		$\downarrow$					
4^22^9	$\to$	$4^13^12^9$	$\to$	3^22^9					
$\downarrow$		$\downarrow$							
4^12^{12}	$\to$	3^12^{12}							
$\downarrow$									
2^{15}									

4. An example derived from the $2^7/\!/2^3$ design: collapsing C

The $2^7/\!/2^3$ design may be rearranged to a $2^44^1/\!/2^3$ design in order to investigate 4 factors each at 2 levels and one factor at 4 levels. It is shown in table 2.24.

In columns 4, 5, and 6 on the left hand side, enclosed by a dotted line, we find four different combinations of levels (111, 001, 010, and 100). These may be considered as four distinct levels of a single variable. It is therefore the inverse of collapsing B. When they are replaced by levels of the new variable X_4' (0, 1, 2, 3) we obtain the design on the right hand side. This design could therefore be used to analyse compatibility with 4 diluents and 4 other classes of excipient, each at 2 levels, by carrying out only 8 experiments. The design is orthogonal.

The choice of the 3 columns is not arbitrary. Any column may be selected for the first 2, but then the third column must be the combination of the other 2 (see chapter 3). For this reason it is not possible to carry out collapsing C a second time on this matrix to obtain a $4^22^1//2^3$ design. Applying collapsing A to the $4^12^4//2^3$ design gives the 3^12^4 design.

Table 2.24 Combining 3 Variables at 2 Levels to Treat 1 Variable at 4 Levels by Collapsing C

No.	$2^7//2^3$								$2^44^1//2^3$				
	X_1	X_2	X_3	X_4	X_5	X_6	X_7		X_1	X_2	X_3	X_4'	X_7
1	0	0	0	1	1	1	0		0	0	0	0	0
2	1	0	0	0	0	1	1		1	0	0	1	1
3	0	1	0	0	1	0	1		0	1	0	2	1
4	1	1	0	1	0	0	0		1	1	0	3	0
5	0	0	1	1	0	0	1		0	0	1	3	1
6	1	0	1	0	1	0	0		1	0	1	2	0
7	0	1	1	0	0	1	0		0	1	1	1	0
8	1	1	1	1	1	1	1		1	1	1	0	1

This appears to be the most interesting example of collapsing C. The technique can be applied to the $2^{15}//2^4$ design but the resulting matrices, except for the $8^12^8//2^4$, are identical to those obtained by collapsing A on the $4^5//4^2$ design. The only other potentially useful possibility appears to be the $9^13^9//3^3$ design with 27 experiments, derived by collapsing C on the $3^{13}//3^3$ design.

B. Special Asymmetrical Designs

There are a few asymmetrical designs which are not derived from the symmetrical designs (by collapsing B or C) and are orthogonal. Two of these for 18 runs are given in appendix II. The first can be used to screen up to 7 factors at 3 levels plus 2 factors at 2 levels (3^72^2). It can be modified to a design for screening a single factor at 6 levels and 6 factors at 3 levels (6^13^6). Designs also exist for 24 experiments (3^42^{12} and 12^12^{12}), 36 experiments (12^13^{12}), 48 experiments (12^14^{12}), and 50 experiments ($5^{10}4^12^4$), but which are unlikely to be useful here.

They may be modified by collapsing just as we saw for the symmetrical factorial designs.

C. Asymmetrical Screening Designs of Any Size (D-Optimal Designs)

1. Limitations in screening designs

The interest, but also the limitations, of the designs of experiments for screening that we have used up until now, are now evident. The two level designs are the most useful. However as soon as we wish to increase the number of levels of a variable we may have difficulties in finding a standard design. The classical designs for screening cover a relatively small proportion of the possible cases, especially if we are limited in the number of experiments that we are able to do. The symmetrical designs at 3, 4, and 5 levels may frequently be adapted to the asymmetrical case but that here also there are limitations. It is not possible, for example, to collapse a $4^5//4^2$ experimental design if one of the variables takes 5 or more levels.

2. Full factorials and the candidate experimental design

Any of the problems that we have looked at in this chapter can be investigated using a full factorial design – that is by doing an experiment at each possible combination of levels.

The symmetrical designs described in section IV were termed fractional factorial. For k variables, each at s (qualitative) levels the number of possible combinations of levels is s^k, these combinations making up the full factorial design. The fractional factorial designs are in fact sub-sets of $s^{k'}$ experiments from the full factorial design. Similarly, the Hadamard design is also a part of a full factorial design.

The full factorial design is a *candidate design* from which the necessary experiments may be extracted to give the optimum design for the purposes of the experiment.

We consider yet again the excipient screening problem in tables 2.8 and 2.25. Here 4 diluents, 4 lubricants, 3 levels of binder (including no binder), 2 disintegrants, glidant and no glidant (2 levels), capsule and no capsule (2 levels), are to be studied. There are 12 independent terms in the model equation, so there need to be at least 12 experiments in the design. We have already seen that, if we are prepared to do as many as 16 experiments, an excellent design may be derived from the symmetrical design by collapsing. This kind of solution does not exist for 12 experiments.

The full factorial design consists of all possible combinations of the variables' levels; that is $4 \times 4 \times 3 \times 2 \times 2 \times 2 = 384$ experiments. The 16 experiment design is a part of the complete design. If it was decided to minimize the number of experiments to determine the effect of each excipient on the stability of the drug molecule, we would have to extract 12 experiments from the full factorial design using one of the available "tools" for doing so, selecting the "best" experiments – those that will give the "best possible" design. Such as design is shown in table 2.25.

Table 2.25 A D-optimal Screening Design of 12 Experiments for Drug-Excipient
Compatibility Testing

	Diluent	Lubricant	Binder	Disintegrant	Glidant	Gelatine capsule
1	lactose	Mg stearate	HPMC	SSG	silica	capsule
2	lactose	gly.behen.	none	starch	none	capsule
3	lactose	stearic acid	PVP	SSG	none	none
4	lactose	HCO	none	starch	silica	none
5	phosphate	stearic acid	HPMC	starch	silica	none
6	phosphate	HCO	PVP	SSG	none	capsule
7	cellulose	Mg stearate	PVP	starch	none	capsule
8	cellulose	gly.behen.	HPMC	SSG	none	none
9	cellulose	stearic acid	none	starch	silica	capsule
10	mannitol	Mg stearate	none	starch	none	none
11	mannitol	HCO	HPMC	SSG	none	capsule
12	mannitol	gly.behen.	PVP	SSG	silica	capsule

3. Extracting experiments from the candidate design

In screening, we look for a design with a small number of experiments which gives precise estimates of the parameter coefficients, independent of one another. The desired number of runs is extracted from the candidate design to give a design which minimizes the standard error of the estimates of the coefficients. A computer program is necessary. Such programs are found in most experimental design and data analysis packages and normally involve an *exchange algorithm*. The resulting design is called *D-optimal*. The details of this method, used also for designs other than for screening, are described in more detail in chapter 8.

D-optimal designs are not always orthogonal and very often no orthogonal design exists for the combination of levels of variables, and the number of runs. (The complete factorial design is orthogonal). But their properties are such that correlation of the estimates of the different coefficients, if not minimized, is still low. Frequently, though not always, designs with "round" numbers of experiments, 12, 16, etc., are found to have better properties, in terms of precision and orthogonality, than those with "odd" numbers of experiments. If one tries to find a D-optimal design where a standard design exists, the exchange algorithm should normally "find" the standard design – which is therefore itself D-optimal.

4. Example of excipient screening

The D-optimal design for the drug-excipient compatibility problem which we have outlined above can be compared with the 16 experiment design derived in section V.A. by "collapsing". This design, consisting of only 12 experiments for 12 coefficients, is efficient. It is, however, inferior to the design obtained by collapsing of the 4^4 symmetrical design (16 experiments) in terms of most of the other criteria for the quality of a design.

Unlike many of the designs we have looked at so far the equations cannot be solved easily by simple examination. A computer program will therefore be needed, and least squares regression analysis (described in chapter 4) will be the method of choice in the majority of cases.

VI. CONCLUDING DISCUSSION

A. Summary of Experimental Designs for Screening

We end this chapter by looking again at the extrusion-spheronization example, imagining five different scenarios according to the factors we want to investigate, and the range of states to be tested for each factor, summarised in table 2.26.

(a) If each factor is studied at two levels, a Plackett-Burman design is selected. With 11 variables, the smallest design is of 12 experiments. In coded variables, the design is identical to the one we have already seen for a dissolution method validation. It allows the effect of various processing parameters to be identified, as well as effects of the amount of binder or its nature (but not both), and enables us to investigate the effect of adding a surfactant, polysorbate or sodium dodecyl sulphate, to the formulation.

(b) The symmetrical fractional factorial design of 9 experiments enables testing of only 4 variables, but each at 3 levels. This design is like the one already described for the compatibility problem (tables 2.14-15). For 5 to 7 variables we could use the 18 experiment $2^1 3^7 //18$ design, described in appendix II. For 8 to 13 variables at 3 levels a design based on a $3^{13}//3^3$ design of 27 experiments would be required.

(c) Now let us imagine that the number of levels and variables are more than can be tested using these designs. We wish to test for the effects of the following variables:

- the quantity of water: 3 levels ("dry", "normal", and "wet")
- the binder: 4 levels (2 binders, each at two concentrations – not necessarily the same concentration for each binder)
- the surfactant: 4 levels (3 different surfactants as well as systems without any surfactant)
- the extrusion rate: 3 levels (as in table 2.26)
- spheronization speed: 3 levels (as in table 2.26)

The asymmetrical factorial design for this problem can be obtained by "collapsing A" on a symmetrical $4^5//4^2$ design to give a $4^2 3^3 //4^2$ design, which is therefore unsaturated.

Table 2.26 Designs for Screening Extrusion-Spheronization in Various Scenarios

Pharmaceutical process or formulation variable	Plackett-Burman (Hadamard)	Symmetrical fractional factorial $3^4//3^2$	Asymmetrical fractional factorial $4^2 3^3//4^2$	D-optimal
Formulation				
1. binder (% and/or nature)	2	3	4	3
2. surfactant additive (presence and nature)	2	3	4	4
Mixing				
3. amount of water	2	–	3	2
4. granulation time	2	–	–	2
5. mixer speed	2	–	–	2
6. load	2	–	–	2
Extrusion				
7. speed (slow – medium – fast)	2	3	3	3
8. grill size (0.8 – 1 mm)	2	–	–	–
Spheronization				
9. load (light – normal)	2	–	–	–
10. time	2	–	–	2
11. speed (slow – medium – fast)	2	3	3	3
Number of independent coefficients in the model	12	9	13	15
Number of experiments	12	9	16	≥ 15

(d) Finally we wish to screen all of the listed factors with no restriction on the levels (see the last column of table 2.26). Most of them are set at 2 levels, but certain factors take 3 or 4 levels. No standard design exists and the full factorial design consists of 3456 experiments. 15 or more experiments may be selected by means of an exchange algorithm to give a D-optimal design.

(e) It is interesting to note that if the number of levels taken by one of the factors at 3 levels were reduced from 3 to 2, then a 16-experiment asymmetrical factorial design of good quality could be derived by collapsing A and B of the $4^5//4^2$ symmetrical design (to a $4^1 3^2 2^6//4^2$ design).

B. What Do We Do About the Factors That Are Left Out?

The number of factors to be screened is potentially large, so the experimenter must choose between them. This choice is normally based on expert or prior knowledge of the system (both theoretical and practical), analogy with other systems, or on the results of previous tests. Studies of individual factors reported in the scientific and technical literature may be helpful. It is normally better to include as many factors as possible at this stage, rather than to try and add them later. A finished study may prove useless if the experimenter has to add extra factors.

So the ones that are chosen are those which would seem to have a possible or probable influence. But what about the ones that are eliminated from the study? They are not thought *a priori* to influence the results, but we still need to consider them.

One possibility is to control each at a constant level, selected arbitrarily or for economic or technological reasons. The values of these factors must be incorporated into the experimental plan even if they themselves do not vary. They are still part of the experimental design. And then, even if the hypothesis that they do not affect the response is mistaken, they will not affect our conclusions about the influences of the factors studied, but only our estimate of the constant term.

Or we allow them to vary in an uncontrolled manner, as it could be too difficult or expensive to control such factors as ambient temperature, relative humidity, batch of starting material. It might also be considered unrealistic to control parameters which cannot be held constant under manufacturing conditions. (This is now rather less of a problem in the pharmaceutical industry than for certain other industries.) The factor is allowed to vary randomly from one experiment to another, in which case its effect, if it has one, will be added to the error and increase the estimate of the experimental error. Thus the estimated effects of certain controlled factors may no longer be considered significant, as they no longer exceed a background noise level that has been (artificially) increased. Or the uncontrolled factor may vary very little during the study and subsequent fluctuation may then give rise to some surprises. If the factor should vary *non-randomly*, so that it is correlated with certain of the factors studied, then the study's conclusions may be falsified. Its effect may then be to mask a factor, or on the other hand to cause another factor to be thought significant when in fact it has no effect.

In conclusion, the values of those experimental conditions that are not part of the design should also be measured and recorded, as they may nevertheless affect the result.

It is possible to prevent an uncontrolled factor causing a biased result by doing the experiments in a random order (*randomization*), by ordering experiments to avoid effects correlated with time (time-trends), by *blocking* experiments, and by studying the alias matrix. These various techniques will be discussed in the following chapters, but they apply equally to screening experiments.

The variability of the response(s) due to uncontrolled factors is the essential theme of a later chapter, the use of statistical experimental design in assuring quality. We will examine there what techniques are available to minimize variation in formulations and processes.

C. Reproducibility and Repeated Experiments

One must have some idea of the reproducibility of a technique to draw conclusions from the results of any experimental design, including those from a screening design. This estimate may be *qualitative* in the sense that our experience of similar processes tells us what sort of value to expect for the standard deviation. Such information is valuable, but it can be expected to give only a very rough estimate of the reproducibility, and it is better not to rely on this alone. We may also use data from a preliminary experiment, repeated several times, one that is not part of the design. This is an improvement on the first method, and is also better than estimating the standard deviation from the redundancy of the design – but, even so, the standard deviation can be over- or under-estimated, leading to erroneous conclusions.

1. Use of centre points

It is better to replicate experiments *within* the design, thus estimating the experimental *repeatability*. If all factors are quantitative it is the centre point that is selected. This has the advantages that if the experimental standard deviation changes within the domain one could reasonably hope that the centre point would represent a mean value and also that if the response within the domain is curved, this curvature may be detected.

In a screening experiment it is rare for all variables to be quantitative. If they are not all quantitative then the centre point can no longer exist. If only one variable F is qualitative it would be possible to take the centre-points for all quantitative variables and repeat them at each level of the qualitative variable, F_1, F_2, F_3... This already requires a considerable number of experiments. With more than one qualitative variable it is rarely feasible. There is no entirely satisfactory solution – one might repeat a number of experiments selected randomly or the experimenter can choose the experimental conditions that he thinks the most interesting.

2. Replication of a design

The experimenter often chooses the smallest design consistent with his objectives and resources. Effects are estimated with a certain precision. For example, we have seen with the Plackett-Burman designs (and it is the same for the 2-level factorial designs of the following chapter) that the standard error of estimation of an effect is $\sigma/N^{\frac{1}{2}}$, where σ is the experimental standard deviation (repeatability) and N is the number of experiments in the design.

If this appears insufficient, the design may be replicated n times. The effects are then estimated with a precision of $\sigma/(nN)^{\frac{1}{2}}$, improved by a factor of $\sqrt{n}$. This also has the advantage that the repeatability σ may be estimated. Alternatively N could be increased by employing a larger Plackett-Burman design, or fractional factorial design, possibly carried out in stages, or blocks (1). It may be then be possible to estimate further effects (interactions), as we shall see in chapter 3.

Both approaches are expensive, and it would often be preferable to try and improve the repeatability of the experimental method, rather than to do more experiments. But where this is impracticable (e.g. biological experiments, *in vivo*, where the variability is innate to the system being studied) replication may be the only way to obtain enough precision.

The replicated experiments are independent of one another, and are treated in the data analysis as separate experiments.

3. Replication of experimental measurements

This replication of a design or within a design is not to be confused with the replicated *measurement* of a response. Here, for example, the particle size distribution may be determined for several powder samples from the same batch of granulate, or a number of tablets of the same batch are tested for dissolution, disintegration time or crushing strength. The number of measurements required in each case is often part of a pharmacopoeal specification. The variance of these measurements is the sum (13) of the sampling variance, σ_S^2, and that of the analytical measurement, σ_A^2, but it excludes all variation from the preceding stages of the experiment, in this case, mixing, granulation, tableting etc., variance σ_p^2. Assuming a single measurement on each of n_A samples, the overall variance σ^2 is given by:

$$\sigma^2 = \sigma_P^2 + \frac{\sigma_A^2 + \sigma_S^2}{n_A}$$

This, replicating measurements only improves the sampling and measurement precision within the experiment. The individual measurements are not complete replicates and should not be treated as separate experiments. In general, the mean value of the n_A measurements is treated as a single datum in the statistical analysis. See the following section and also the section in chapter 7 on the *dependence of the residuals on combinations of 2 or more factors* (split-plot designs).

D. Some Considerations in Analysing Screening Data and Planning Future Studies

1. Assumptions in screening

The screening phase allows us to group the factors into those whose effects are clearly important and significant, those we are certain may be neglected, and those whose effects are intermediate. What choices are there available after completing such a study?

We are reminded that setting up a design involves a number of suppositions, some explicit, others understood but not formally specified. The fact that we study a finite number of factors means that all others that are not held constant are assumed to have a negligible effect on the response. The design will give no information on these factors.

No mathematical hypothesis is ever perfectly respected. The experimenter must put forward reasonable assumptions. They can be verified only by further experimentation. A case in point is the assumption that the experimental variance is constant throughout the experimental region. This can be verified (but in fact we cannot prove it, only show that the hypothesis is not unreasonable) by repeating each experiment several times and comparing the dispersions for the different experiments. Repeating large numbers of experiments is not at all compatible with the economy of a screening experiment, and would only be resorted to if strong evidence of excessive variation forced us to take this approach.

Screening presupposes that we accept the (considerable) approximation of the additivity of the different factors and of the absence of interaction. If this hypothesis is thought to be totally invalid, the only solution is to go straight on to the quantitative study of the effects of factors (chapter 3). No screening design will enable us to verify additivity and the interpretation of the results can only be valid if the hypothesis of additivity is itself valid! But it would be most unusual to try to verify this supposition during the screening phase. We recall that screening is generally followed by a more detailed study of the factors and as long as the results found at these stages are not in total contradiction with the results of screening, the present assumptions are not questioned.

2. Proportion of factors found significant or active

If *all or most of the factors* are found to be significant this may be for one or more of the following reasons.
   (a) It is true that all factors are significant within the domain!
   (b) The experimental variance is underestimated, possibly because:
   • The experiments used to determine the standard deviation were carried out consecutively, not reflecting the variation over the whole design, carried out over a longer period.
   • They were not completely replicated. This is often observed if the experimentation takes place in several stages and it is only the last stage that is repeated (*split-plot design*) or even that it is only the

measurement that is repeated. This is a common situation in industrial experiments.

- The conditions for the repeated experiments are not typical of the variation in the rest of the domain.

A thorough analysis of the conditions is then required, both of the screening experiment and of those in which the reproducibility data were obtained, with an analysis of the assumed mathematical model. Further experiments may be needed.

(c) The more significant factors exist, the greater is the probability that the factors interact and bias the estimates of other effects. The only possible solution would be the detailed study of several or all of the factors. Should the number of experiments be prohibitively high, the values of some of the factors could be fixed.

If *no factor* is found to be important this may be for one of the following reasons.

(d) No factor is indeed significant within the experimental region.

(e) There were mistakes in selecting the experimental domain, and its limits should be widened.

Cases (d) and (e) would normally lead to the design being abandoned as the experimenter would normally have introduced the largest possible number of factors with wide limits to the experimental domain.

(f) The experimental variation ("background noise") is too high. It may be that the analytical methods used are imprecise, or non-controlled factors have too much effect.

(g) certain factors are significant, but are not seen because their effects are cancelled out by interactions with other factors. This phenomenon may be observed but it is relatively improbable.

Cases (f) and (g) may be demonstrated if the experimenter knows in advance that certain factors *must* be significant. Case (f) might be detected on comparing the repeatability and the reproducibility of the results. Case (g) can only be treated by constructing a design matrix that allows the study of interactions (chapter 3). This will be very expensive if there are many factors to be studied.

(e) the effect of a quantitative, continuous factor shows a pronounced curvature within the domain. This would be shown by the results of experiments at the centre points.

In the *intermediate case,* non-significant factors may be fixed at a suitable level (representing minimum cost for example), or perhaps left uncontrolled (such as the origin of the starting materials). Factors found to be significant may be kept in, for more detailed study. Alternatively, some of these may be set at their optimum level. This applies particularly to qualitative variables, but only if there is no ambiguity as to the effect of the factor, with no risk of confounding with an interaction between two other very significant factors (see chapter 3).

3. Studies following screening

Two types of study follow screening: the detailed study of the *influence* of the factors (including that of all or a part of the interactions) and also the more detailed study of the *response* in part of the domain. The screening study often leads to the displacement of the centre of the experimental domain. The variation limits of certain factors are also frequently changed, usually being made narrower.

Finally it may well be the case that among the experiments carried out in the screening study we find one or more which give results so satisfactory that the study may be considered as completed!

References

1.   G. E. P. Box, W. G. Hunter and J. S. Hunter, Statistics for Experimenters, John Wiley, N. Y., 1978.
2.   R. L. Plackett and J. P. Burman, The design of optimum multifactorial experiments, *Biometrica*, **33**, 305-325 (1946).
3.   W. Mulholland, P. L. Naish, D. R. Stout, and J. Waterhouse, Experimental design for the ruggedness testing of high performance liquid chromatography methodology, *Chemom. Intell. Lab. Syst.*, **5**, 263-270 (1989).
4.   S. R. Goskonda, G. A. Hileman, and S. M. Upadrasha, Development of matrix controlled release beads by extrusion spheronization technology using a statistical screening design, *Int. J. Pharm.* **100**, 71-79 (1993).
5.   A. A. Karnachi, R. A. Dehon, and M. A Khan, Plackett-Burman screening of micromatrices with polymer mixtures for controlled drug delivery, *Pharmazie*, **50**, 550-553 (1995).
6.   S. V. Sastry, M. D. DeGennaro, I. K. Reddy, and M. A. Khan. Atenolol gastrointestinal therapeutic system. 1. Screening of formulation variables, *Drug Dev. Ind. Pharm.* **23**, 157-165 (1997).
7.   D. Monkhouse and A. Maderich, Whither compatibility testing? *Drug. Dev. Ind. Pharm.*, **15**, 2115-2130 (1989).
8.   R. N. Kacker, E. S. Lagergren, and J. J. Filliben, Taguchi's fixed element arrays are fractional factorials, *J. Qual. Tech.*, **23**, 107-116 (1991).
9.   H. Leuenburger and W. Becker, A factorial design for compatibility studies in preformulation work, *Pharm. Acta. Helv.*, **50**, 88-91 (1975).
10. T. Durig and A. R. Fassihi, Identification of stabilizing and destabilizing effects of excipient-drug interactions in solid dosage form design, *Int. J. Pharm.*, **97**, 161-170 (1993).
11. G. A. Lewis and D. Mathieu, Screening designs for compatibility testing, *Proc. 14th Pharm. Tech. Conf.*, **2**, 432-439 (1995).
12. S. Addelman, Orthogonal main effect plans for asymmetric factorial experiments, *Technometrics*, **4**, 21-46 (1962).
13. J. C. Miller and J. N. Miller, Statistics for Analytical Chemistry, Ellis Horwood, Chichester, U. K., 1984.

3

FACTOR INFLUENCE STUDIES

Applications of Full and Fractional Factorial Designs at 2 Levels

I. INTRODUCTION

A. The Place of a Factor Study in Development

We have now covered the standard methods used for screening studies, which, carried out early in a project's lifetime, consume only a small part of the resources of time, money, materials, availability of equipment, etc... allocated to it. We may therefore suppose that, having completed such a study, we are left with rather fewer factors and are thus ready to carry out a detailed quantitative study of the influence of these factors. In fact, a separate screening study is only justified if there are many factors and not all are expected to be influential.

B. Recognising Situations Requiring a Factor-Influence Study

These are analogous in many ways with the screening situation:

- As with screening, the factors may be qualitative or quantitative. *Quantitative* factors, such as spheronization time, take few distinct levels, and are generally limited to only 2 levels, upper and lower. Exceptionally, they may be set at 3 equidistant levels. *Qualitative* factors, such as the nature of an excipient, can be tested at any number of levels.
- The experimental region (domain) is described as "cubic" as it is defined by the lower and upper level of each factor.
- The designs used are the same kind as for screening.

However there are some important differences:

- Fewer factors are normally studied.
- All factors are likely or supposed to be significant or active.
- The experimental domain is usually less extensive than for screening: but this rule is not absolute. The limits for some factors may even be enlarged.
- Additional *interaction* terms are added to the model, either directly or in stages (see below), the result being a *synergistic* model. This is the most important difference from screening.
- The experimenter will try to understand and to explain what is happening mechanistically, perhaps even in physical-chemical terms, trying in particular to account for interactions. The mathematical model continues to have a descriptive role, but it is also used to help interpret the observed phenomena.
- The factor-influence study is frequently linked to optimization of the process or formulation being studied, as described in the succeeding chapters.

As always, these remarks are general and it is the exception which proves the rule. Also the experimenter should note the limitations of these studies and what he must *not* expect from them.

The models are not predictive. They are constructed for the purpose of measuring the change in the response from one extreme of a factor to another and for determining interactions, and not for mapping a response over the domain. Experience has shown that the use of the synergistic model for prediction is rarely satisfactory for interpolation as much as for extrapolation.

In other words, we strongly advise against using the methods of this chapter for optimizing a response, or for modelling it within the zone of interest. In chapter 5 we describe far better methods. However the experiments conducted in this phase of the study may often be re-used at the optimization, or response surface modelling phase.

This is why no distinction is drawn here between qualitative and quantitative factors; quantitative factors are treated in the same way as qualitative factors (and not the reverse – a mistake made by many users of factorial designs).

Here the experimenter is often less interested in the global significance of the model than in each coefficient's individual significance. The appearance of an interaction term in the model is not thought of as a mathematical "artifact". Its presence was allowed for when setting up the design; it was looked for, and its mathematical existence (in the sense that it is statistically significant) often demonstrates the existence of a real physical phenomenon, a synergy or antagonism between two factors. We will, however, take a very different approach to this in a later chapter on "response surface methodology" .

C. Standard Approaches to a Factor-Influence Study

Even though for this kind of study both quantitative and qualitative factors (especially the latter) may be set at more than 2 levels, the number of coefficients in the model equation, and therefore the number of experiments needed to determine them, both increase sharply with the number of levels once we have decided to study interactions between variables. We will therefore also limit both kinds of variables to 2 levels, in this chapter. This is an artificial limitation and might sometimes prove to be too restrictive, especially in the case of qualitative factors.

We will begin by demonstrating the form and meaning of the mathematical models used in factor-influence studies, using a simple 2 factor example. We will go on to look at the most widely used designs, mainly factorial and fractional factorial designs at 2 levels, but also Rechtschaffner and ¾-factorial designs and demonstrate how fractional factorial designs will often allow experimentation to be carried out sequentially.

In conclusion, we note that after completion of a factor study, the work will often continue by empirical modelling of the responses within the experimental domain. The experiments already carried out may be re-incorporated in the design for this next stage.

D. Interactions

In contrast to the screening designs, we assume for the factor study that the effect of a factor may well depend on the level of other factors. There may be *interactions* between the variables.

For example, the effect of the presence or not of an excipient on the formulation could be greater or less depending on whether certain other excipients are present. So that if a drug substance is unstable when formulated, it could still be the case that the presence of a pair of excipients might have a stabilising effect that is greater than that which could be predicted from the individual effects of those excipients. There is a *synergy* between the two factors.

II. FULL FACTORIAL DESIGNS AT 2 LEVELS

A. The 2^2 Design - Example of Extrusion-Spheronization

The general form of the synergistic model will be demonstrated using a full factorial design for 2 factors.

1. Objectives – experimental domain

We continue with the extrusion-spheronization project, discussed in chapter 2, section III. Here the effects of a large number of factors on the yield of pellets of the correct size were examined. Amongst those found to have quite large, and statistically significant, effects were the speed of the spheronizer ·and the spheronization time.

In view of these results we might wish to examine the influence on the yield of these two factors in more detail, at the same time keeping the values of the remaining process and formulation factors constant. The ranges, shown in table 3.1, are identical to those in the previous set of experiments.

2. Experimental design – experimental plan – responses

All combinations of the extreme values in table 3.1 can be tested in 4 experiments, as shown in table 3.2. This is called a (full) 2^2 factorial design (1).

Table 3.1 Experimental Domain for Extrusion-Spheronization Study

Factor	Associated variable	Lower level (coded -1)	Upper level (coded +1)
spheronization time (min)	X_1	2	5
spheronizer speed (rpm)	X_2	700	1100

Table 3.2 2^2 Full Factorial Design for Extrusion-Spheronization Study

No.	X_1	X_2	Spheronization time (min)	Spheronizer speed (rpm)	Yield (%)
1	-1	-1	2 min	700 rpm	68.3
2	+1	-1	5 min	700 rpm	63.1
3	-1	+1	2 min	1100 rpm	62.5
4	+1	+1	5 min	1100 rpm	42.1

3. Mathematical model

Up to this point we have assumed that the system is adequately described by the first-order or additive (linear) model:

$$y = \beta_0 + \beta_1 x_1 + \beta_2 x_2 + \varepsilon \tag{3.1}$$

If this were indeed the case, and the values of β_1 and β_2 were -4.1% and -6%, as previously determined in the screening experiment, the results of the 4 experiments would be as shown in figure 3.1a.

In the screening study we were concerned either with qualitative variables or with the values of a response at certain discrete levels of quantitative variables, such as here. We did not attempt to interpolate between the upper and lower levels of each quantitative variable to estimate the response over the whole of the experimental domain. Whether or not such an interpolation can be justified depends on our knowledge of the system, and to some extent on the aims of the project. Here the variables are quantitative and normally continuous so trying to predict the response at values between -1 and +1 at least makes theoretical sense. However, as previously stated, factor studies are not designed for modelling or predicting responses and extreme caution is advised in using such response surfaces as the ones in figure 3.1b and 3.2b without proper validation of the model. They are included only to illustrate the form of the surface.

If equation 3.1 holds, then the response surface of the dependence of the yield has the planar form shown in figure 3.1b. The deviations between the experimental results and the predictions of the model, at each corner of the design space, would be expected to be small, and possibly due only to random fluctuations in the experimental conditions.

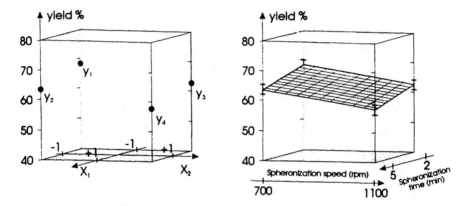

Figure 3.1 2^2 design with *non-interacting* variables.

Increasing the spheronization time from 2 to 5 minutes in this system would therefore have the same effect on the yield whatever the rotation speed of the spheronizer (within the range of the experiment and within the limits of its repeatability).

The actual results, shown in table 3.2 are rather different from what has been predicted. Without doing the statistical tests requiring a knowledge of the repeatability we suspect that either the process is poorly controlled, or the model does not describe the system accurately. We will assume the second hypothesis to be the more probable. We calculate the values of the coefficients b_1 and b_2 as before:

$$b_1 = \frac{1}{4}(-y_1 + y_2 - y_3 + y_4) = -6.4\%$$

$$b_2 = \frac{1}{4}(-y_1 - y_2 + y_3 + y_4) = -6.7\%$$

The calculation of these estimators b_1 and b_2 of the main effects β_1 and β_2 can be interpreted in a different way. If we consider only experiments 1 and 2 in the above design, only X_1 varies, from level -1 to +1, and X_2 remains constant at level -1. Any change in the response (yield of pellets) can only be a result of the change in X_1. Let us define a *partial* effect $b_1^{(-1)}$ for the variable X_1 as the change in response corresponding to a change of 1 unit in X_1 as X_2 is held constant at level -1. Another partial effect $b_1^{(+1)}$ is defined for X_1, where X_2 is held constant at level +1.

$$b_1^{(-1)} = \tfrac{1}{2}(y_2 - y_1) = \tfrac{1}{2}\,(63.1 - 68.3) = -2.6$$

$$b_1^{(+1)} = \tfrac{1}{2}(y_4 - y_3) = \tfrac{1}{2}\,(42.1 - 62.5) = -10.2$$

These are two different estimates of the effect of the variable X_1 on the response, calculated from independent experimental results. They should be equal, allowing for experimental errors. This is evidently not the case, the difference between the two values is too great to be attributed solely to random variation. The effect of the variable X_1 is not an intrinsic property of the factor "spheronization time", but on the contrary the effect of the spheronization time depends on the value of the spheronizer speed, variable X_2. There is an *interaction effect* β_{12} between the two variables X_1 and X_2. This is estimated by b_{12}, defined as follows:

$$
\begin{aligned}
b_{12} &= \tfrac{1}{2} \times [b_1^{(+1)} - b_1^{(-1)}] \\
&= \tfrac{1}{2} \times [\tfrac{1}{2}(y_4 - y_3) - \tfrac{1}{2}(y_2 - y_1)] \\
&= \tfrac{1}{4} \times [+y_1 - y_2 - y_3 + y_4] \quad = -3.8\%
\end{aligned}
$$

This interaction effect will usually be referred to simply as an *interaction*. It is the apparent change of the effect b_1 of X_1, on changing X_2 by 1 unit. Depending on whether it is positive or negative the phenomenon may be described as *synergism* or *antagonism*.

We can examine the effect on b_2 of changing the level of X_1, using the same reasoning, calculating the interaction effect b_{21}. The same formula is found for b_{21}

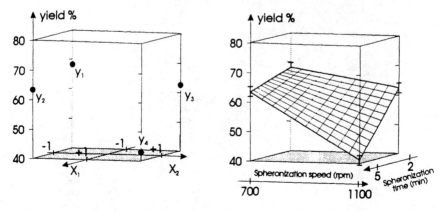

Figure 3.2   Results of 2^2 design (extrusion-spheronization) showing interaction between X_1 and X_2.

as for b_{12}, so β_{12} represents the interaction of X_1 on the effect of X_2 as well as the interaction of X_2 on the effect of X_1.

The main effect β_1 may be estimated from the mean of the two partial effects:

$$b_1 = \tfrac{1}{2}(b_1^{(+1)} + b_1^{(-1)}) = \tfrac{1}{4}(-y_1 + y_2 - y_3 + y_4) = -6.4\%$$

The additive linear model 3.1 is inadequate for describing the phenomenon. The model is completed by adding the variable X_1X_2:

$$y = \beta_0 + \beta_1 x_1 + \beta_2 x_2 + \beta_{12} x_1 x_2 + \varepsilon \tag{3.2}$$

The results of table 3.2 are shown graphically in figure 3.2a, with the response surface corresponding to equation 3.2 in figure 3.2b.

4. The model matrix: X

The model 3.2 is rewritten as:

$$y = \beta_0 x_0 + \beta_1 x_1 + \beta_2 x_2 + \beta_{12} x_1 x_2 + \varepsilon$$

where X_0 is a "pseudo-variable" associated with the constant term β_0 and therefore equal to +1. For each of the 4 experiments, we may replace each of the variables by its numerical value (-1 or +1), thus obtaining:

$$
\begin{aligned}
y_1 &= \beta_0 - \beta_1 - \beta_2 + \beta_{12} + \varepsilon_1 \\
y_2 &= \beta_0 + \beta_1 - \beta_2 - \beta_{12} + \varepsilon_2 \\
y_3 &= \beta_0 - \beta_1 + \beta_2 - \beta_{12} + \varepsilon_3 \\
y_4 &= \beta_0 + \beta_1 + \beta_2 + \beta_{12} + \varepsilon_4
\end{aligned}
$$

These may be written in matrix form (see chapter 4 and appendix I):

$$
\begin{pmatrix} y_1 \\ y_2 \\ y_3 \\ y_4 \end{pmatrix}
=
\begin{pmatrix}
+1 & -1 & -1 & +1 \\
+1 & +1 & -1 & -1 \\
+1 & -1 & +1 & -1 \\
+1 & +1 & +1 & +1
\end{pmatrix}
\times
\begin{pmatrix} \beta_0 \\ \beta_1 \\ \beta_2 \\ \beta_{12} \end{pmatrix}
+
\begin{pmatrix} \varepsilon_1 \\ \varepsilon_2 \\ \varepsilon_3 \\ \varepsilon_4 \end{pmatrix}
\tag{3.3}
$$

or

$$\mathbf{Y} = \mathbf{X}\boldsymbol{\beta} + \boldsymbol{\varepsilon}$$

Y is the vector (column matrix) of the experimental response, X is known as the *effects matrix* or *model matrix* (see below), β is the vector of the effects of the variables and ε is the vector of the experimental errors. From now on we will usually represent the model matrix as a table, as we have done for the design, as in table 3.3, below. Here it consists of 4 lines, each corresponding to one of the 4 experiments of the design, and 4 columns, each corresponding to one of the 4 variables of the model. The experimental design is enclosed in double lines.

Table 3.3 Model Matrix, **X**, of a Complete 2^2 Factorial Design Matrix

X_0	X_1	X_2	X_1X_2
+1	-1	-1	+1
+1	+1	-1	-1
+1	-1	+1	-1
+1	+1	+1	+1

In the case of a 2-level factorial design *the different columns of the model (effects) matrix correspond to the linear combinations for calculating the corresponding effects.*

5. Interactions described graphically

We show here one way to visualize a first-order interaction (between 2 factors). The 4 experiments in figure 3.2 are *projected* onto the (X_1, y) plane in figure 3.3. The points 1 and 2 corresponding to $X_2 = -1$ (spheronizer speed = 700 rpm) are joined by a line, as are the points 3 and 4 corresponding to $X_2 = -1$ (spheronizer speed = 1100 rpm). The lines are not parallel, thus showing that there is an interaction β_{12}. An equivalent diagram may be drawn in the (X_2, y) plane, showing the effect of changing the spheronizer speed at the two different spheronization times.

This way of representing an interaction is common in the literature. It has the advantage that the effects and the existence of an interaction may be seen at a glance. It has one major drawback: it is the suggestion that an interpolation is possible between the experimental points. For qualitative variables this does not present any danger as there is no meaning to an interpolation between levels "-1" and "+1". For quantitative variables, such an interpolation could be possible, but it is not recommended in a factor study.

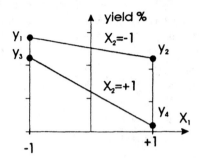

Figure 3.3 Simple graphical representation of an interaction.

B. Full Factorial Design for 3 Factors (2^3) - Formulation of an Oral Solution

1. The synergistic model for more than 2 variables

When the number of factors k exceeds 2 the complete synergistic model contains all the terms corresponding to the interactions of the factors, taken 2 by 2, then 3 by 3, up to k by k. There are 8 terms in the model for $k = 3$:

$$y = \beta_0 + \beta_1 x_1 + \beta_2 x_2 + \beta_3 x_3 + \beta_{12} x_1 x_2 + \beta_{13} x_1 x_3 + \beta_{23} x_2 x_3 + \beta_{123} x_1 x_2 x_3 + \varepsilon$$

and 16 coefficients for $k = 4$:

$$y = \beta_0 + \beta_1 x_1 + \beta_2 x_2 + \beta_3 x_3 + \beta_4 x_4 + \beta_{12} x_1 x_2 + \beta_{13} x_1 x_3 + \beta_{14} x_1 x_4 + \beta_{23} x_2 x_3$$
$$+ \beta_{24} x_2 x_4 + \beta_{34} x_3 x_4 + \beta_{123} x_1 x_2 x_3 + \beta_{124} x_1 x_2 x_4 + \beta_{134} x_1 x_3 x_4 + \beta_{234} x_2 x_3 x_4$$
$$+ \beta_{1234} x_1 x_2 x_3 x_4 + \varepsilon$$

The higher order coefficients may be considered as correction terms for the lower order coefficients:

- β_1 represents the mean change in the response when the variable X_1 changes by 1 unit.
- β_{12} represents the mean change in the effect β_1 when the variable X_2 changes by 1 unit.
- β_{123} represents the mean change in the effect β_{12} when the variable X_3 changes by 1 unit.

We will illustrate this using a study of the influence of 3 factors on the solubility of a drug.

2. Defining the problem – objectives

Senderak, Bonsignore and Mungan (2) have described the formulation of an oral solution of a very slightly water-soluble drug, using a non-ionic surfactant. After a preliminary series of experiments, also carried out using a factorial design, they selected a suitable experimental region and studied the effects of polysorbate 80, propylene glycol and invert sugar concentrations on the cloud point and the turbidity of the solution. They used a *composite* design, which, as we shall see in chapter 5, consists of a factorial design with additional experiments. We *select* the 8 data given in their paper for the runs corresponding to the full 2^3 factorial design. The experimental domain is given in table 3.4.

Table 3.4 Experimental Domain for the Formulation of an Oral Solution

Factor	Associated variable	Lower level (coded -1)	Upper level (coded +1)
Polysorbate 80 (%)	X_1	3.7	4.3
Propylene glycol (%)	X_2	17	23
Sucrose invert medium (%)	X_3	49	61

For simplicity we will refer to the "sucrose invert medium" as sucrose.

3. Experimental design, plan and responses

If each of the 3 factors is fixed at 2 levels, and we do experiments at all possible combinations of those levels, the result is a 2^3 full factorial design, of 8 experiments. The design, plan and results (turbidity measurements only) are listed in table 3.5a, in the standard order.

A design is said to be in the standard order when all variables are at level -1 for the first experiment and the first variable changes level each experiment, the second changes every two experiments, the third every four experiments, etc. Although the design is often *written down* in the standard order, this is *not* usually the order in which the experiments are carried out.

The complete synergistic model is proposed:

$$y = \beta_0 + \beta_1 x_1 + \beta_2 x_2 + \beta_3 x_3 + \beta_{12} x_1 x_2 + \beta_{13} x_1 x_3 + \beta_{23} x_2 x_3 + \beta_{123} x_1 x_2 x_3 + \varepsilon \quad (3.4)$$

We can write the equation for each of the 8 experiments either as before, or in matrix form. The model matrix **X** is as shown in table 3.5b.

Table 3.5 2^3 Full Factorial Design for the Formulation of an Oral Solution
(a) Experimental Design, Plan and Results (Turbidity)
[from reference (2), with permission]

No.	X_1	X_2	X_3	polysorbate 80 (%)	propylene glycol (%)	sucrose invert medium (mL)	turbidity y (ppm)
1	-1	-1	-1	3.7	17	49	3.1
2	+1	-1	-1	4.3	17	49	2.8
3	-1	+1	-1	3.7	23	49	3.9
4	+1	+1	-1	4.3	23	49	3.1
5	-1	-1	+1	3.7	17	61	6.0
6	+1	-1	+1	4.3	17	61	3.4
7	-1	+1	+1	3.7	23	61	3.5
8	+1	+1	+1	4.3	23	61	1.8

(b) Model Matrix **X** for the 2^3 Design and Complete Synergistic Model

No.	X_0	X_1	X_2	X_3	X_1X_2	X_1X_3	X_2X_3	$X_1X_2X_3$
1	+1	-1	-1	-1	+1	+1	+1	-1
2	+1	+1	-1	-1	-1	-1	+1	+1
3	+1	-1	+1	-1	-1	+1	-1	+1
4	+1	+1	+1	-1	+1	-1	-1	-1
5	+1	-1	-1	+1	+1	-1	-1	+1
6	+1	+1	-1	+1	-1	+1	-1	-1
7	+1	-1	+1	+1	-1	-1	+1	-1
8	+1	+1	+1	+1	+1	+1	+1	+1

The column for each of the interaction coefficients is derived from the product of the corresponding columns in the experimental design matrix. Thus the interaction column X_1X_3 (also noted simply as **13**) is obtained by multiplying the elements in the column X_1 by those in the column X_3. A column X_0 representing the constant term in the model is introduced. It consists of +1 elements only.

4. Calculation of the effects

The coefficients in the model equation 3.4 may be estimated as before, as linear combinations or **contrasts** of the experimental results, taking the columns of the effects matrix as described in section III.A.5 of chapter 2. Alternatively, they may be estimated by multi-linear regression (see chapter 4). The latter method is more usual, but in the case of factorial designs both methods are mathematically equivalent.

The calculated effects are listed below and shown graphically in figure 3.4. These data allow us to determine the effects of changing the medium on the response variable (in this case the turbidity).

$b_0 =$ 3.450	$b_1 =$ **-0.675**	$b_{12} =$ 0.050	$b_{123} =$ 0.175
	$b_2 =$ **-0.375**	$b_{13} =$ **-0.400**	
	$b_3 =$ 0.225	$b_{23} =$ **-0.650**	

We see that b_1 and b_{23} are the most important effects, followed by b_2 and b_{13}. Absolute values of the remaining effects are 2 to 3 times less important.

Having carried out as many experiments as there are coefficients in the model equation and not having any independent measurements or estimation of the experimental variance we cannot do any of the standard statistical tests for testing the significance of the coefficients. However all factorial matrices, complete or fractional, have the fundamental property that all coefficients are estimated with equal precision and, like the screening matrices, all the coefficients have the same unit, that of the response variable. This is because they are calculated as contrasts of the experimental response data, and they are coefficients of the dimensionless coded variables X_i. They can therefore be compared directly with one another.

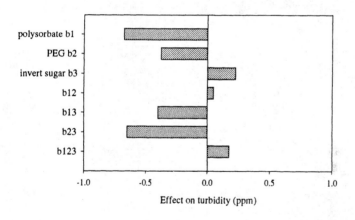

Figure 3.4 2^3 design: main and interaction effects.

But the fact that there are interactions prevents us from interpreting the main effects directly. We showed in paragraph II.A.3 that a factor can no longer be said to have a single effect when it interacts with another factor. It is therefore wrong to conclude simply that increasing the level of surfactant or of propylene glycol leads to a decreased turbidity. The analysis should be carried out directly on the interactions, as shown below, using the first order interaction diagrams.

5. First order interaction diagrams

The factorial design 2^3 may be represented by a cube, the 8 experiments being
situated at each corner (figure 3.5a). To study the interaction β_{23} between the
variables X_2 and X_3, we project the corners of the cube on the (X_2, X_3) plane to give
a square, and calculate the mean value of the responses at each corner of the
square, as shown in figure 3.5b.

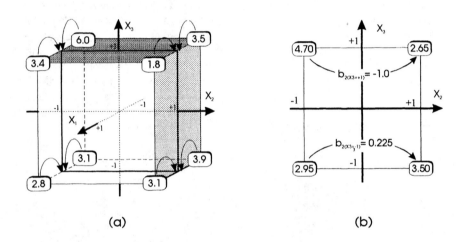

(a)   (b)

Figure 3.5 (a) Projection of a cube to give an interaction diagram.
(b) Interaction b_{23} between propylene glycol (X_2) and sucrose (X_3).

Figure 3.5b shows that when the sucrose is fixed at its lower level (49 mL;
$X_3 = -1$), varying the concentration of propylene glycol has little effect – the mean
response increases from 2.95 to 3.50. On the other hand when the concentration of
sucrose is higher $(X_3 = +1)$ then the effect of increasing the concentration of
propylene glycol is to decrease markedly the turbidity (from 4.70 to 2.65).

It therefore seems that interpreting b_2 alone gives only a mean value of the
effect of X_2, for an untested mean value $X_3 = 0$ of the sucrose invert medium. (This
mean value would not even exist for a qualitative variable.) The interaction of
propylene glycol with the effect of sucrose is still more visible as there is a change
of sign: with 17% propylene glycol, increasing the quantity of sucrose *increases* the
turbidity, whereas with 23% propylene glycol increasing sucrose leads to a slightly
decreased turbidity.

The b_{13} interaction is of less importance, but should be considered
nevertheless (figure 3.6). Changing the sucrose concentration has much less effect

on the turbidity when polysorbate 80 is at 4.3% than when it is at the low level (3.7%).

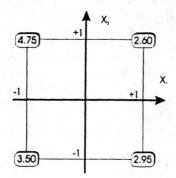

Figure 3.6 Interaction diagram (b_{13}) for polysorbate 80 (X_1) and sucrose (X_3).

This analysis enables us to identify active main interaction effects. It would be unwise to try to use equation 3.4 to predict values of the turbidity or the solubility within the experimental domain and expect to obtain reliable predictions, at least without doing further experiments to test the reliability of the model.

One way of testing the model is to perform experiments at the centre of the domain. We consider here the results of 3 replicates (polysorbate 80 = 4.0%, propylene glycol = 20%, sucrose invert medium = 55% by volume). A mean value of 3.0 ppm is obtained for the turbidity. This is slightly different from the mean value of the factorial experiment data, but further analysis is necessary before deciding whether or not the model is sufficient for predictive purposes in the experimental region. This problem is therefore developed further in chapter 5.

C. Full Factorial Design for 4 Factors (2^4)– a Mixture Example

1. Defining the problem – objectives

Vérain and coworkers (3) described the formulation of an effervescent paracetamol tablet dosed at 500 mg, containing saccharose and sorbitol as diluents. Other components were anhydrous citric acid, sodium or potassium bicarbonate, PVP, and sodium benzoate. The tablets were characterised by measurement of a number of responses, in particular the friability, the volume of carbon dioxide produced per tablet when it is put in water, and the time over which the tablet effervesced. The objective was to study the effects of 4 factors, the quantities of sorbitol and of citric acid per tablet, the nature of the bicarbonate (whether sodium or potassium bicarbonate), and the effect of different tableting forces on these responses. The total tablet mass was held constant by adjusting the proportion of saccharose (as "filler"). Other component levels (paracetamol, PVP, sodium benzoate) were held constant.

This is, strictly speaking, a *mixture problem* as the responses depend on the relative proportions of each of the constituents. Unlike the classical problems studied up to now, where each factor may be varied independently of all others, here the sum of all 8 constituents must add up to unity. It follows that the designs we have used up until now cannot normally be applied to such a case. We note that there is also a process variable (the degree of compression).

Two chapters at the end of the book are devoted to the mixture problem, and there we will also treat problems combining mixture and process variables. Here we will show how a mixture problem can in certain circumstances be treated in the same way as the general case for independent variables and factorial designs be used.

2. Experimental domain

The authors first fixed the quantities (and thus the proportions) of three of the constituents, paracetamol (500 mg, that is 15.5%), PVP (1 mg, 0.03%) and sodium benzoate (90 mg, 2.8%). The total tablet mass was 3.22 g. There were 5 *variable* components, making up 82.42% of the tablet mass.

In fact, sodium and potassium bicarbonate are not 2 separate constituents, as the authors did not intend making tablets containing both salts at the same time. There was therefore only a single constituent, "bicarbonate". Its *nature*, whether sodium or potassium salt is therefore considered as an independent *qualitative* variable.

The amount of bicarbonate was maintained at a constant molar ratio – 3 moles of bicarbonate to one mole of anhydrous citric acid in the mixture. So citric acid and bicarbonate together can be taken as a single component.

The problem therefore becomes one of a mixture of 3 constituents that can each theoretically vary between 0 and 82.42% of the tablet mass: the saccharose, citric acid plus bicarbonate, and the sorbitol. The experimental domain may be represented by the ternary diagram of figure 3.7a.

More realistically there are restrictions on the quantities or relative amounts of the different excipients. If we were to fix lower limits for sorbitol at 100 mg (3%), for the citric acid/sodium bicarbonate mixture at 46.6% (slightly greater when the bicarbonate is potassium) and for the saccharose 300 mg (9%) the resulting reduced experimental domain would be that shown in figure 3.7b. The experimental domain remains triangular and it would be possible to treat this problem using standard mixture designs (see chapter 8). The simplest design would involve experiments at each of the vertices of the triangular zone of interest.

In addition to these lower limits, an upper limit of 300 mg per tablet (9.3%) was imposed on the sorbitol, and an upper limit of 65.2% for the citric acid/sodium bicarbonate mixture. The unshaded zone in figure 3.7c shows the resulting experimental domain.

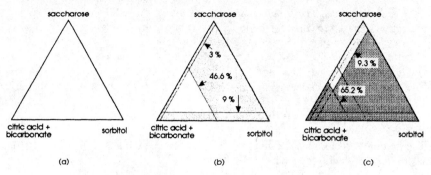

Figure 3.7 Factor space for paracetamol effervescent tablet.

Table 3.6 Formulation of an Effervescent Paracetamol Tablet

Substances	Dose/tablet	% composition
paracetamol	500 mg	15.5%
PVP	1 mg	0.03%
sodium benzoate	90 mg	2.8%
sorbitol	100 - 300 mg	81.67%
anhydrous citric acid	710 - 996 mg	
sodium or potassium bicarbonate	3 moles for 1 mole of citric acid	(total of 4 components)
saccharose	QS 3220 mg	

Table 3.7 Experimental Domain for the Formulation of an Effervescent Tablet

Factor	Associated variable	Lower level (coded -1)	Upper level (coded +1)
sorbitol (mg/tablet)	X_1	100	300
anhydrous citric acid (mmol/tablet)	X_2	3.38	4.74
nature of the bicarbonate	X_3	sodium	potassium
compression force (kg/cm^2)	X_4	775	1150

The proportion of saccharose was allowed to vary, so that the tablets would be of constant mass. The design space thus becomes a parallelogram, each side of the parallelogram representing the replacement of saccharose by another excipient. As the saccharose is a filler, completing the formulation to 100%, the proportions of sorbitol and the citric acid/bicarbonate mixture may be set independently of one another at any selected value, and may thus be considered as independent variables. The tablet formula is summarised in table 3.6, and the levels are given in table 3.7.

3. Experimental design and plan and experimental responses

The complete factorial 2^4 design and the plan for experimentation are shown in table 3.8, along with the experimental results.

4. Mathematical model

The (complete factorial) design allows the calculation of all the main and interaction effects of the 4 variables. We therefore postulate the model:

$$y = \beta_0 + \beta_1 x_1 + \beta_2 x_2 + \beta_3 x_3 + \beta_4 x_4 + \beta_{12} x_1 x_2 + \beta_{13} x_1 x_3 + \beta_{14} x_1 x_4 +$$
$$\beta_{23} x_2 x_3 + \beta_{24} x_2 x_4 + \beta_{34} x_3 x_4 + \beta_{123} x_1 x_2 x_3 + \beta_{124} x_1 x_2 x_4 +$$
$$\beta_{134} x_1 x_3 x_4 + \beta_{234} x_2 x_3 x_4 + \beta_{1234} x_1 x_2 x_3 x_4 + \varepsilon \qquad (3.5)$$

5. Calculation of the effects

As before, the coefficients may be estimated either by linear combinations (contrasts) corresponding to the 16 columns of the model matrix, or by multi-linear regression. These estimates are listed in table 3.9 and plotted in figure 3.8.

As in the previous example, we have estimated as many coefficients as there are experiments, so we cannot calculate the statistical significance of the coefficients by the standard statistical tests. But we have at our disposal certain techniques, outlined in the following section, that can help us choose what effects to analyse and retain for further study.

We first compare numerically the values obtained, as we did for the previous example. Effects which *appear* to be significant are shown in bold type in table 3.9. We will not use the term (*statistically*) *significant* but describe rather such effects as *active*. The responses will be examined in turn.

Friability: The nature of the bicarbonate (b_3) and the compression force (b_4) have effects with similar orders of magnitude. The main effect of the citric acid (b_2) and the interaction b_{24} (citric acid × compression force) are 3 or 4 times smaller. The largest interaction b_{23} also appears to be inactive.

Effervescence time: The level of sorbitol has no effect, but the other three main effects b_2, b_3, b_4, those corresponding to the citric acid level, the kind of bicarbonate used, and the compression force applied all seem to be important.

Table 3.8 2^4 Factorial Design for the Formulation of an Effervescent Tablet [adapted from reference (3) with permission]

	X_1	X_2	X_3	X_4	Sorbitol (mg)	Citric acid (mmol)	Bicarbonate	Compression (kg.cm^{-2})	Friability (%)	Effervesc. time (s)	Volume of CO_2 (mL)
1	-	-	-	-	100	3.38	sodium	775	1.50	126	195
2	+	-	-	-	300	3.38	sodium	775	1.35	132	197
3	-	+	-	-	100	4.74	sodium	775	1.63	98	280
4	+	+	-	-	300	4.74	sodium	775	1.10	93	276
5	-	-	+	-	100	3.38	potassium	775	1.90	82	196
6	+	-	+	-	300	3.38	potassium	775	1.89	84	199
7	-	+	+	-	100	4.74	potassium	775	2.50	70	280
8	+	+	+	-	300	4.74	potassium	775	2.05	75	280
9	-	-	-	+	100	3.38	sodium	1150	0.54	145	199
10	+	-	-	+	300	3.38	sodium	1150	0.30	159	200
11	-	+	-	+	100	4.74	sodium	1150	0.38	115	283
12	+	+	-	+	300	4.74	sodium	1150	0.87	119	281
13	-	-	+	+	100	3.38	potassium	1150	1.26	112	205
14	+	-	+	+	300	3.38	potassium	1150	1.14	113	195
15	-	+	+	+	100	4.74	potassium	1150	2.00	85	285
16	+	+	+	+	300	4.74	potassium	1150	1.67	90	283

Table 3.9 Effervescent Tablet: Estimates of the Model Coefficients

		Friability (%)	Effervescence time (s)	Volume CO_2 (mL/tablet)
constant	b_0	1.38	106	239.6
sorbitol	b_1	-0.08	2	-0.8
citric acid	b_2	0.14	**-13**	**41.4**
nature of bicarbonate	b_3	**0.42**	**-17**	0.8
compression	b_4	**-0.36**	**11**	1.8
(first-order interactions)	b_{12}	-0.02	-1	-0.3
	b_{13}	-0.03	0	-0.4
	b_{14}	0.06	1	-0.9
	b_{23}	0.11	**4**	0.3
	b_{24}	0.06	-2	0.3
	b_{34}	0.08	0	-0.1
(second-order interactions)	b_{123}	-0.06	2	0.9
	b_{124}	0.08	0	0.9
	b_{134}	-0.06	-1	-1.0
	b_{234}	0.00	-2	0.1
(general interaction)	b_{1234}	-0.06	0	0.5

It is possible that the interaction b_{23} (= 4) is also active. All other effects are much less important and therefore probably inactive.

Volume of carbon dioxide evolved: It is clear that only the amount of citric acid (and bicarbonate as these are in a fixed molar ratio) has an effect on this response.

The concentration of sorbitol has very little effect on any of the responses. We may select levels for the remaining variables, that will give an *improved*, but not optimized, formulation. This could in some cases mean a compromise choice, as an increase in the level of a factor may lead to an improvement in one response but have a disadvantageous effect on another one.

D. Identifying Active Factors in Saturated Designs

Of the three responses measured in this design, it is the interpretation of the effervescence time that seems to be the least clear. We therefore use it to illustrate the various methods for identifying active factors.

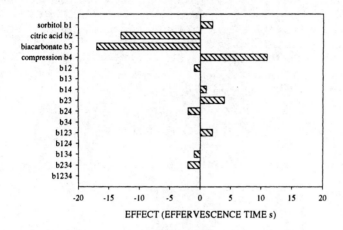

Figure 3.8 Estimated coefficients for effervescent paracetamol tablet.

When the number of experiments exceeds the number of effects calculated, or when the experimental variance can be estimated independently (for example by repeating certain experiments) the significance of the coefficients may be estimated and the analysis of variance of the regression may be carried out. The methods that are shown below are for saturated or near saturated designs, where the experiments are too costly or too long to carry out and the main criterion in choosing a design is its R-efficiency.

The use of these methods assumes that all coefficients are estimated with the same precision. Now this is true for the designs with variables at 2 levels, discussed up to this point: Plackett-Burman designs, factorial designs, as well as those discussed later in this chapter, fractional factorial designs, Rechtschaffner, and ¾ designs. It is not so for many other experimental designs, described from chapter 5 onwards. These methods can only be applied in such cases after normalising the coefficients, dividing each by a factor related to its standard deviation. This is the square root of the corresponding diagonal element c^{ii} in the dispersion matrix (chapter 4). We will see in the section on linear regression in chapter 4 how the standard deviation of the coefficients may be obtained.

1. Normal and half-normal plots

It is to be expected that the large effects, whether positive or negative, are the statistically significant ones. If none of the effects was active it would be expected that they would be all normally distributed about zero. A cumulative plot on "probability paper", or "probit analysis", would give an approximately straight line.

Let the effects, sorted in increasing absolute magnitude, and omitting the constant term, be E_i. (The constant term is omitted in the case of all methods described in this section.) There are therefore 15 effects. The total probability being unity, the probability interval associated with each effect is $1/15 = 0.067$. We associate the first, smallest effect with a cumulative probability of half that amount, 0.033. (Since each datum defines the probability interval, it will be shared between two intervals.) Then for the next smallest, the estimated probability of an effect being less or equal to its value E_2 will be $0.033 + 0.067 = 0.1$, for the next smallest, E_3, the probability will be $0.1 + 0.067$, and so on, as shown in table 3.10.

Table 3.10 Coefficients in Order of Increasing Absolute Magnitude

Coefficient		Value	Cumulative probability of absolute value
E_1	b_{1234}	0.000	0.033
E_2	b_{34}	0.000	0.100
E_3	b_{124}	0.125	0.167
E_4	b_{13}	-0.375	0.233
E_5	b_{12}	-0.875	0.300
E_6	b_{14}	1.000	0.367
E_7	b_{134}	-1.125	0.433
E_8	b_{234}	-1.625	0.500
E_9	b_{123}	1.750	0.567
E_{10}	b_{24}	-2.000	0.633
E_{11}	b_1	2.000	0.700
E_{12}	b_{23}	4.125	0.767
E_{13}	b_4	11.125	0.833
E_{14}	b_2	-13.000	0.900
E_{15}	b_3	-17.250	0.967

The absolute values of the coefficients are then plotted on a linear scale and the cumulative probabilities on a "probability scale". This is known as a *half-normal plot* (4). The result is shown in figure 3.9. Most of the points appear to lie on a straight line but the last 4 points deviate from it. This suggests that:

- The effect of the sorbitol b_1 is probably not active.
- The other main effects are active, as is the interaction b_{23} between the amount of citric acid and the type of bicarbonate used.
- All other interactions may be ignored.

An approximate value of the standard deviation can also be obtained. It is read off on the straight line as the absolute value of a coefficient corresponding to 0.72 probability. In this way we obtain a standard deviation of 2.6 s for the time over which a tablet effervesces.

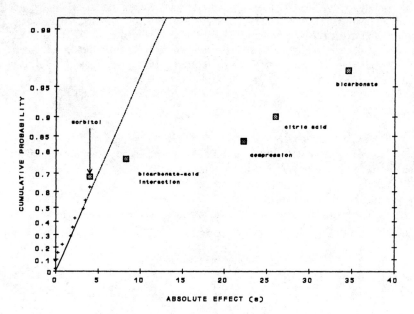

Figure 3.9 Half-normal plot of coefficients (paracetamol effervescent tablet).

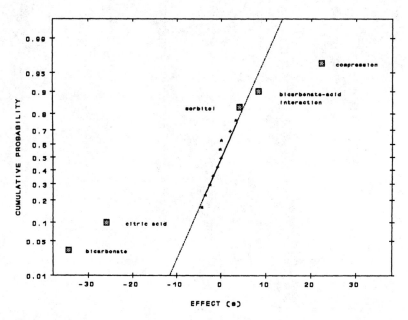

Figure 3.10 Normal plot of coefficients (paracetamol effervescent tablet).

An alternative approach is the *normal plot*, where the coefficients are arranged in the order of their actual algebraic values rather than their absolute ones. They are plotted on a 2-way probability scale. This is shown in figure 3.10. Most of the points lie on a straight line passing close to zero on the ordinate scale and 50% probability on the coordinate. The points for b_4 and b_{23} deviate at the positive end of the graph and those for b_2 and b_3 at the negative end.

2. Lenth's method (6)

As for the previous method the effects (except for the constant term) are set out in order of increasing absolute values. An initial estimate s_b of the standard deviation of the coefficients is obtained by multiplying the median of the absolute values coefficients by 1.5. Any coefficient whose absolute value exceeds 2.5 s_b is eliminated from the list. The process is then repeated until no further effect is eliminated. Then if p_r is the number of coefficients remaining in the list at the end, the signification limit L may be calculated as:

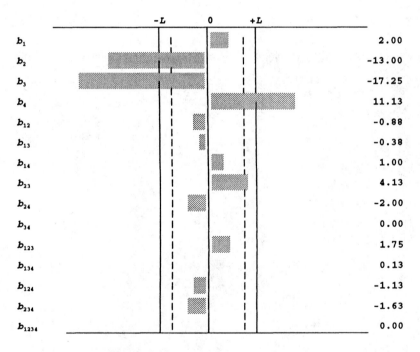

Figure 3.11 Analysis by Lenth's method.

$$L = t_{(0.025,\,v)} \times s_b$$

where $v = (p_r/3)_{round}$ and $t_{(0.025,\,v)}$ is the value of Student's t for a probability 0.025 and v degrees of freedom.

The list of coefficients in increasing absolute value is given in table 3.10. The median value is 1.625, corresponding to the 8th out of the 15 coefficients. Thus the first estimate of s_b is $s_b^{(1)} = 1.625 \times 1.5 = 2.437$. Coefficients with absolute values greater than $2.5 \times 2.437 = 6.094$ are eliminated from the list – that is b_3, b_2 and b_4.

The median of the new list of coefficients is now the mid-point of the 6th and 7th coefficients. Thus the new estimate of the pseudo-standard deviation is $s_b^{(2)} = \frac{1}{2}(1.000 + 1.125) \times 1.5 = 1.594$. Coefficients greater than $2.5 \times 1.594 = 3.984$ are eliminated from the list. The only one is $b_{23} = 4$. At the third stage, the median value is 1.000, $s_b^{(3)} = 1.5$ and the new limit is 3.75. There are no more coefficients to leave out. There remain $p_r = 11$ coefficients, and $v = (11/3)_{round} = 4$. A value of $t = 2.78$ is taken from a table of Student's t.

The significance limit is therefore $L = 2.78 \times 1.5 = 4.2$ seconds. It corresponds to the outer limits in figure 3.11. Any coefficient exceeding this limit is considered *active*. Another confidence interval may be calculated by a similar method – corresponding to the inner limit of figure 3.11. All coefficients not exceeding this limit are considered as *inactive*. Coefficients between the two limits, as is the case for b_{23}, may be considered either as active or inactive.

3. Pareto charts

The effects may be represented graphically in the form of a Pareto chart (4). Three kinds of representations are possible:
- A Pareto chart of the absolute value of each effect (figure 3.12a).
- A Pareto chart of the normalised square of each effect l_i, (figure 3.12b).
- A chart of the cumulative function c_i of l_i (figure 3.12c).

The functions l_i and c_i are defined as follows:

$$l_i = 100 \times \frac{E_i^2}{\sum\limits_{j=2}^{p} E_j^2} \qquad\qquad c_i = \sum\limits_{j=2}^{i} l_j$$

where E_i are defined in table 3.10, so that here also i is such that the l_i are ordered in increasing values and p is the number of coefficients. The 3 first coefficients add up together to nearly 95% of the total of the sums of squares of the coefficients. According to these methods the 4th coefficient, b_{23}, does not seem to be active.

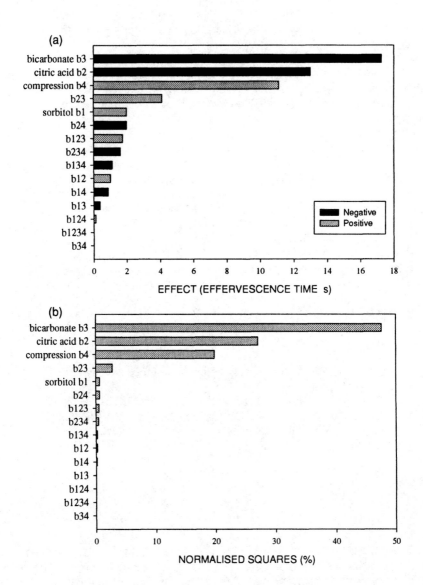

Figure 3.12   Pareto charts for paracetamol effervescent tablet data.
(a) Absolute values of the coefficient estimates, **(b)** Normalised squares of the coefficient estimates.

(c)

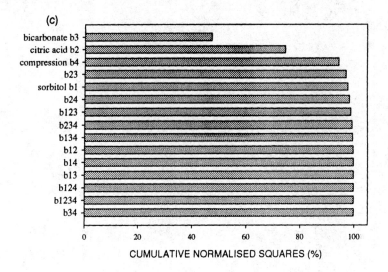

Figure 3.12c   Pareto chart: cumulative plot of normalised squares.

4. Bayesian analysis of the coefficients

This is an *a posteriori* calculation of the probability that each of the effects is active (7). It is based on the following hypotheses.

- If one studies a large number of effects, only some of them are active (significant). Let α be the probability *a priori* that a given effect is active. α is usually chosen to be between 0.1 and 0.45; that is we suppose that between 10% and 45% of the effects are active.
- Inactive effects are assumed to be normally distributed, with mean value zero, and constant variance.
- Active effects are also assumed to be normally distributed, but a greater variance. Let k be the ratio of the variance of the active effects to that of the inactive ones. k is assumed to lie between 5 and 15.

The probabilities of each effect are calculated *a posteriori* for different values of k and α, and the maximum and minimum probabilities are plotted, as solid boxes, as shown in figure 3.13, for the effervescence time. The results are in agreement with the other methods, in that the main effects b_2, b_3, b_4 are active, with minimum probability 100%. The maximum probability of no effect being active (column N.S.) is close to zero. The status of the b_{23} interaction is less clear as the minimum probability is only 24%. Nevertheless we would probably consider that the interaction between the level of citric acid + bicarbonate and the nature of the bicarbonate salt should be taken into account.

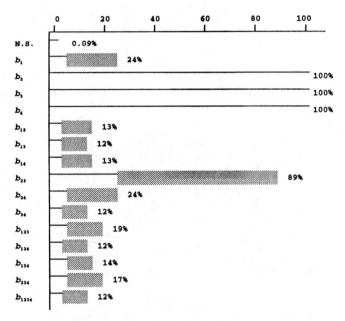

Figure 3.13   Bayesian analysis of coefficients (paracetamol tablet data).

At the end of the above study a suitable simplified model might be proposed for the 3 responses:

$$y = \beta_0 + \beta_2 x_2 + \beta_3 x_3 + \beta_4 x_4 + \beta_{23} x_2 x_3 + \varepsilon$$

Estimates of the statistical significance of the coefficients can and frequently should be obtained by other means – in particular by replicated experiments (usually centre points) followed by multi-linear regression of the data, and analysis of variance, as developed in chapter 4. The methods we have described above are complementary to those statistical methods and are especially useful for saturated designs of 12 to 16 or more experiments. For designs of only 8 experiments, the results of these analyses should be examined with caution.

E. General Forms of Full Factorial Designs and their Mathematical Models

1. The synergistic model

The complete synergistic model for k factors contains all the terms corresponding to the interactions of the factors, taken 2 by 2, then 3 by 3, up to k by k. We have, therefore:

$$y = \beta_0 + \beta_1 x_1 + ... + \beta_k x_k + \beta_{12} x_1 x_2 + ... + \beta_{k-1,k} x_{k-1} x_k + ... + \beta_{12..k} x_1 x_2 ... x_k + \varepsilon \quad (3.6)$$

The coefficient β_0 is the *constant term*. The coefficient β_1 is called the main effect of the variable (or the factor) X_1. The coefficient β_{12} is a *2-factor interaction* or a *first-order interaction* effect. The coefficient β_{123} is called a *3-factor interaction* or a *second-order interaction* effect. The coefficient $\beta_{12..k}$ is called the *general interaction* effect. Thus β_{123} is the general interaction effect for a 3 factor model and β_{1234} is the general interaction effect for a 4 factor model.

A complete synergistic model consists of:

1 constant term β_0,

k main effects β_i,

$k(k-1)/2!$ first-order interaction effects β_{ij},

$k(k-1)(k-2)/3!$ second-order interaction effects β_{ijk},

...

$k!/(d!(k-d)!)$ interaction effects between d factors.

...

1 general interaction effect between all k factors $\beta_{12..k}$.

The full model for k factors contains 2^k coefficients in all.

2. Complete design of k factors at 2 levels: 2^k

A complete factorial design of k factors each at 2 levels (represented by the symbols "-1" and "+1") consists of all possible combinations of these 2 levels, that is 2^k combinations corresponding to $N = 2^k$ distinct experiments. The same symbol 2^k will be used in the remainder of the book for the design itself.

Table 3.11 Complete factorial designs at 2 levels

No.	X_1	X_2	X_3		X_{k-1}	X_k
1	-	-	-	...	-	-
2	+	-	-	...	-	-
3	-	+	-	...	-	-
4	+	+	-	...	-	-
5	-	-	+	...	-	-
...	...	...	...	...	...	...
2^{k-1}	+	+	+	...	+	-
$2^{k-1}+1$	-	-	-	...	-	+
...	...	...	...	...	...	...
2^k-1	-	+	+	...	+	+
2^k	+	+	+	...	+	+

We have seen that a design is in the standard order when all variables are at level -1 for the first experiment, and the first variable changes level each experiment, the second every 2 experiments, the third every 4 experiments etc.

To simplify, we write the levels simply as "-" and "+". The general form of a complete factorial design at 2 levels 2^k is shown in table 3.11 where the different matrices enclosed by dotted lines represent the designs 2^1, 2^2, ..., 2^{k-1}, 2^k.

III. FRACTIONAL FACTORIAL DESIGNS

In the previous section a complete factorial 2^4 design at 2 levels and 4 factors was used to calculate 16 effects. This is a fairly large number of effects, but it required an equally large number of experiments. The experimenter may well hesitate before beginning such a study, as it involves looking at higher order interactions (β_{123}, etc.) when he might not even be totally sure of the influence of the main effect. (This particular consideration is well illustrated by the results for the effervescent paracetamol tablet.)

It is generally true that, except in certain special cases, the probability that a given interaction effect is significant decreases as the order of the interaction increases. Interactions of order 2 and above are usually assumed negligible.

Fractional factorial designs allow us to carry out only a fraction (half, quarter, eighth, etc.) of the full factorial design, obtaining the items of information that are *a priori* the most important and at the same time allowing us to add all or some of the missing experiments later on if the interpretation reveals certain ambiguities. There are several ways of constructing them which give equivalent results, provided they are applied correctly. According to the circumstances, one method may be simpler than another. We will examine two such methods in this chapter.

Fractional factorial designs are especially important when there are 5 or more variables, as the full factorial designs comprise large, or very large, numbers of experiments with frequently many non-significant higher order effects. However, for simplicity we will begin by looking at fractions of the 2^4 full factorial design and only afterwards apply the theory to larger designs.

A. Partition of the (Full) Factorial Design

1. The starting point: the 2^4 full factorial design

We take the full factorial design with the number of variables required and select half the experiments. Each half fraction can in turn be divided into quarter fractions. Whereas for a given number of variables at 2 levels there is only one possible full factorial design, there can be many possible fractional factorial designs and they are not at all equivalent to one another. We saw in paragraph II.C.3 that the 2^4 design consists of 16 experiments representing all possible combinations of the levels "-1" and "+1". For clarity, and because there is no possible ambiguity, we will identify the variables X_1, X_2, X_3, and X_4 as **1**, **2**, **3**, and **4**. The standard order design is enclosed in double lines in table 3.8.

Table 3.12 shows the model (effects) matrix for the full synergistic equation 3.5, and the 2^4 design of table 3.8. Column X_0 represents the pseudo-variable associated with the constant term β_0 in the synergistic mathematical model. Columns **12**, **13**, ... , **1234** correspond to the variables X_1X_2, X_1X_3, ..., $X_1X_2X_3X_4$, which are the variables associated with the interaction effects β_{12}, β_{13}, ..., β_{1234} in the model equation 3.5. The contrasts (linear combinations of the experimental values of the response y_i) corresponding to each column allow the associated effect to be calculated (after dividing by the number of experiments $N = 16$).

2. Partition of the design: the effect of a bad choice

We split the full factorial design in two, in the simplest possible way, and observe what information can be obtained and what information is lost in the process. Consider what the situation would be if only the second half of the design, had been carried out. The reader may examine the resulting design and model matrix, that is experiments 9 to 16, by covering up rows 1-8 of table 3.12.

We first of all see that column **4** (factor X_4) is now set at the level "+1" for all experiments. It is thus impossible to obtain b_4, the estimate of β_4. On the contrary, the mean value for the 8 experiments is an estimate of the sum of the constant term and of the effect β_4, because the effect of X_4, if it exists, is the same in all 8 experiments.

$$E(b_0) = \beta_0 + \beta_4$$

Therefore, this partition is not useful, as only 3 of the main effects can be estimated. Also, we see that columns **1** and **14** are identical so that when the variable X_1 is at level -1, the variable X_1X_4 equals -1. It is impossible to detect the individual effects of these variables. We measure only their sum. We can therefore write:

$$b_1 = (-y_9 + y_{10} - y_{11} + y_{12} - y_{13} + y_{14} - y_{15} + y_{16})/8$$
and
$$E(b_1) = \beta_1 + \beta_{14}$$

Columns **2** and **24** are also identical and so are **3** and **34**. Estimates of these variables are therefore biased. Using the same reasoning as above we obtain for all of the effects, confounded amongst themselves:

$$
\begin{aligned}
E(b_0) &= \beta_0 + \beta_4 & \qquad E(b_{12}) &= \beta_{12} + \beta_{124} \\
E(b_1) &= \beta_1 + \beta_{14} & \qquad E(b_{13}) &= \beta_{13} + \beta_{134} \\
E(b_2) &= \beta_2 + \beta_{24} & \qquad E(b_{23}) &= \beta_{23} + \beta_{234} \\
E(b_3) &= \beta_3 + \beta_{34} & \qquad E(b_{123}) &= \beta_{123} + \beta_{1234}
\end{aligned}
$$

To summarise, the 16 experiments were thus separated into 2 blocks of 8, according to the sign of column **4**. As we saw in this example the properties of the resulting matrix are a function of the choice of column for the partition. We must choose the column for partitioning according to our previously fixed objectives, the properties that we are looking for in the design, and the effects that we are seeking to estimate. It is clear that one would not normally partition on a column that corresponds to a main effect, as we have done here.

3. A good partitioning of the 2^4 design

It is the choice of the column for the partition that determines the properties of the fractional factorial design, so the column must be chosen according to the objectives. These may be taken to be the *calculation of the constant term and the 4 main effects without them being confounded with any first-order (2-factor) interactions*.

Only 8 independent effects may be estimated from the data for 8 experiments. This leaves only 3 terms, apart from the main effects and constant term. The model contains a total of 6 first order interactions, so these may not be estimated independently. In general, the best way to obtain a half fraction of the design is to partition the design on the last column of the model matrix, the one which corresponds to the highest order interaction that is possible (**1234** in this case). The resulting design, and its associated model (effects) matrix, is shown in table 3.13.

As there are 16 effects but only 8 experiments, we would expect the effects to be confounded with each other, as these are only 8 independent combinations of terms. Thus the columns **1** and **234** are the same; so are **12** and **34**. This shows us that when we are estimating β_1 we in fact obtain an estimate of $\beta_1 + \beta_{234}$. The effects are confounded or aliased as follows:

$$E(b_1) = \beta_1 + \beta_{234} \qquad\qquad E(b_0) = \beta_0 + \beta_{1234}$$
$$E(b_2) = \beta_2 + \beta_{134} \qquad\qquad E(b_{12}) = \beta_{12} + \beta_{34}$$
$$E(b_3) = \beta_3 + \beta_{124} \qquad\qquad E(b_{13}) = \beta_{13} + \beta_{24}$$
$$E(b_4) = \beta_4 + \beta_{123} \qquad\qquad E(b_{23}) = \beta_{23} + \beta_{14}$$

Let those linear combinations, or *contrasts*, that allow coefficients involving interactions between 2 factors to be calculated, be written as l_{12}, l_{13}, l_{23}. These are the last three.

As an approximation we might neglect all second-order interactions (interactions between 3 factors). In this case we obtain estimates of the main effects. The hypothesis:

$$\beta_{1234} = 0 \quad \text{leads to} \quad E(b_0) = \beta_0$$
$$\beta_{234} = 0 \qquad\qquad\quad E(b_1) = \beta_1 \quad \text{etc.}$$

But we cannot make any equivalent assumptions about the second-order terms, confounded 2 by 2. There is in the general case no reason at all for eliminating one rather than another. We wrote "E(b_{12})", "E(b_{13})", "E(b_{23})", but this is not really correct and is certainly misleading. It might well be better to write:

$$E(b_{12+34}) = \beta_{12} + \beta_{34}$$

as better indicating our state of incertitude. This is the nomenclature we will use in much of this chapter.

An unambiguous interpretation of the linear combinations l_{12}, l_{13}, l_{23} is impossible. If we find the values of any of these to be significant, we have three choices.

(a) Carry out the remaining 8 experiments in the full design.
(b) We may assume that only those interactions are active, where one or both main effects are large.
(c) We may use previous practical or theoretical knowledge of the system to guess which of the two aliased interactions is important.

Consider a problem of drug excipient compatibility (see chapter 2) where we wish to test a new drug formulated in a hard gelatine capsule. The aim is to estimate the effect of changing the concentration of the drug substance in the mixture (X_1), the effect of wet granulation (X_2), the effect of changing diluents (X_3) and the effect of changing the lubricant (X_4). All these variables are qualitative. They are allowed to take 2 levels, -1 and +1. This problem may be studied using the design shown in table 3.13.

Experience tells us that of these factors, the effect of granulation is often the most important. In cases of instability, it is the granulated mixtures which tend to be less stable. It may well happen that all the non-granulated mixtures are relatively stable, without great differences between them, whereas the granulated mixtures may show considerable differences due to the excipients present. We thus expect the coefficients of the interaction terms involving the granulation, β_{12}, β_{23}, β_{24}, to be active. The remaining interactions can be ignored.

If we were to see that the main effect of granulation was small compared with the effects of changing diluent and lubricant we would probably revise our opinion.

4. The concept of a generator

The effect used to partition the design is called the **generator**. The above fractional factorial design was got by taking those experiments for which the element in column **1234** is "+1". Its generator is thus $G = +1234$ and the design obtained by choosing experiments with element "-1" has the generator $G = -1234$.

A half factorial design has a single generator.

Table 3.12 Model Matrix for a 2^4 Factorial Design

No	I	1	2	3	4	12	13	14	23	24	34	123	124	134	234	1234
1	+	-	-	-	-	+	+	+	+	+	+	-	-	-	-	+
2	+	+	-	-	-	-	-	-	+	+	+	+	+	+	-	-
3	+	-	+	-	-	-	+	+	-	-	+	+	+	-	+	-
4	+	+	+	-	-	+	-	-	-	-	+	-	-	+	+	+
5	+	-	-	+	-	+	-	+	-	+	-	+	-	+	+	-
6	+	+	-	+	-	-	+	-	-	+	-	-	+	-	+	+
7	+	-	+	+	-	-	-	+	+	-	-	-	+	+	-	+
8	+	+	+	+	-	+	+	-	+	-	-	+	-	-	-	-
9	+	-	-	-	+	+	+	-	+	-	-	-	+	+	+	-
10	+	+	-	-	+	-	-	+	+	-	-	+	-	-	+	+
11	+	-	+	-	+	-	+	-	-	+	-	+	-	+	-	+
12	+	+	+	-	+	+	-	+	-	+	-	-	+	-	-	-
13	+	-	-	+	+	+	-	-	-	-	+	+	+	-	-	+
14	+	+	-	+	+	-	+	+	-	-	+	-	-	+	-	-
15	+	-	+	+	+	-	-	-	+	+	+	-	-	-	+	-
16	+	+	+	+	+	+	+	+	+	+	+	+	+	+	+	+

Matrix of the 2^{4-1} Factorial Design on the 1234 Column

No.	I	1	2	3	4	12	13	14	23	24	34	123	124	134	234	1234
1	+	-	-	-	-	+	+	+	+	+	+	-	-	-	-	+
4	+	+	+	-	-	+	-	-	-	-	+	-	-	+	+	+
6	+	+	-	+	-	-	+	-	-	+	-	-	+	-	+	+
7	+	-	+	+	-	-	-	+	+	-	-	-	+	+	-	+
10	+	+	-	-	+	-	-	+	+	-	-	+	-	-	+	+
11	+	-	+	-	+	-	+	-	-	+	-	+	-	+	-	+
13	+	-	-	+	+	+	-	-	-	-	+	+	+	-	-	+
16	+	+	+	+	+	+	+	+	+	+	+	+	+	+	+	+

Table 3.14 2^{4-1} Design for Effervescent Tablet [data taken from reference (3), with permission]

	X_1	X_2	X_3	X_4	Sorbitol (mg)	Citric acid (mmol)	Bicarbonate	Compression (kg/cm^2)	Friability (%)	Effervescence time (s)	Volume of CO$_2$ (mL)
1	-	-	-	-	100	3.38	sodium	775	1.50	126	195
4	+	+	-	-	300	4.74	sodium	775	1.10	93	276
6	+	-	+	-	300	3.38	potassium	775	1.89	84	199
7	-	+	+	-	100	4.74	potassium	775	2.50	70	280
10	+	-	-	+	300	3.38	sodium	1150	0.30	159	200
11	-	+	-	+	100	4.74	sodium	1150	0.38	115	283
13	-	-	+	+	100	3.38	potassium	1150	1.26	112	205
16	+	+	+	+	300	4.74	potassium	1150	1.67	90	283

5. Notation for a half factorial design

The fractional factorial design of table 3.13 is a half fraction of a complete design. The number of experiments N is defined by:

$$N = 2^4/2 = 2^{4-1}$$

This notation 2^{4-1} for the half factorial design for 4 variables at 2 levels indicates the fractional nature of this design.

It should be noted that the design is a complete factorial design for 3 factors (see the columns for factors X_2, X_3, X_4). We will make use of this fact for the second method of constructing fractional designs.

6. Example of a 2^{4-1} design: paracetamol tablet

As an example we will again use the data for the effervescent tablet of section II.C. These were for a complete 2^4 design. We take the half fraction partitioned on the column **1234** as described above, (that is with generator **1234**). We thus imagine that only experiments 1, 4, 6, 7, 10, 11, 13 and 16 were carried out - those where all elements of column **1234** are equal to +1. The experimental design, plan, and response data are given in table 3.14. The coefficients may be calculated using the 8 different columns of the model matrix. They are listed in table 3.15

Comparison with the results for the full factorial design (table 3.9) shows similar estimates for the main effects. The conclusions are at least qualitatively the same for the two designs. So 8 experiments are sufficient to obtain the major part of the information.

Table 3.15 Estimation of Effects from the 2^{4-1} Design

		Friability %	Effervescence time (s)	Volume of gas emitted (mL)
Constant	b_0	1.33	106	240
Sorbitol	b_1	-0.09	0	-0.6
Acid	b_2	-0.09	-14	**40.4**
Carbonate	b_3	**0.51**	-17	-1.6
Compression	b_4	**-0.42**	13	2.6
	b_{12+34}	0.06	-1	-0.4
	b_{13+24}	-0.03	2	0.1
	b_{14+23}	0.17	5	-0.6

Interaction effects between the factors are on the whole of little importance, unless 2 confounded interaction effects are equal and opposite. For example, we have seen that "b_{13}" ("b_{13+24}" in table 3.15) is an estimate of the sum of two interactions, $\beta_{13} + \beta_{24}$. So although the interaction effect is small, we cannot totally exclude the possibility that one of the interactions partially cancels out the other.

Without doing any significance testing (which we cannot do as there are no repeated experiments) we can see that the "b_{14+23}" effect for the effervescence time is relatively large, though still smaller than the main effects. This interaction may be attributed either to the sorbitol-compression interaction (β_{14}) or to an interaction between the level of citric acid and the type of carbonate used (β_{23}). The fact that the main effects b_2 and b_3 are large whereas b_1 is negligible leads us to expect that b_{14+23} is an estimate of the citric acid–carbonate interaction β_{23} (in this case and for this response). From a pharmaceutical or physico-chemical point of view we would come to the same conclusion – which is confirmed by the data for the full factorial design given in section II.C.

Therefore, on the basis of 8 experiments, the main effects of all 4 variables can be estimated without interference from interaction effects. The experiments also demonstrate that interaction effects are on the whole small. One could well decide to stop here. On the other hand, if large interaction effects were found, it might be necessary to carry out the remaining half of the design to identify the interactions without ambiguity.

A similar treatment may be used for 5 factors. However, here the half fraction design 2^{5-1} of 16 experiments, partitioned on the highest order interaction **12345**, allows all the main effects *and first-order interactions* to be estimated. They are not confounded with one another, only with higher order interactions. Thus, for example, **12** is aliased with **345**. So it is this half fraction design which is normally used. An example is the study by Dick, Klassen, and Amidon (9), on the influence of formulation and processing conditions on a tableting process. Similarly, Lindberg and Jönsson (10) used the 2^{6-1} design of 32 experiments partitioned on **123456** to investigate the effect of 6 factors on the wet granulation of a lactose-starch mixture.

B. Double Partition of a Complete Factorial Design

1. Quarter fraction of a 2^4 design

When a large number of factors (more than 4) is being studied then the number of experiments in the full factorial design increases so rapidly that such a design may become prohibitively expensive. Even the half-design may be too large, at least for an initial study. One would like to reduce the number of experiments still further, in fractionating the full factorial design yet again. This involves separating the experiments according to the signs symbolising the levels for *two* of the columns in the model matrix.

It is difficult to demonstrate this procedure on a 2^5 design because of the page size available. We will therefore first of all *demonstrate* this approach using the 2^4 design. We again take the **1234** column and also one of the remaining highest order columns, **234**, and select all experiments with the + level in both columns **1234** and **234** {+ ; +}. Otherwise we might have chosen the combinations {- ; -}, {+ ; -} or {- ; +} each of which would have given another group of 4 different experiments, each group with equivalent properties.

The model matrix for the design of 4 experiments is given in table 3.16 Since the partition was carried out by selecting experiments so that the columns **1234** and **234** are at level +1, **1234**, and **234** are generators. We may therefore write:

$$G_1 \equiv 234 \quad \text{and} \quad G_2 \equiv 1234$$

A column which contains only +1 elements is called **I** (*Identity*). This is used to identify the column corresponding to the constant term, in the model matrix. Thus **234** ≡ **1234** ≡ **I**. Since the product of a column with itself contains only +1 elements it is clear that the product of column **1234** and column **234** is column **1**. Column **1** is also a generator.

There are 2 independent generators as there are 2 partitions, and a total of 3 (2^2) generators, as **I** is excluded from their number. *The defining relation* of a fractional factorial design matrix consists of all its generators. For the design in table 3.16 it is:

$$I \equiv 234 \equiv 1234 \equiv 1$$

There are 4 experiments so only 4 effects may be estimated:

$$E(b_0) = \beta_0 + \beta_1 + \beta_{234} + \beta_{1234}$$
$$E(b_2) = \beta_2 + \beta_{12} + \beta_{34} + \beta_{134}$$
$$E(b_3) = \beta_3 + \beta_{13} + \beta_{24} + \beta_{124}$$
$$E(b_4) = \beta_4 + \beta_{14} + \beta_{23} + \beta_{123}$$

The confounding pattern above may be deduced from the defining relation. It shows directly that the effects represented by **1**, **234** and **1234** cannot be distinguished from the constant term. The second line (for b_2) is derived as follows: multiply the defining relation by **2**. We have $2 \times I \equiv 2 \times 234 \equiv 2 \times 1234 \equiv 2 \times 1$. A column in the design when multiplied by itself gives a column of "+1" elements. Thus any column appearing twice in the above expression has no effect on the final result and **22** may be eliminated from the product to give:

$$2 \equiv 34 \equiv 134 \equiv 12$$

which is another way of writing $E[b_2] = \beta_2 + \beta_{12} + \beta_{34} + \beta_{134}$.

This design, written as 2^{4-2}, is a complete factorial (in the standard order) for the two factors X_3 and X_4. Used on its own it is insufficient for estimating the 4 main effects plus the constant term. We will use it later in constructing the ¾ design.

Table 3.16 Double Partition of the 2^4 Factorial Design on the **234** and **1234** Columns

	I	1	2	3	4	12	13	14	23	24	34	123	124	134	234	1234
4	+	+	+	-	-	+	-	-	-	-	+	-	-	+	+	+
6	+	+	-	+	-	-	+	-	-	+	-	-	+	-	+	+
10	+	+	-	-	+	-	-	+	+	-	-	+	-	-	+	+
16	+	+	+	+	+	+	+	+	+	+	+	+	+	+	+	+

2. Quarter fraction of a 2^5 design

We take the 2^5 complete design of 32 experiments. The model consists of a constant term, 5 main effects, 10 first-order interactions, 10 second-order interactions etc. The design normally used is the half fraction design (16 experiments) partitioned on the highest order interaction **12345**, allowing estimation all the main and first-order interaction effects.

Although this $2^{5 \cdot 1}$ fractional factorial design is of good quality, it is not suitable for further partitioning. If we want to continue fractionating the design we would normally consider an alternative partition, for reasons which will be explained later, in section D.2. We first partition on the **1245** column - that is we divide the matrix in half - on the one hand the experiments where the column **1245** in the model matrix is +1, and on the other hand where it is -1. We then divide them again according to whether the column **234** is equal to +1 or -1. These columns **234** and **1245** are the *independent generators*. The product of these (**234** × **1245**) is also a generator, **135**.

The design is split into 4 blocks, each of 8 experiments, as is shown diagrammatically in figure 3.14. There are thus 4 generators in all, which make up the defining relation of each of the four blocks. These defining relations are:

Block 1: $\quad$ I $\equiv$ 1245 $\equiv$ 234 $\equiv$ 135	I $\equiv$ G_1 $\equiv$ G_2 $\equiv$ $G_1 G_2$
Block 2: $\quad$ I $\equiv$ 1245 $\equiv$ -234 $\equiv$ -135	I $\equiv$ G_1 $\equiv$ $-G_2$ $\equiv$ $-G_1 G_2$
Block 3: $\quad$ I $\equiv$ -1245 $\equiv$ 234 $\equiv$ -135 $\quad$ or	I $\equiv$ $-G_1$ $\equiv$ G_2 $\equiv$ $-G_1 G_2$
Block 4: $\quad$ I $\equiv$ -1245 $\equiv$ -234 $\equiv$ 135	I $\equiv$ $-G_1$ $\equiv$ $-G_2$ $\equiv$ $G_1 G_2$

These four blocks are equivalent to one another.

The defining relation of each block is obtained by multiplying the independent generators to give the one remaining generator. For example:

$$G1 \times G2 \equiv 1245 \times 234 \equiv 1\cancel{22}3\cancel{44}5 \equiv 135$$

on eliminating columns occurring twice.

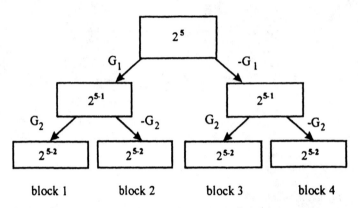

block 1   block 2   block 3   block 4

Figure 3.14   Double partition of a 2^5 factorial design.

3. Example of a 2^{5-2} design: extrusion spheronization

Defining the problem – choice of experimental domain
The effect of five parameters in the extrusion spheronization process was studied by Chariot *et al.* (8) for a placebo formulation. In an initial study they estimated the main effects of five variables on the percentage yield of pellets of the correct size. This can be done most efficiently using a reduced factorial design of 8 experiments. The factors studied and their levels are shown in table 3.17.

Table 3.17 Extrusion Spheronization: Factors Studied

Factor	Associated variable	Lower level (-1)	Upper level (+1)
Spheronization time (min)	X_1	2	5
Spheronizer speed (rpm)	X_2	650	1350
Rate of extrusion (rpm)	X_3	15	59
Spheronizer load (kg)	X_4	1	4
Extrusion screen size (mm)	X_5	0.8	1.5

Mathematical model

The synergistic model for 5 variables contains 32 coefficients. With 8 experiments we may calculate only 8 independent linear combinations. We could on the other hand postulate a model containing only first degree terms:

$$y = \beta_0 + \beta_1 x_1 + \beta_2 x_2 + \beta_3 x_3 + \beta_4 x_4 + \beta_5 x_5 + \varepsilon$$

and if this model were correct the estimates of the five coefficients would be unbiased. This model assumes that the effect of changing the value of a factor does not depend in any way on the values of any other factor (provided of course that these other factors remain constant).

Experimental design – experimental plan – results

The design was constructed using the independent generators **234** and **1245**. Thus the 8 experiments are selected where both these columns in the model matrix are at level +1. (The reasons for not choosing the highest order interaction as a generator here will be discussed later, in section III.D).

The experimental design is set up as follows.

(a) The first 3 columns X_1, X_2, X_3 are written down in the standard order.

(b) Since **234** is a generator, multiplying column **4** by column **23** gives "+" elements only ("I"). So columns **23** and **4** are identical and the column for X_4 is constructed by multiplying columns **2** and **3**.

(c) Column **5** is similarly constructed by multiplying columns **1** and **3**, since **135** is a generator.

The resulting experiment design is shown in table 3.18 in terms of the coded variables, X_i and the natural variables for each factor. The table also shows experimental values of the yield of pellets for each run.

Table 3.18 2^{5-2} Design for Extrusion-Spheronization: Coded and Natural Variables and Experimental Yields (adapted from reference (8), by courtesy of Marcel Dekker Inc.)

No.	X_1	X_2	X_3	X_4	X_5	Spher. time	Spher. speed	Extrus. speed	Spher. load	Extrus. screen	Yield pellets
				23	13	min	rpm	rpm	kg	mm	%
1	-	-	-	+	+	2	650	15	4	1.5	11.1
2	+	-	-	+	-	5	650	15	4	0.8	92.8
3	-	+	-	-	+	2	1350	15	1	1.5	19.7
4	+	+	-	-	-	5	1350	15	1	0.8	55.5
5	-	-	+	-	-	2	650	59	1	0.8	75.5
6	+	-	+	-	+	5	650	59	1	1.5	45.4
7	-	+	+	+	-	2	1350	59	4	0.8	46.5
8	+	+	+	+	+	5	1350	59	4	1.5	55.0

Estimation of the effects

The pattern of confounded effects is deduced from the defining relation for the matrix:

$$I \equiv 234 \equiv 135 \equiv 1245$$

Thus for the main effect of X_1 we obtain on multiplying each generator by 1:

$$1 \times I \equiv 1234 \equiv \mathbf{11}35 \equiv \mathbf{11}245$$

so β_1 is confounded with β_{35}, β_{245} and β_{1234}.

We can treat all the effects in the same way, multiplying the defining relation by 2, 3, 4, 5, etc. The effects are as follows, ranged in increasing order of interaction, with second and third order interactions displaced to the right. See later, in section III.C.1, for a further discussion.

$$
\begin{aligned}
E(b_0) &= \beta_0 & &+ \beta_{234} + \beta_{135} + \beta_{1245} \\
E(b_1) &= \beta_1 + \beta_{35} & &+ \beta_{245} + \beta_{1234} \\
E(b_2) &= \beta_2 + \beta_{34} & &+ \beta_{145} + \beta_{1235} \\
E(b_3) &= \beta_3 + \beta_{24} + \beta_{15} &&+ \beta_{12345} \\
E(b_4) &= \beta_4 + \beta_{23} & &+ \beta_{125} + \beta_{1345} \\
E(b_5) &= \beta_5 + \beta_{13} & &+ \beta_{124} + \beta_{2345} \\
E(b_{12+45}) &= \beta_{12} + \beta_{45} & &+ \beta_{134} + \beta_{235} \\
E(b_{14+25}) &= \beta_{14} + \beta_{25} & &+ \beta_{123} + \beta_{345}
\end{aligned}
$$

We recall that the contrast (or linear combination):

$$l_1 = (-y_1 + y_2 - y_3 + y_4 - y_5 + y_6 - y_7 + y_8)/8 = 95.9/8 = 12.0$$

may be thought of as the estimate b_1 of β_1, only if the interaction coefficients β_{35}, β_{245}, and β_{1234} are negligible. It would normally be reasonable to ignore the second and third order interactions, but not the two factor interaction β_{35}. Therefore we write it below as b_{1+35}.

The estimates of the coefficients, calculated by linear combinations of the eight data as described in chapter 2, are shown in table 3.19.

Some factors seem to have a considerable effect on the yield. The spheronization time is important, a longer spheronization time giving an improved yield. However, we have no way of knowing if this estimate refers to the spheronization time only, or to the interaction between the rate of extrusion and the extrusion screen size, or to both. The estimated effect of the diameter of the extrusion screen is also important, but this also may contain other aliased terms, and we have no way of knowing their relative importance.

Table 3.19 Estimates of the Effects from the Response Data of Table 3.18

Factor	Coefficient	Yield of pellets (%)
constant	b_0	**50.2**
spheronization time	b_{1+35}	**12.0**
spheronizer speed	b_{2+34}	-6.0
extrusion speed	$b_{3+24+15}$	5.4
spheronizer load	b_{4+23}	1.2
extrusion screen	b_{5+13}	**-17.4**
	b_{12+45}	-0.9
	b_{14+25}	**10.6**

We do see that there is at least one important interaction (b_{14+25} = 10.6%), but since this corresponds to the sum of two aliased interactions terms, we cannot draw any definite conclusions as to its physical meaning. The authors concluded that the interaction between spheronization time and spheronizer load (β_{14}) was probably more important than that between the spheronization speed and the extrusion screen (β_{25}).

Whatever the conclusions reached, we recall that the main effects may not be analysed separately and the interaction diagrams must be used (see figure 3.17).

Unbiased estimates of the main effects were obtained after carrying out a further series of 8 experiments, as we will see in a later section (III.D) of this chapter. These demonstrated that the above estimates of the main effects were quite good ones, in spite of the bias caused by aliasing.

C. Generalisation: the Construction and Properties of Fractional Factorial Designs

1. Generators: defining relation

The general nomenclature of these designs is as follows: a fractional factorial design, formed from a full design of k factors, each taking 2 distinct levels, is written as 2^{k-r}. This shows that:

- it consists of $N = 2^{k-r}$ experiments,
- it is a $(½)^r$ fraction of the complete factorial design 2^k,
- it can also be constructed from a complete factorial $2^{k'}$ where $k' = k - r$,
- its defining relation contains 2^r terms, of which $2^r - 1$ are generators, themselves defined by r independent generators G_1, G_2, ..., G_r,
- it results from r successive partitions of the complete factorial design,

according to the r independent generators as shown in figure 3.14 for the quarter-fractions of the 2^5 design.

All the properties depend solely on the totality of the independent generators, which themselves define the fractional design. We will therefore show how the generators of a $2^{k \cdot r}$ design can be found. Consider first the case of an experimental design that is already known or written down. We will envisage another situation in a later section (C.6), where we explain how a design is obtained starting from known generators.

2. Identifying generators of an existing design

We choose $k' = k - r$ factors (columns) that can be rearranged to give the standard order of a complete $2^{k'}$ design. The r remaining columns must of necessity be identified with certain products of the first k' columns, 2 by 2, or 3 by 3 etc. Take for example the 2^{6-3} design of table 3.20.

Table 3.20 A 2^{6-3} Design

N°	X_1	X_2	X_3	X_4	X_5	X_6
1	-	-	-	+	-	+
2	-	+	-	+	+	-
3	+	+	-	-	-	+
4	+	-	-	-	+	-
5	-	+	+	-	-	-
6	-	-	+	-	+	+
7	+	-	+	+	-	-
8	+	+	+	+	+	+

We see that taking the 3 columns X_5, X_1 and X_3 we have a complete 2^3 factorial design in the standard order. In general the order of the rows should be adjusted to the standard order to reveal the design's structure. A systematic examination of the remaining columns shows that the columns X_2, X_4 and X_6 are identical to the products of columns $X_1X_3X_5$, X_1X_3, and X_3X_5. So we may write for all of the experiments of the design:

$$X_2 \equiv X_1X_3X_5 \qquad X_4 \equiv X_1X_3 \qquad X_6 \equiv X_3X_5$$

or more simply, using the notation introduced in section III.A.1.

$$2 \equiv 135 \qquad\qquad 4 \equiv 13 \qquad\qquad 6 \equiv 35$$

The design is defined by 3 independent generators ($2^{k \cdot r}$ design with $r = 3$). We may work out what these 3 generators are from the previous line. A generator corresponds to a column of "+" signs in the model matrix. Multiplying a column by itself will result in a column of "+" signs. And if column 2 (variable X_2) is identical to column 135 (variable $X_1X_3X_5$), the product of the two columns 1235 (variable $X_1X_2X_3X_5$) must also be a generator. We can therefore write down the first four terms in the defining relation of the matrix.

$$I \equiv 1235 \equiv 134 \equiv 356$$
$$\quad\quad G1 \quad\quad G2 \quad\quad G3$$

We shall see that there is a total of $2^r = 2^3 = 8$ terms in the defining relation. The product of 2 generators is also a generator, as the product of two columns of "+" signs is inevitably also a column of "+" signs.

Take, for example, the product of 1235 and 134. The result after rearrangement is 1123345. Now since the product of a column by itself (11 or 33) is a column "+" (written as I), 245 must also give a column "+". Column 245 is therefore a generator but not independent, as it is obtained as the product of 2 other generators.

Using the same argument, all products of the independent generators are also generators.

$$
\begin{aligned}
&G1 \times G2 && = 245 \\
&G1 \times G3 && = 1235 \times 356 && = 12\ \cancel{33}\ \cancel{55}\ 6 && = 126 \\
&G2 \times G3 && = 134 \times 356 && = 1\cancel{33}\ 456 && = 1456 \\
&G1 \times G2 \times G3 = 1235 \times 134 \times 356 && = \cancel{11}\ 23\ \cancel{33}\ 4\ \cancel{55}\ 6 && = 2346
\end{aligned}
$$

The complete defining relation is:

$I \equiv 1235 \equiv 134 \equiv 356 \equiv \quad 245 \quad \equiv \quad 126 \quad \equiv \quad 1456 \quad \equiv \quad\quad 2346$
$\quad\quad G1 \quad\quad G2 \quad\quad G3 \quad\quad G1 \times G2 \quad G1 \times G3 \quad G2 \times G3 \quad G1 \times G2 \times G3$

3. Confounded (aliased) effects

It is this defining relation that is used to determine which effects are aliased with one another. We have seen in paragraph III.A.2 that when columns in the model matrix are equal we cannot calculate those individual effects but only their sum. Thus, the column X_0 which consists of "+" signs is used to estimate the constant term but is also confounded with all effects corresponding to the generators.

$$E(b_0) = \beta_0 + \beta_{1235} + \beta_{134} + \beta_{356} + \beta_{245} + \beta_{126} + \beta_{1456} + \beta_{2346}$$

Also if a column is multiplied by a generator it remains unchanged. Take for example the column corresponding to the variable X_1. Thus $1 \times 1235 \equiv 1$ but also $1 \times 1235 \equiv (11)235 \equiv 235$. We multiply column 1 by each generator in turn:

$$1 \times I \equiv 1 \times 1235 \equiv 1 \times 134 \equiv 1 \times 356 \equiv 1 \times 245 \equiv 1 \times 126 \equiv 1 \times 1456$$
$$\equiv 1 \times 2346$$

which becomes, after simplification:

$$1 \equiv 235 \equiv 34 \equiv 1356 \equiv 1245 \equiv 26 \equiv 456 \equiv 12346$$

As these columns are identical the corresponding effects are confounded:

$$E(b_1) = \beta_1 + \beta_{235} + \beta_{34} + \beta_{1356} + \beta_{1245} + \beta_{26} + \beta_{456} + \beta_{12346}$$

The process can be repeated for all the factors:

$$E(b_2) = \beta_2 + \beta_{135} + \beta_{1234} + \beta_{2356} + \beta_{45} + \beta_{16} + \beta_{12456} + \beta_{346}$$
$$E(b_3) = \beta_3 + \beta_{125} + \beta_{14} + \beta_{56} + \beta_{2345} + \beta_{1236} + \beta_{13456} + \beta_{246}$$
$$E(b_4) = \beta_4 + \beta_{12345} + \beta_{13} + \beta_{3456} + \beta_{25} + \beta_{1246} + \beta_{156} + \beta_{236}$$
$$E(b_5) = \beta_5 + \beta_{123} + \beta_{1345} + \beta_{36} + \beta_{24} + \beta_{1256} + \beta_{146} + \beta_{23456}$$
$$E(b_6) = \beta_6 + \beta_{12356} + \beta_{1346} + \beta_{35} + \beta_{2456} + \beta_{12} + \beta_{145} + \beta_{234}$$

A further item of information can be obtained, corresponding to an 8th linear combination of the experimental data. The interaction effect β_{15} is not found in any of the above equations. We calculate as before:

$$15 \times I \equiv 15 \times 1235 \equiv 15 \times 134 \equiv 15 \times 356 \equiv 15 \times 245 \equiv 15 \times 126$$
$$\equiv 15 \times 1456 \equiv 15 \times 2346$$

which gives:

$$E(b_{15}) = \beta_{15} + \beta_{23} + \beta_{345} + \beta_{136} + \beta_{124} + \beta_{256} + \beta_{46} + \beta_{123456}$$

β_{15} is thus confounded with two other first-order interaction effects, as well as a number of higher order interactions. In addition the estimators for the main effects show that these also are each confounded with a number of interactions of different degrees. In practice it would probably be assumed that all second and higher order interactions could be neglected. This still leaves each main effect confounded with 2 first-order interactions, and this fact must be taken into account in the interpretation of the results. This may mean that certain estimates of the main effects will be viewed with some caution. The above design is essentially used for screening, as described in chapter 2.

4. Resolution of a fractional factorial design

With fractional factorial designs the estimates of the effects are always aliased. We need to know how serious is the confounding. If two main effects are confounded then the matrix is not very useful. If a main effect is confounded only with second order interactions then it is very likely that the interaction could be neglected. The *resolution* of a fractional factorial design is the property which determines, for example, if it will be possible to estimate all main effects without them being aliased with first-order interactions and whether all first-order interaction effects can be determined separately.

It is defined as the *length of the shortest generator*. For the 2^{6-3} design given above, the shortest generators are of 3 elements. The design is of resolution III (this is often written as R_{III}). It allows main effects to be estimated independently of one another but they are each confounded with first-order interactions.

The 2^{4-1} design described previously (section III.A) has a single generator of 4 elements, and it is therefore of resolution IV (R_{IV}). We have seen that all main effects could be estimated independently, and they were not aliased with first-order, but only with second-order interactions. However the estimates of the first-order interactions were aliased with one another. All main effects and first-order interactions of a resolution V (R_V) design can be estimated separately.

Thus many different fractional factorial designs may be constructed from a single full factorial design, and that the properties of the resulting design depend on the choice of generators. In general, and without prior knowledge of the existence of certain interactions and not others, the chosen design would be the one with the highest possible resolution. This enables us to confound main effects (the most probable) with only the highest order interactions, those which *a priori* are the least probable. We will however later on set out an alternative approach, useful if certain interactions are expected.

If the design is a half fraction (2^{k-1}) we obtain the design of highest possible resolution if we choose the highest order interaction as generator – that is the product of all the factors. There is only one generator.

If the design is a quarter fraction (2^{k-2}) there is no simple method for choosing the two independent generators, as one must consider also the length of the product of the 2 generators.

5. Aberration of a design

The main criterion for quality in the choice of a fractional factorial design is that it should be of maximum resolution. There could however be several fractional designs, each of the same resolution, but with apparent differences in quality. We take for example two 16 experiment designs for 6 factors (2^{6-2}), both of resolution III. Their defining relations are:

$$I \equiv 123 \equiv 456 \equiv 123456$$

$$I \equiv 123 \equiv 3456 \equiv 12456$$

In the first matrix all main effects are aliased with a first-order interaction. In the second, only 3 main effects, **1**, **2**, and **3**, are so aliased. The second design appears to be better than the first.

When several designs are of equal resolution, the best design is that with the fewest independent generators of minimum length. This is called a *minimum aberration design* (11).

As far as the above example is concerned, there is also a 2^{6-2} design of resolution IV, with independent generators **1235** and **1246** which is normally to be preferred to either of the two designs given, neither of which appears in the table of optimum fractional factorial designs (table 3.21).

Table 3.21 Fractional Factorial Designs of Maximum Resolution and Minimum Aberration

k Resolution	5	6	7	8	9	10	11
III	2^{5-2} 124 135	2^{6-3} 124 135 236	2^{7-4} 124 135 236 1237		2^{9-5} 1235 1246 1347 2348 12349	2^{10-6} 1235 2346 1347 1248 12349 12.10	2^{11-7} 1235 2346 1347 1248 12349 12.10 13.11
IV		2^{6-2} 1235 1246	2^{7-3} 1235 1246 1347	2^{8-4} 1235 1246 1347 2348	2^{9-4} 12346 12357 12458 13459	2^{10-5} 12346 12357 12458 13459 2345.10	2^{11-6} 1236 2347 3458 1349 145.10 245.11
V	2^{5-1} 12345	*	*	2^{8-2} 12347 12568	*	2^{10-3} 12378 12469 2345.10	2^{11-4} 12378 23459 1346.10 1234567.11

* No resolution V design exists. Use either resolution IV or VI/VII design.

6. Optimum matrices of maximum resolution and minimum aberration

Table 3.21 gives designs of maximum resolution and minimum aberration for a number of factors between 5 and 11, which covers most possible problems. Certain squares in the table are empty. There is, for example, no design of resolution V for 6, 7, or 9 factors, where it is necessary to use either a resolution IV or VI design.

The independent generators are arranged in increasing order. Where the number of the column number is 10 or more it is separated by a dot. So **12.10** corresponds to the product of X_1, X_2, and X_{10}. Of the half fraction designs (2^{k-1}), which are of resolution k, only 2^{5-1} is given in this table.

Up until now we have shown how generators may be obtained from a known design. The problem in using this table is the opposite one: how is the design to be constructed from the data in table 3.21?

Take for example the 2^{9-4} design of resolution IV. The independent generators are **12346, 12357, 12458,** and **13459.** There are 9-4 = 5 independent factors. So the complete 2^5 design of 32 experiments is constructed from the first 5 factors. We now need expressions for constructing columns 6-9. The fact that **12346** is a generator shows that column **6** is constructed from the product of columns **1234.** Similarly because **12357** is a generator, column **7** is the product of columns **1235.** Column **8** is constructed from **1245,** and column **9** from **1345.**

It is clear that equivalent designs can be obtained by permutating the symbols representing factors in the generating function – for example, changing **1** for **9** throughout, and **9** for **1.**

D. Continuation or Complement to a Fractional Factorial Design

Fractional factorial designs allow main effects and some interaction effects to be estimated at minimal cost, but the result is not without ambiguity – main effects may be confounded with first-order interactions (designs of resolution III) or first-order interaction effects confounded with one another (resolution IV).

After analysing the results, the experimenter may well wish for clarification and he will very likely need to continue the experiment with a second design. This is usually of the same number of experiments as the original, the design remaining fractional factorial overall.

If a half-fraction factorial, such as the 2^{4-1} design for 4 factors, gives insufficient information, it is possible to do the remaining experiments of the full factorial design. This kind of complementary design is trivial, and will not be discussed further.

1. Complementary design

Let us consider a design fractionated first into two blocks using G_1 and $-G_1$ as generators. Each block is then fractionated again, with another generator G_2 and $-G_2$, as shown in figure 3.14 for the 2^5 design. The defining relations of the 4 blocks, each a 2^{k-2} design, are, as we have already seen:

Block 1:	$I \equiv G_1 \equiv G_2 \equiv G_1G_2$
Block 2:	$I \equiv G_1 \equiv -G_2 \equiv -G_1G_2$
Block 3:	$I \equiv -G_1 \equiv G_2 \equiv -G_1G_2$
Block 4:	$I \equiv -G_1 \equiv -G_2 \equiv G_1G_2$

Any two blocks allows us to construct a 2^{k-1} design, of which the single generator is the one that has not changed sign in going from one block to the other. For example if we complete the experiments in block 1 by those of block 2 the generator of the new half fraction design is G_1. Completing block 1 by block 3 will give a design with generator G_2 and adding block 4 to block 1 will give a design with generator G_1G_2. Thus, starting with block 1, we have three possible choice of generator, depending on which block is chosen for the second stage.

The new design, with twice as many experiments, allows twice as many effects to be calculated. Since its defining relation contains only one generator the coefficients are confounded with only one other coefficient. The nature of this confounding depends on the generator and thus on the choice of associated block.

2. Continuation of the extrusion-spheronization study

Definition of a complementary design – results
In the extrusion-spheronization study described above (8) we recollect that the estimations of the 5 main effects contained aliased terms. These are listed below, on the left hand side (neglecting all interactions between more than two factors). It is clear that if we could carry out a second series of experiments with a different confounding pattern, with the estimations listed on the right hand side, then on combining the two series of experiments we would be able to separate main effects from first-order interactions.

$E(b_0)$	$= \beta_0$	$E(b'_0)$	$= \beta_0$
$E(b_1)$	$= \beta_1 + \beta_{35}$	$E(b'_1)$	$= \beta_1 - \beta_{35}$
$E(b_2)$	$= \beta_2 + \beta_{34}$	$E(b'_2)$	$= \beta_2 - \beta_{34}$
$E(b_3)$	$= \beta_3 + \beta_{24} + \beta_{15}$	$E(b'_3)$	$= \beta_3 - (\beta_{24} + \beta_{15})$
$E(b_4)$	$= \beta_4 + \beta_{23}$	$E(b'_4)$	$= \beta_4 - \beta_{23}$
$E(b_5)$	$= \beta_5 + \beta_{13}$	$E(b'_5)$	$= \beta_5 - \beta_{13}$

The defining relation of the first design being $I \equiv 234 \equiv 135 \equiv 1245$, such a complementary design must have a defining relation $I \equiv -234 \equiv -135 \equiv 1245$. In the original design the main effects are confounded with first-order interactions because of the presence of generators of length 3. If the signs of those generators with *three* terms are reversed in the new design, then they will disappear from the defining relation of the design of the two blocks added together, but leaving the longest generator. Therefore the new defining relation is $\mathbf{I} \equiv \mathbf{1245}$. It is of resolution IV and allows us to resolve main effects free of all first-order interactions and confounded only with second-order interactions that we assume negligible. The new design is given in table 3.22.

Table 3.22 Complementary $2^{5\text{-}2}$ Design for the Extrusion-Spheronization Study ($2^{5\text{-}2}$ Design) of Table 3.18 [adapted from reference (8)]

Associated variable	X_1	X_2	X_3	X_4	X_5	y
	Spher. time	Spher. speed	Extrus. speed	Spher. load	Extrus. screen	Yield of pellets
No.	(min)	(rpm)	(rpm)	(kg)	(mm)	(%)
1	2	650	15	1	0.8	62.5
2	5	650	15	1	1.5	34.9
3	2	1350	15	4	0.8	56.9
4	5	1350	15	4	1.5	29.0
5	2	650	59	4	1.5	1.2
6	5	650	59	4	0.8	78.7
7	2	1350	59	1	1.5	10.2
8	5	1350	59	1	0.8	47.0

The complementary design is set up as follows.
- Write the first 3 columns as a 2^3 design in the standard order.
- We now require X_4. Column **234** corresponds to the product of X_4 and X_2X_3. If **-234** is to be a generator it must contain negative signs only. Thus we set X_4 to $-X_2X_3$.
- Similarly X_5 must be of opposite sign to X_1X_3 if **-135** is to be a generator. The two designs together make what is known as a "fold-over" pair.

Calculation of the effects

This second design or block may be treated the same as the first one was (see section III.B). The new estimates of the effects are shown in table 3.23 in the right hand column along with the equivalent results calculated from the first block (in the third column).

Without further analysis, these results add little to what we have already seen. Nevertheless, we notice that the estimates of the effects are of the same order, with three important exceptions, the inversion of the coefficient supposed to represent the effect of the extrusion speed, considerable changes in the mixed interaction effect $b'_{14\text{-}25}$, and also a difference in the constant term b'_0 (figure 3.15).

Table 3.23 Effects Calculated from the Second 2^{5-2} Design

Factor	Coefficient	Calculated effect – yield of pellets (%)	
	(2nd block)	1st block	2nd block
constant	b'_0	50.2	40.0
spheronization time	b'_{1-35}	12.0	7.4
spheronization speed	b'_{2-34}	-6.0	-4.3
extrusion speed	$b'_{3-24-15}$	5.4	-5.8
spheronization load	b'_{4-23}	1.2	1.4
extrusion screen	b'_{5-13}	-17.4	-21.2
	b'_{12-45}	-0.9	-5.1
	b'_{14-25}	10.6	5.1

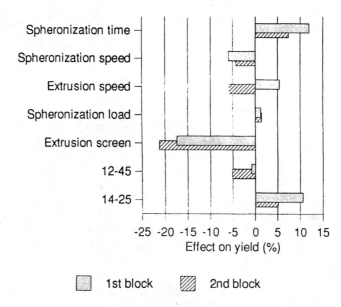

Figure 3.15 Coefficient sets estimated from the two blocks.

We can obtain a better estimate from the mean of the two effects, which eliminates the terms that change sign. For example:

$$b_1 = (b_{1+35} + b'_{1-35})/2 = 9.7$$

Additional effects may be calculated from the differences of the two estimates:

$$b_{35} = (b_{1+35} - b'_{1-35})/2 = 2.3$$

The results for the main effects, first-order interaction effects, and a few purely second-order confounded interactions are listed in table 3.24 and shown graphically in figure 3.16. Identical results can be obtained by fusion of the two matrices to a single 2^{5-1} design with defining relation $I \equiv 1245$. The design, as we recall of resolution IV, enables the main effects to be calculated without ambiguity, unaliased with any second-order interaction terms. The effects may be calculated either from the contrasts defined by the columns of the model matrix, or by multi-linear regression.

Table 3.24 Estimated Effects Calculated from the Two 2^{5-2} Designs

Factor	Coefficient	Yield of pellets (%)
constant	b_0	45.1
spheronization time	b_1	**9.7**
spheronization speed	b_2	**-5.1**
extrusion speed	b_3	-0.2
spheronization load	b_4	1.3
extrusion screen	b_5	**-19.3**
	b_{12+45}	-3.0
	b_{14+25}	**7.8**
	b_{15+24}	**5.6**
	b_{13}	1.9
	b_{23}	-0.1
	b_{34}	-0.9
	b_{35}	2.3
	$b_{123+345}$	2.8
	$b_{134+235}$	2.1
	$b_{234+135}$	**5.1**

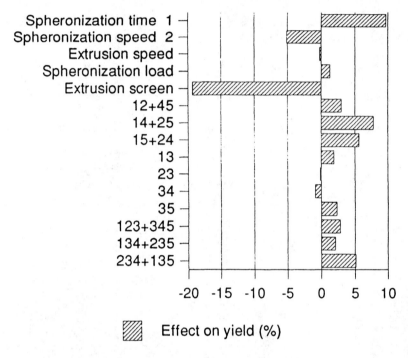

Figure 3.16 Estimated effects for 2^{5-1} design (extrusion-spheronization).

Interpretation of interactions

The authors of the paper concluded that three factors had pharmaceutically significant effects, the spheronization time β_1, the spheronization speed β_2 and the extrusion screen size β_5. In addition there were important interactions. These represented aliased terms. They considered that the interaction term b_{14+25} was to be attributed mainly to β_{14}, an interaction between spheronizer load and spheronization time. The value of this coefficient was positive.

The second interaction b_{15+24} was attributed to the spheronization speed and the load, β_{24}. The third unresolved interaction b_{12+45} is much less likely to be important. This interpretation is used in drawing the interaction diagrams of figure 3.17. A certain degree of caution is required however.

It may also be noticed that $b_{234+135}$ is rather high for a second order interaction effect. It represents in fact a difference between the two blocks of 8 experiments. This is discussed in more detail in the section on time trends and blocking (IV).

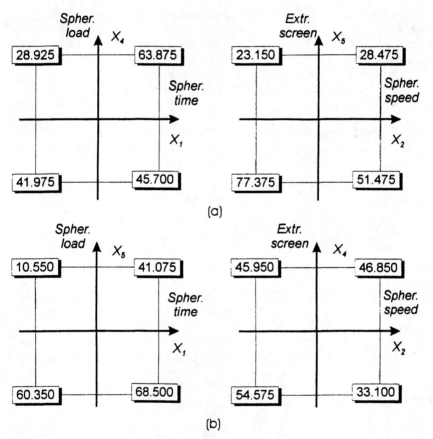

Figure 3.17 Interaction diagrams for extrusion-spheronization showing the aliased pairs of interactions. (a) b_{14+25} (spheronization time × load aliased with spheronization speed × extrusion screen) (b) b_{15+24} (spheronization time × extrusion screen aliased with spheronization speed × load). For the explanation of the interaction diagram see figure 3.5. The values of the real variables corresponding to the coded variable settings -1 and +1 are given in table 3.17.

Although the authors arrived at the above conclusions on the basis of pharmaceutical considerations it is worth noting that different conclusions may be reached on comparing the relative values of the main effects. In the case of b_{15+24} the main effects b_1 and b_5 are larger than b_2 and b_4. So on that basis we would expect b_{15+24} to represent β_{15} rather than β_{24}. Further experiments would be needed to confirm this.

It is worth asking at this point whether the above choice of design(s) was the best possible one. The $2^{5\text{-}1}$ design that results from joining the two designs is only of resolution IV whereas a design of resolution V may be constructed by taking the general interaction 12345 as generator (as in reference 9). If this design had been carried out we would not have had the above ambiguity over the interactions, as all first-order interactions could have been estimated, unconfounded with one another.

However, if it were necessary to carry out the experiment in two steps (and it was not certain that the second stage would be carried out) it would be necessary to partition the $2^{5\text{-}1}$ design, and in the case of the resolution V design this would cause problems. Since 12345 is a generator, each third-order interaction is aliased with a main effect (for example 1234 with 5) and each second-order interaction is aliased with a first-order interaction (e.g. 245 with 13). So we can only partition on a first-order effect and if we do this the two main effects of the interaction are confounded (resolution II).

The stepwise approach is thus not possible if we wish to obtain the resolution V design at the end.

3. Optimum strategy for a succession of fractional factorial designs

Fractional factorial designs can lead to biased estimations of the effects if certain interactions are wrongly assumed to be negligible for the purpose of setting up the design. This can prevent analysis of the design and lead to more information being required in the form of a second experimental design. It is here that we come across a problem. We need to weigh up the properties of the first block of experiments against the properties of the final design. In the previous case we are forced to take an inferior final design if we are to have individual blocks that can be analysed.

Here we examine a different case. We wish to study 6 factors, with 16 experiments. We would normally select a design of maximum resolution and minimum aberration. Reference to table 3.21 shows us that the defining relation of such a design is:

$$I \equiv 1235 \equiv 1246 \equiv 3456$$

and it is of resolution IV. Now none of the main effects is biased by first-order interactions, only by second-order interactions. However, *all* the first-order interactions are confounded with one another (in pairs: e.g. 13 with 25, or in one case threes: e.g. 12 with 35 and 46). Whatever complementary design we choose the result will be resolution IV and the defining relation will contain *one* of the above generators. (We can choose which that is to be by selecting the appropriate block for the second design). Of the 15 first-order interaction effects 9 will be resolved and 6 will remain aliased.

On the other hand, consider the 2^{6-2} design of resolution III:

$$I \equiv 135 \equiv 123456 \equiv 246$$

Each of the main effects is confounded with a first-order interaction effect. The design is therefore less efficient than the one shown above. However a complementary design can be found of defining relation $I \equiv -135 \equiv 123456 \equiv -246$, and this leads to a 2^{6-1} design of resolution VI and defining relation $I \equiv 123456$.

In general the optimality of the overall design is more important than the optimality of the individual steps. There is however no hard and fast rule and it is the responsibility of the experimenter to weigh up the various considerations according to the circumstances.

4. Adding another block to the 2^{5-1} design to give a ¾ fraction of a factorial design

Figure 3.14 summarised the partition of a 2^5 design into 4 blocks. Defining relations were written down for the 4 blocks 2^{5-2}. We have shown that by taking any 2 blocks we can estimate 16 out of the 32 effects in the full synergistic model. If having done so we take a further block we may estimate 8 further effects, that is 24 in all.

Each block of 8 experiments can be analysed separately as we saw for the extrusion-spheronization example, and aliasing terms eliminated by combining pairs of estimates. We examine further the extrusion-spheronization example. Since the overall design 2^{5-1} design was of resolution IV this resulted in certain first-order interaction terms being confounded. We may want to clarify the position with respect to the confounded terms in the model. The blocks are:

Block 1	$I \equiv 234 \equiv 135 \equiv 1245$
Block 2	$I \equiv 234 \equiv -135 \equiv -1245$
Block 3	$I \equiv -234 \equiv 135 \equiv -1245$
Block 4	$I \equiv -234 \equiv -135 \equiv 1245$

We used block 1 first of all, followed by block 4. Let us see what happens on adding block 2.

Each block can be used to calculate 8 effects, using contrasts of the response data, represented by $l_0, l_1, ..., l_7$. These are estimates of the constant term, the main effects, and 2 interaction terms, each aliased with various interactions. The aliasing or confounding pattern will be different according to the block being analysed. The blocks themselves are identified by a superscript, for example $l_i^{(4)}$ for the linear combinations from block 4.

They are listed in table 3.25. Interactions between more than 3 variables have been left out.

Table 3.25 Contrasts of the 3 Blocks in a ¾ × 2^{5-2} Design

i	Block 1 $l_i^{(1)}$	Block 4 $l_i^{(4)}$	Block 2 $l_i^{(2)}$
0	$\beta_0 + \beta_{234} + \beta_{135}$	$\beta_0 - \beta_{234} - \beta_{135}$	$\beta_0 + \beta_{234} - \beta_{135}$
1	$\beta_1 + \beta_{35} + \beta_{245}$	$\beta_1 - \beta_{35} + \beta_{245}$	$\beta_1 - \beta_{35} - \beta_{245}$
2	$\beta_2 + \beta_{34} + \beta_{145}$	$\beta_2 - \beta_{34} + \beta_{145}$	$\beta_2 + \beta_{34} - \beta_{145}$
3	$\beta_3 + \beta_{24} + \beta_{15}$	$\beta_3 - \beta_{24} - \beta_{15}$	$\beta_3 + \beta_{24} - \beta_{15}$
4	$\beta_4 + \beta_{23} + \beta_{125}$	$\beta_4 - \beta_{23} + \beta_{125}$	$\beta_4 + \beta_{23} - \beta_{125}$
5	$\beta_5 + \beta_{13} + \beta_{124}$	$\beta_5 - \beta_{13} + \beta_{124}$	$\beta_5 - \beta_{13} - \beta_{124}$
6	$\beta_{12} + \beta_{45} + \beta_{134} + \beta_{235}$	$\beta_{12} + \beta_{45} - \beta_{134} - \beta_{235}$	$\beta_{12} - \beta_{45} + \beta_{134} - \beta_{235}$
7	$\beta_{14} + \beta_{25} + \beta_{123} + \beta_{345}$	$\beta_{14} + \beta_{25} - \beta_{123} - \beta_{345}$	$\beta_{14} - \beta_{25} + \beta_{123} - \beta_{345}$

In particular the previously aliased interactions may be estimated.

$$b_{12} = \frac{1}{2}(l_6^{(1)} + l_6^{(2)}) \qquad\qquad E(b_{12}) = \beta_{12} + \beta_{134}$$

$$b_{45} = \frac{1}{2}(l_6^{(1)} - l_6^{(2)}) \qquad\qquad E(b_{45}) = \beta_{45} + \beta_{235}$$

This design of 3 blocks, appropriately known as a ¾ design (12, 13, 14), can be used as we have seen in a stepwise or sequential approach using factorial design.

They may under certain circumstances be used instead of the factorial design as an *initial* design. An example is the 4 factor ¾ design of 12 experiments, described at the end of this chapter, which allows estimation of all main effects and first-order interactions in the synergistic model.

E. Factorial Designs Corresponding to a Particular Model

1. Synergistic models with only certain interaction terms

We have seen how a fractional factorial design may be set up by selecting all the experiments from a complete design where one or more columns in the model matrix have a certain structure. Thus certain effects are confounded or aliased. This is not a problem if one can manage to confound effects which are highly probable with those that are likely to be insignificant, main effects confounded with second-order interactions for example. We will now demonstrate another way of dealing with the problem, and of setting up the design.

Suppose we wish to study 5 factors, characterised by 5 variables at 2 levels -1, +1 written as X_1 to X_5 and we are prepared to risk neglecting most of the interactions between the factors, except those between X_2 and X_3 and between X_3 and X_5. The incomplete synergistic model postulated is:

$$y = \beta_0 + \beta_1 x_1 + \beta_2 x_2 + \beta_3 x_3 + \beta_4 x_4 + \beta_5 x_5 + \beta_{23} x_2 x_3 + \beta_{35} x_3 x_5 + \varepsilon \qquad (3.7)$$

There are 8 coefficients in the model. The design must therefore contain not less than 8 experiments. Is it possible to construct a fractional factorial design of only 8 (2^3) experiments allowing the resolution of the problem?

We are reminded first of all that a fractional factorial design 2^{k-r} of k factors contains a complete 2^{k-r} design of $k-r$ factors. The question may therefore be asked in another way: can we take a complete 2^3 factorial design and use it to construct a 2^{5-2} design with 5 factors? If the answer is "yes", which 3 factors do we choose for the initial design and how can we extend it to the fractional design?

Assume first of all that we have solved the problem! We call the 3 selected variables A, B, and C. The full factorial design is therefore the design whose model matrix as shown below in table 3.26. AB, AC, BC, ABC represent the interactions.

Table 3.26 Model (Effects) Matrix of a 2^3 Design

No.	X_0	A	B	C	AB	AC	BC	ABC
1	+	-	-	-	+	+	+	-
2	+	+	-	-	-	-	+	+
3	+	-	+	-	-	+	-	+
4	+	+	+	-	+	-	-	-
5	+	-	-	+	+	-	-	+
6	+	+	-	+	-	+	-	-
7	+	-	+	+	-	-	+	-
8	+	+	+	+	+	+	+	+

The second step is to establish a link or correspondence between the effects that we *wish* to calculate (**1, 2, 3, 4, 5, 23, 35**) and those that we *can* calculate (A, B, C, AB, AC, BC, ABC) with this design.

Now we assume that the invented effects A, B, and C correspond to the real *variables* X_2, X_3, *and* X_5 (those that figure amongst the interactions we have postulated). Then the products AB and BC correspond to the real variables $X_2 X_3$ and $X_3 X_5$. The contrasts corresponding to each column allow calculation of each effect. For example:

$$b_3 = (-y_1 - y_2 + y_3 + y_4 - y_5 - y_6 + y_7 + y_8)/8 \qquad : \text{column B}$$
$$b_5 = (-y_1 - y_2 - y_3 - y_4 + y_5 + y_6 + y_7 + y_8)/8 \qquad : \text{column C}$$
$$b_{35} = (+ y_1 + y_2 - y_3 - y_4 - y_5 - y_6 + y_7 + y_8)/8 \qquad : \text{column BC}$$

Columns AC and ABC correspond to interactions X_2X_5 and $X_2X_3X_5$ which were not postulated in the model, and are therefore assumed to be negligible. We can use these columns for factors X_1 and X_4. This, with a simple rearrangement of the order of columns, gives us the design and model matrix of table 3.27.

The construction of this design gives the independent generators automatically. X_1 being set identical to X_2X_5 the product of those two columns $X_1X_2X_5$ contains only "+" signs, and **125** is a generator. In the same way, column $X_2X_3X_5$ is used for X_4 and therefore the second independent generator is **2345**. The defining relation is:

$$\mathbf{I} \equiv \mathbf{125} \equiv \mathbf{134} \equiv \mathbf{2345}$$

Table 3.27 Model Matrix for Equation 3.7

N°	X_0	AC X_1	A X_2	B X_3	ABC X_4	C X_5	AB X_2X_3	BC X_3X_5
1	+	+	-	-	-	-	+	+
2	+	-	+	-	+	-	-	+
3	+	+	-	+	+	-	-	-
4	+	-	+	+	-	-	+	-
5	+	-	-	-	+	+	+	-
6	+	+	+	-	-	+	-	-
7	+	-	-	+	-	+	-	+
8	+	+	+	+	+	+	+	+

This method is widely applicable: one sets up a full factorial model matrix, with enough experiments to calculate all of the desired effect. The real factors and the postulated interactions are then made to correspond to these imaginary factors and interactions. However this does not give a solution in every case. A well known example is the problem of 4 factors X_1, X_2, X_3, X_4 with the interactions X_1X_2 and X_3X_4. The model is:

$$y = \beta_0 + \beta_1 x_1 + \beta_2 x_2 + \beta_3 x_3 + \beta_4 x_4 + \beta_{12} x_1 x_2 + \beta_{34} x_3 x_4 + \varepsilon$$

As there are only 7 coefficients one could well be hopeful of solving the problem with a 2^{4-1} half fraction design. Unfortunately no such design exists. The only factorial design that will give estimates of the interactions sought is the full factorial design of 16 experiments, and its R-efficiency is low (44%). D-optimal or ¾ designs may offer a more economical alternative.

2. Screening designs

This method also allows the construction of screening designs for k factors each at 2 levels: the complete factorial design for k' factors is set up, $2^{k'}$, with k' chosen so that $2^{k'} \geq k + 1$. The remaining $k-k'$ factors are assigned to the interactions columns in the model matrix. For example a design for screening 7 factors is based on a 2^3 full factorial design, as shown in table 3.28. Factors X_1, X_2 and X_3 are assigned to the corresponding columns of the full factorial design, and X_4, X_5, X_6 and X_7 are assigned to the interactions. The result is a Hadamard design of 8 experiments (in a different order from the design given by the cyclic construction method of Plackett and Burman).

Table 3.28 Design for the Model (3.7)

N°	A X_1	B X_2	C X_3	AB X_4	AC X_5	BC X_6	ABC X_7
1	-	-	-	+	+	+	-
2	+	-	-	-	-	+	+
3	-	+	-	-	+	-	+
4	+	+	-	+	-	-	-
5	-	-	+	+	-	-	+
6	+	-	+	-	+	-	-
7	-	+	+	-	-	+	-
8	+	+	+	+	+	+	+

Similarly we may construct the Hadamard design of 4 experimental runs, adding a third variable X_3 to the 2^2 design. X_3 is confounded with the interaction $X_1 X_2$ (AB). This gives the 2^{3-1} fractional factorial design with defining relation:

$$I \equiv 123$$

No.	A X_1	B X_2	AB X_3
1	-1	-1	+1
2	+1	-1	-1
3	-1	+1	-1
4	+1	+1	+1

The confounding pattern is:

$$E(b_1) = \beta_1 + \beta_{23}$$
$$E(b_2) = \beta_2 + \beta_{13}$$
$$E(b_3) = \beta_3 + \beta_{12}$$
$$E(b_0) = \beta_0 + \beta_{123}$$

Therefore l_0, l_1, l_2 and l_3, are unbiased estimates only so long as the interaction effects can be neglected.

IV. TIME TRENDS AND BLOCKING

Up until now we have supposed that all of the experiments of a design are carried out in a consistent fashion under constant conditions. This means that the results do not depend on the order in which the experiments were carried out, nor the distance that separates them – either in time or in space. The results depend on the experimental conditions, on the levels X_i of the factors and an experimental error of constant variance.

However several untoward effects may arise and be superimposed on those we are studying, perturb the results (in some cases seriously) and affect the conclusions. We will consider here how to allow for two such phenomena.

A. Effect of Time (Time Trend)

Time may affect the response, in the sense that it may drift. This can be related to a number of effects, such as aging of reagents, changes in the apparatus (warming up), increasing experience of the operator, temperature or other atmospheric changes or arrangement in an oven. That most frequently considered is a linear effect of time. Suppose a complete factorial design of 3 factors and 8 experiments is carried out in the standard order and a drift is superimposed on the response y_i that would have been measured for each experiment in its absence. Let the drift per experiment be τ. Therefore the actual response measured for the ith experiment is:

$$y'_i = y_i + (i - 1)\tau$$

Table 3.29 describes the resulting situation.

Table 3.29 Full Factorial Design with Linear Time Trend

Standard order	X_1	X_2	X_3	Measured response	Corrected order
1	-	-	-	$y'_1 = y_1$	7
2	+	-	-	$y'_2 = y_2 + \tau$	4
3	-	+	-	$y'_3 = y_3 + 2\tau$	6
4	+	+	-	$y'_4 = y_4 + 3\tau$	1
5	-	-	+	$y'_5 = y_5 + 4\tau$	2
6	+	-	+	$y'_6 = y_6 + 5\tau$	5
7	-	+	+	$y'_7 = y_7 + 6\tau$	3
8	+	+	+	$y'_8 = y_8 + 7\tau$	8

All the estimates of the main effects b'_i are more or less biased with respect to the values they would have taken without the time trend. The most extreme example is b_3 where:

$$b'_3 = b_3 + 2\tau$$

It is easy to show on the other hand that the estimations of the interaction terms are not biased.

If the interactions may be assumed negligible the time trend may be allowed for by carrying out the experiments in a different order, the "corrected" order of the right hand column of table 3.29. With this new order the time trend terms cancel out in the contrasts for estimating the main effects. However, now the interactions are biased and so this solution is limited to the rare cases where the interactions may be neglected safely.

Another solution, by far the more general one, is to carry out the experiments in a *random order*. If there are the same number of experiments in the design as there are parameters in the postulated model, the effect of time is to perturb the different estimations in a random fashion. If the number of experiments exceeds the number of parameters to be estimated then the experimental error estimated by multi-linear regression (see chapter 4) includes the effect of time and is therefore overestimated with respect to its true value.

B. Block Effects

A block is a group of experiments, part of the total design, that may be considered homogenous. We suppose that the experimental results within the block are not affected by the type of time trend phenomenon or drift described above, or if such a trend does exist it is relatively small and allowed for by random ordering of the experiments inside the block. On the other hand, the results may vary from one block to another.

This happens in two kinds of circumstances. The first may be involuntary, when the experiments are carried out sequentially, in two or several stages. This may be for reasons of economy, when a preliminary fractional factorial experiment has been carried out and a complementary design added some time later, to clarify the ambiguities in estimations of the main and interaction effects. The extrusion-spheronization experiment where a 2^{5-2} design was followed by a second 2^{5-2} foldover design is an example of this.

The second case is one where the number of experiments in the design is too large for it to be carried out under constant conditions. This may be because of time available, or the need to use two or more machines, more than one operator, different batches of raw material, because no one batch is sufficient for carrying out all the experiments. A particular case is in crossover experiments on animals where the design is too large for a single animal to undergo all possible treatment (see chapter 4).

In both cases we assume that the effects of factors in the two blocks are equivalent. The solution is to add one or several *qualitative* factors called **blocking factors** which are considered to be constant within a given block. We are assuming that these blocking factors will take account for the most part of the uncontrolled factors in the experiment, but that these uncontrolled factors can be assumed constant within a block.

If there are two blocks we take one blocking factor with values ±1. If the design consists of 3 blocks the blocking factor could take 3 levels, but it is simpler to take two blocking factors, each allowed to take levels ±1. This last solution applies also where there are 4 blocks. The arrangement, or blocking structure is such that if there are block effects, they will not bias our estimations of the main effects, nor of the interactions that we are trying to measure. On the other hand we generally assume that there is no interaction between the block effects and the main effects that we are studying – that is that the main effects and interactions do not change from one block to another. If this is not the case, then the block effects and their interactions must be taken into account just like any other, when constructing the design.

Consider for example the 2^{5-1} design obtained by joining the two fractional factorial designs of the extrusion-spheronization example described in sections III.B and III.D. Each of the 2^{5-2} designs may be considered as a separate block. We add a blocking factor X_6 set at -1 for the first block and +1 for the second (table 3.30).

The first part of the design had the defining relation:

$$I \equiv 234 \equiv 135 \equiv 1245$$

We see that **-6** is also a generator. So the new defining relation for the first part of the design, a 2^{6-3} design is:

$$I \equiv 234 \equiv 135 \equiv 1245 \equiv -6 \equiv -2346 \equiv -1356 \equiv -12456.$$

Similarly, for the second part of the design, **6** is a generator, so we combine this with the defining relation already established for the 2^{5-2} foldover design in section III.D.2, to give:

$$I \equiv -234 \equiv -135 \equiv 1245 \equiv 6 \equiv -2346 \equiv -1356 \equiv 12456.$$

Three of these generators are common to the two expressions and it is these that are found in the defining relation for the overall 2^{6-2} design.

$$I \equiv 1245 \equiv -2346 \equiv -1356$$

We now see why the third order effect $b_{234+135}$ was found to be large in the extrusion spheronization experiment. The defining relation shows that these second-order interactions **234** and **135** are confounded with the block effect **-6**.

Table 3.30 Blocking of Extrusion-Spheronization Experimental Design

No.	X_1	X_2	X_3	X_4	X_5	X_6 (block)
1	-	-	-	+	+	-
2	+	-	-	+	-	-
3	-	+	-	-	+	-
4	+	+	-	-	-	-
5	-	-	+	-	-	-
6	+	-	+	-	+	-
7	-	+	+	+	-	-
8	+	+	+	+	+	-
9	-	-	-	-	-	+
10	+	-	-	-	+	+
11	-	+	-	+	-	+
12	+	+	-	+	+	+
13	-	-	+	+	+	+
14	+	-	+	+	-	+
15	-	+	+	-	+	+
16	+	+	+	-	-	+

The block effect can be calculated from the contrast for b_6:

$$b_6 = (-y_1 -y_2 -y_3 -y_4 -y_5 -y_6 -y_7 -y_8+y_9+y_{10}+y_{11}+y_{12}+y_{13}+y_{14}+y_{15}+y_{16})/16$$

is the same as half the difference between the means of the two blocks (see table 3.23):

$$b_6 = \tfrac{1}{2}(b_0^{(2)} - b_0^{(1)}) = \tfrac{1}{2}(40.0 - 50.2) = -5.1$$

Evidently, the certain interval of time between carrying out the 2 blocks led to a difference between them in one or more of the uncontrolled factors. The difference was detected by blocking, but it did not falsify the analysis of the effects of the controlled factors.

Finally, we note that whole designs may be replicated as blocks, the results being treated by analysis of variance (see chapter 4).

V. OTHER DESIGNS FOR FACTOR STUDIES

A. ¾ Designs

1. The ¾ design for 4 factors (12 experiments)

Consider the synergistic model equation for 4 factors, consisting of the constant term, the 4 main effects and the 6 first-order interactions. These can be estimated using the ¾ of 2^4 design shown in table 3.31. This consists of 12 experiments, constructed as described in paragraph III.B.1, from 3 blocks of 4 experiments, defined by the following relations:

Block 1	I ≡	24 ≡	123 ≡	134
Block 2	I ≡	-24 ≡	123 ≡	-134
Block 3	I ≡	24 ≡	-123 ≡	-134

In spite of not being completely orthogonal, the design is perfectly usable (12, 13, 14). The inflation factors (see section IV of chapter 4) are acceptable (between 1.33 and 1.5) and the variances of all of the coefficients are equal to $\sigma^2/8$. In addition the very low redundancy and high R-efficiency (equal to 92%) makes this design very interesting in circumstances where economy is called for. It is less efficient than the factorial design with regard to precision, as the standard deviation of the estimates of the effects is the same as for a reduced factorial 2^{4-1} design of 8 experiments.

We may take the results of the 2^4 factorial study of an effervescent table formulation reported earlier, and select the data corresponding to the 12 experiments of table 3.31. Estimates of the coefficients obtained either by contrasts or by the usual method of multi-linear regression are very close to those estimated from the full design (shown in table 3.9).

Table 3.31 ¾ of a Full Factorial 2^4 Design

No.	X_1	X_2	X_3	X_4
1	-	-	+	-
2	+	-	-	-
3	-	+	-	+
4	+	+	+	+
5	-	-	+	+
6	+	-	-	+
7	-	+	-	-
8	+	+	+	-
9	-	-	-	-
10	+	-	+	-
11	-	+	+	+
12	+	+	-	+

2. Other ¾ designs of resolution V

We have seen previously that a 2^{5-1} design of resolution IV may be augmented to a ¾ of 2^5 design of resolution V, enabling estimation of main and first-order interaction effects. There is no resolution V ¾ of 2^{6-1} design of 24 runs. ¾ designs with 48 runs, both of resolution V, exist for 7 and for 8 factors (13).

3. Use of the ¾ design for a specific model

This type of design may also be used in cases where the experimenter proposes a specific model (compare section III.E.1). Suppose that 8 factors must be studied and the existence of 3 interaction effects, β_{12}, β_{13}, and β_{45} is suspected. We rename the 4 variables in the design of table 3.31, A, B, C, and D (table 3.32). We know that we may estimate these 4 main effects plus the effects corresponding to the interactions AB, AC, AD, BC, BD, and CD. We may also estimate ABC. As previously, we establish the correspondence between these variables and those whose effects we actually want to study (**1, 2, 3, 4, 5, 6, 7, 8, 12, 13, 45**). If variables X_1, X_2, and X_3 are made to correspond to A, B, and C, then the interaction effects **12** and **13** will correspond to AB and AC. X_4 and X_5 are in turn set to correspond to BC and BD, and then the interaction effect **45** will correspond to CD, which is also available. It only remains to set X_6, X_7, and X_8. These are made to correspond to the remaining factors D, AD, and ABC, which results in the design of table 3.32.

Table 3.32 ¾ of a Factorial 2^{8-4} Design

	A	B	C	D	AD	BC	BD	ABC
No.	X_1	X_2	X_3	X_6	X_7	X_4	X_5	X_8
1	-	-	+	-	+	-	+	+
2	+	-	-	-	-	+	+	+
3	-	+	-	+	-	-	+	+
4	+	+	+	+	+	+	+	+
5	-	-	+	+	-	-	-	+
6	+	-	-	+	+	+	-	+
7	-	+	-	-	+	-	-	+
8	+	+	+	-	-	+	-	+
9	-	-	-	-	+	+	+	-
10	+	-	+	-	-	-	+	-
11	-	+	+	+	-	+	+	-
12	+	+	-	+	+	-	+	-

B. Rechtschaffner Designs

These designs are little used, or at least their use is seldom reported in the literature, but they are sufficiently interesting to the pharmaceutical scientist to merit a brief description. They may be used where the synergistic model is limited to main effects and first-order interactions. Their R-efficiencies are 100%; there are only as many experiments as effects to estimate. They are constructed as follows (15):

- an experiment with all elements "-1"
- k experiments with 1 element "-1" and all others "+1"; all possible permutations of this
- $k(k-1)/2$ experiments with 2 elements "+1" and all others "-1"; all possible permutations of this.

See table 3.33 for the design for 6 factors.

They are not balanced and do not have the factorial designs' property of orthogonality (except the 5-factor Rechtschaffner design, identical to the 2^{5-1} factorial design of resolution V). With this single exception, the estimations of the effects are not completely independent. Their main advantage is that they are saturated.

Table 3.33 Rechtschaffner Design for 6 Factors

No.	X_1	X_2	X_3	X_4	X_5	X_6
1	-	-	-	-	-	-
2	-	+	+	+	+	+
3	+	-	+	+	+	+
4	+	+	-	+	+	+
5	+	+	+	-	+	+
6	+	+	+	+	-	+
7	+	+	+	+	+	-
8	+	+	-	-	-	-
9	+	-	+	-	-	-
10	+	-	-	+	-	-
11	+	-	-	-	+	-
12	+	-	-	-	-	+
13	-	+	+	-	-	-
14	-	+	-	+	-	-
15	-	+	-	-	+	-
16	-	+	-	-	-	+
17	-	-	+	+	-	-
18	-	-	+	-	+	-
19	-	-	+	-	-	+
20	-	-	-	+	+	-
21	-	-	-	+	-	+
22	-	-	-	-	+	+

Table 3.22 showed that the only factorial designs allowing estimation of all first-order interactions are 2^4, 2^{5-1}, 2^{6-1}, 2^{7-1}, 2^{8-2}, 2^{10-3}, and 2^{11-4}. Rechtschaffner designs may be used instead, for 4, 6 or 7 factors, to give less expensive estimates of the effects. For $k = 8$ a D-optimal design is probably better. See also the discussion of these designs in terms of variance inflation factors (VIF) in chapter 4, section IV.

Comparison of the different designs for 7 factors is instructive. If only first-order interactions are considered then the number of coefficients $p = 28$. Three possible solutions are given in table 3.34. The Rechtschaffner design is far more efficient than the full factorial design, as we have already seen, but also more efficient than the ¼ design, with a standard deviation of estimation of the effects only 25% higher than that of the ¼ design. The method of linear combinations cannot be used to estimate the effects from a Rechtschaffner design; multilinear least squares regression must be employed, as described in chapter 4.

Table 3.34 Comparison of Some Reduced Designs for 7 Factors

Design		R-eff	σ_b^2	σ_b
Fractional factorial (2^{7-1})	$N = 64$	45%	$\sigma^2/64 = 0.0156\sigma^2$	0.125σ
¾-design (¾ $\times 2^{7-1}$)	$N = 48$	60%	$\sigma^2/32 = 0.031\sigma^2$	0.177σ
Rechtschaffner	$N = 29$	100%	$0.05\sigma^2$	0.224σ

C. D-Optimal Designs

The whole of chapter 8 is devoted to these designs, which have a number of advantages but whose construction requires special computer programs. We give an example here of a D-optimal design (table 3.35) constructed as an alternative to the ¾ $\times 2^4$ design (table 3.31) and the 4-factor Rechtschaffner design (see above).

It is set up by selecting the best 12 experiments from the full factorial 2^4 design, those which provide most information, and give the most precise estimates of the coefficients.

Table 3.35 D-Optimal Design to Obtain the 4 Main Effects and all First-Order Interactions

No.	X_1	X_2	X_3	X_4	No.	X_1	X_2	X_3	X_4
1	-	-	-	-	7	-	+	+	-
2	+	-	-	-	8	-	-	-	+
3	+	+	-	-	9	-	+	-	+
4	+	-	+	-	10	+	+	-	+
5	-	-	+	-	11	+	-	+	+
6	+	+	+	-	12	-	+	+	+

In the same way as for the ¾ $\times 2^4$ design, we can select the data for these 12 experiments from the experimental results of table 3.8 (2^4 effervescent tablet factor study) and estimate the coefficients by multi-linear regression. (The linear combinations method is not applicable here.) The results, given in table 3.36, are almost identical to those found with the full design and reported in table 3.9.

Table 3.36 Estimations of Main Effects for the Effervescent Tablet Example, from the D-optimal Design of Table 3.35 and the Experimental Data (Effervescence Time) of Table 3.8

$b_0 = 105$ $b_1 = 3$ $b_3 = -18$ $b_{23} = 6$
$b_2 = -12$ $b_4 = 11$ All other interactions are negligible.

D. Conclusion

A number of the different designs available for factor-influence studies are given in table 3.37. We have tried there to summarise the recommendations of this section of the chapter and only the designs which seem to us the most frequently useful are included. Other minimum aberration designs are listed in table 3.22.

Table 3.37 Summary of the Best Designs for Factor Influence Studies

No of factors	Fractional factorial designs				Complete factorial designs		Other designs of resolution V (R_V)	
	R_{IV}		$R_{V/VI}$					
	Design	N	Design	N	Design	N	Design	N
2					2^2	4		
3					2^3	8		
4	2^{4-1}	8			2^4	16	Three-quarter	12
							D-optimal	12
							Rechtschaffner	11
5			2^{5-1}	16				
6	2^{6-2}	16	2^{6-1}	32			Rechtschaffner	22
7	2^{7-3}	16					Rechtschaffner	29
8	2^{8-4}	16	2^{8-2}	64			D-optimal	37
9	2^{9-4}	32					D-optimal	50

D-optimal designs are discussed more fully in the final section of chapter 8. All the designs of resolution V and higher for 7 or more factors require a large number of experiments and the reader's attention is drawn, in particular, to the resolution IV designs which may provide a viable alternative, though further experimentation may be required where interactions are found active.

References

1. D. C. Montgomery, Design and Analysis of Experiments, J. Wiley, N.Y., 1984.
2. E. Senderak, H. Bonsignore, and D. Mungan, Response surface methodology as an approach to the optimization of an oral solution, *Drug Dev. Ind. Pharm.*, **19**, 405-424 (1993).
3. A. Abebe, D. Chulia, J. P. Richer, M. Tendero, A. Verain, and P. Ozil, Formulation methodology: application to an effervescent form, *Proc. 5th Internat. Conf. Pharm. Tech.*, APGI (1989).
4. N. Draper and H. Smith, Applied Regression Analysis, 2nd edition, Wiley-Interscience, 177-182 (1981).
5. C. Daniel, Use of half-normal plots in interpreting two-level factorial

experiments, *Technometrics*, **1**, 311-342 (1959).

6. R. V. Lenth, Quick and easy analysis of unreplicated factorials, *Technometrics*, **31**(4), 469-473 (1989).

7. G. E. P. Box and R. D. Meyer, An analysis for unreplicated fractional factorials, *Technometrics*, **28**(1), 11-18 (1986).

8. M. Chariot, J. Francès, G. A. Lewis, D. Mathieu, R. Phan-Tan-Luu, and H. N. E. Stevens, A factorial approach to process variables of extrusion spheronization of wet powder masses, *Drug Dev. Ind. Pharm.*, **13**, 1639-1651 (1987).

9. C. F. Dick, R. A. Klassen, and G. E. Amidon, Determination of the sensitivity of a tablet formulation to variations in excipient levels and processing conditions using optimization techniques, *Int. J. Pharm.*, **38**, 23-31 (1987).

10. N. O. Lindberg and C. Jönsson, The granulation of lactose and starch in a recording high-speed mixer, Diosna P25, *Drug Dev. Ind. Pharm.*, **11**, 387-403 (1985).

11. A. Fries and W. G. Hunter, Minimum aberration 2^{k-p} designs, *Technometrics*, **22**, 601-608 (1980).

12. P. W. M. John, Statistical Design and Analysis of Experiments, MacMillan Ed., N. Y., 163-170, 1971

13. P. W. M. John, Three-quarter replicates of 2^4 and 2^5 designs, *Biometrics*, **17**, 319-321 (1961).

14. P. W. M. John, Three-quarter replicates of 2^n designs, *Biometrics*, **18**, 172-184 (1962).

15. R. L. Rechtschaffner, Saturated fractions of 2^n and 3^n factorial designs, *Technometrics*, **9**, 569-575 (1967).

Further Reading

•   G. E. P. Box, W. G. Hunter, and J. S. Hunter, Statistics for Experimenters, chapters 10 to 13, J. Wiley, N.Y., 1978.

•   F. Giroud, J. M. Pernot, H. Brun, and B. Pouyet, Optimization of microencapsulation of acrylic adhesives, *J. Microencapsul.*, **12**, 389-400 (1995).

•   R. Carlson, A. Nordahl, T. Barth, and R. Myklebust, An approach to evaluating screening experiments when several responses are measured, *Chemom. Intell. Lab. Syst.* **12**, 237-255 (1992).

•   N.-O. Lindberg and T. Lundstedt. Application of multivariant analysis in pharmaceutical development work, *Drug Dev. Ind. Pharm.* **21**, 987-1007 (1995).

4

STATISTICAL AND MATHEMATICAL TOOLS

Introduction to Multi-Linear Regression and Analysis of Variance

I. INTRODUCTION

Whether it is to understand or interpret a phenomenon, as in the factor study (chapter 3), or to predict results under different conditions by response surface modelling (as we will see in the next chapter), we require a mathematical model that is close enough in its behaviour to that of the real system. The models we use are polynomials of coded (normalised) variables, representing factors that are all transformed to the same scale and with constant coefficients. It is the unknown value of each coefficient that must be estimated with the best possible precision by experiments whose position in the experimental factor space is chosen according to the form of the mathematical model postulated. For this to be possible, the number of *distinct* experiments (that is not counting replications) must be at least equal to the number of coefficients in the model.

If the number of distinct experiments, N, equals the number of coefficients, the design is **saturated**. The calculation is therefore the resolution of a series of N equations with N unknowns.

However, each experimental result is associated with an error called the experimental error. Even if we suppose that our mathematical model does correspond to the reality (and it can only be a more or less good approximation), the values obtained for the coefficients by resolution of the system are only estimations, more or less close to the "true" values - those which would have been obtained in the absence of this experimental error. Now it is almost as important generally to know the precision of the estimates of the coefficients as to know their values and for this it is necessary to perform more experiments than there are coefficients: $N > p$. The extra experiments may be distinct or some or all of them may be replications. In both cases, the number of distinct experiments must exceed p.

Calculating estimations of the coefficients is no longer so simple and cannot be done manually. In fact, the method of linear combinations we have employed up to now for estimating the effects is very little used in practice. *Least squares multi-linear regression* already referred to in chapter 2 and requiring a computer program, is used for treating most designs and models covered in the rest of the book. The techniques described in this chapter therefore complete the sections on screening matrices and factor influence studies and are a necessary preparation for those on the use of second-order models, response surface methodology, and optimal designs. There are a large number of excellent books and articles devoted to the subject of multilinear regression analysis (1, 2). Here we will merely introduce the notation and those principles that are strictly necessary when considering the design of experiments.

II. LEAST SQUARES REGRESSION

A. Single Variable: Example of a First-Order Model

The principle will first be illustrated with some data for a response depending on a single variable, shown in figure 4.1. We assume a first-order relationship:

$$y = \beta_0 + \beta_1 x_1 + \varepsilon \qquad (4.1)$$

The variable X_1 is coded so that the points are centred around zero. This is done for simplicity, and because it is the case for most examples in this book. It is not necessary to the argument.

There is no straight line which passes through all the points. We will assume for the time being that this is because of experimental error. However it would be relatively easy to draw a straight line which passes fairly close to all the points. An objective method is required for determining the best straight line. Let the relationship for a straight line passing close to all the points be:

$$\hat{y} = b_0 + b_1x_1 \tag{4.2}$$

$\hat{y}$ ("y hat") is the calculated value of y, as shown by the double line in figure 4.1. Then, for each point corresponding to a value of x_1, we may calculate the difference between the experimental value of the response, y and the calculated value, $\hat{y}$. These are the *residuals,* shown by the vertical lines. The objective of least squares regression is to obtain the most probable value of the model coefficients, and hence the best straight line, by minimizing the sum of the squares of the residuals. The line of equation 4.2, shown in figure 4.1, is the fitted relationship using the estimates of the model coefficients calculated according to this criterion, that is by least squares regression.

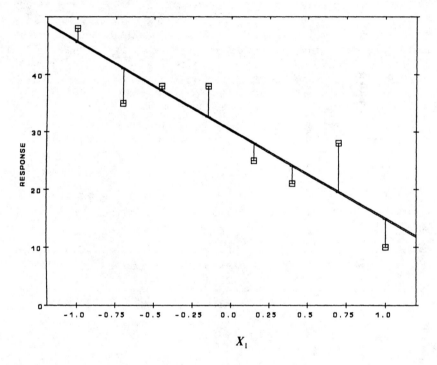

Figure 4.1 First-order dependence on a single variable x_1, showing regression line and residuals.

B. Two Variables: Example of Solubility in Surfactant Mixture

We introduce a simple example that will be used again and also extended in chapter 5 to illustrate the predictive use of models. The data will be used here to demonstrate least squares regression, and the succeeding statistical tests.

1. Definition of the problem

Bile salts such as sodium cholate, taurocholate and glycocholate form micelles in solution. Long chain phosphotidyl choline may be added and the phospholipid molecules, although for geometric reasons they do not form micelles on their own, can be incorporated into the bile acid micelles. These so-called mixed micelles are relatively well tolerated when injected intravenously. They have been used to solubilize water-insoluble vitamins and other slightly soluble pharmaceuticals, such as diazepam.

2. Experimental domain

We look at an experiment to investigate the solubilizing power of a mixed micelle system consisting of sodium cholate and lecithin on an experimental drug. This is evidently a mixture problem where the three components are the bile salt, lecithin and water and the response is the equilibrium solubility. The concentration of bile salt in such systems is normally allowed to vary between 2.5% and 7.5%. The phospholipid concentration is usually defined as its ratio with respect to the bile salt – from 0.25 to 1.25 molecules of lecithin to 1 molecule of bile salt. The percentage of water in the system varies from 78% to 97%. Since water is in considerable excess, the quantities (in percentage, concentration or molar ratio) of the two other constituents may be fixed independently of each other. Water content becomes the "slack variable". Taking rather narrower limits to the above, and expressing bile salt concentration in mole L^{-1}, we may define an experimental domain as in table 4.1.

Table 4.1 Experimental Domain: Solubility of an Experimental Drug in Mixed Micelles

Factor	Associated variable	Lower level (coded -1)	Upper level (coded +1)
Concentration of bile salt (M)	X_1	0.075	0.125
Lecithin-cholate molar ratio	X_2	0.6 : 1	1.4 : 1

3. Mathematical model

Considering that the domain is somewhat restricted we might assume as a first approximation that the solubility might be described satisfactorily by the synergistic model, equation 4.3:

$$y = \beta_0 + \beta_1 x_1 + \beta_2 x_2 + \beta_{12} x_1 x_2 + \varepsilon \tag{4.3}$$

We have already seen that this model may be estimated using a 2^2 factorial design.

4. Experimental design

Additional experiments are needed for there to be enough data ($N > p$) for a statistical analysis. We add an experiment to the design at the centre of the domain, which is the point furthest from the positions of the experiments of the factorial design. This will allow us to verify, at least partially, the mathematical model's validity. Therefore the solubility was determined in a mixed micelle containing 0.1M bile salt ($X_1 = 0$) and an equimolar lecithin to bile salt ratio ($X_2 = 0$). Also each distinct experiment was done twice. These replications, which are complete and carried out under identical experimental conditions, allow us to estimate the repeatability, *without any external influence,* and thus to have a good idea of the dispersion of the results and the extent of the experimental error. The 10 experiments are carried out in random order. The experiment design, plan, and the response data are all listed in table 4.2 in the standard order.

Table 4.2 2^2 Full Factorial Design for the Solubility of a Drug in Mixed Micelles

No.	X_1 X_2	Concentration of bile salt M	Lecithin-cholate molar ratio	Experimental solubility mg/mL	
1	-1 -1	0.075	0.6 : 1	6.58	6.30
2	+1 -1	0.125	0.6 : 1	10.18	9.90
3	-1 +1	0.075	1.4 : 1	9.41	10.03
4	+1 +1	0.125	1.4 : 1	14.15	14.75
5	0 0	0.100	1:1	11.70	11.04

C. Multi-Linear Regression by the Least Squares Method

1. Writing the model in matrix form

In postulating a synergistic mathematical model in the previous section, we supposed that each experimental result j may be written in that form, the variables x_i taking values corresponding to the experimental conditions x_{ji}:

$$y_1 = \beta_0 x_{10} + \beta_1 x_{11} + \beta_2 x_{12} + \beta_{12} x_{11} x_{12} + \varepsilon_1$$
...
$$y_j = \beta_0 x_{j0} + \beta_1 x_{j1} + \beta_2 x_{j2} + \beta_{12} x_{j1} x_{j2} + \varepsilon_j \tag{4.4a}$$
...

$$y_N = \beta_0 x_{N0} + \beta_1 x_{N1} + \beta_2 x_{N1} + \beta_{12} x_{N1} x_{N2} + \varepsilon_N$$

where:
- y_j is the value of the response for the jth experiment.
- β_i is one of the p coefficients of the postulated model.
- x_{ji} is the value of the ith variable for the jth experiment. As is usual in expressions of matrix components the row number (j) comes first, and the column number after.
- x_{j0} is a "pseudo-variable", constant and equal to 1, added simply to make the first element of the equation homogeneous with the others when equations are written in matrix form.
- ε_j is the (unknown) experimental error in the jth experiment.

Take for example the 5 distinct experiments of table 4.2. We have:

$$\begin{aligned}
y_1 &= \beta_0 - \beta_1 - \beta_2 + \beta_{12} + \varepsilon_1 \\
y_2 &= \beta_0 + \beta_1 - \beta_2 - \beta_{12} + \varepsilon_2 \\
y_3 &= \beta_0 - \beta_1 + \beta_2 - \beta_{12} + \varepsilon_3 \\
y_4 &= \beta_0 + \beta_1 + \beta_2 + \beta_{12} + \varepsilon_4 \\
y_5 &= \beta_0 + \varepsilon_5
\end{aligned}$$
(4.4b)

and a similar set of equations for the 5 replications.

The set of 5 distinct equations 4.4b may be written in matrix form:

$$\mathbf{Y} = \mathbf{X}\beta + \mathbf{\varepsilon}$$
(4.4c)

where:

$$\mathbf{Y} = \begin{bmatrix} y_1 \\ y_2 \\ y_3 \\ y_4 \\ y_5 \end{bmatrix} \quad \mathbf{X} = \begin{bmatrix} x_{10} & x_{11} & x_{12} & x_{11}x_{12} \\ x_{20} & x_{21} & x_{22} & x_{21}x_{22} \\ x_{30} & x_{31} & x_{32} & x_{31}x_{32} \\ x_{40} & x_{41} & x_{42} & x_{41}x_{42} \\ x_{50} & x_{51} & x_{52} & x_{51}x_{52} \end{bmatrix} \quad \beta = \begin{bmatrix} \beta_0 \\ \beta_1 \\ \beta_2 \\ \beta_{12} \end{bmatrix} \quad \mathbf{\varepsilon} = \begin{bmatrix} \varepsilon_1 \\ \varepsilon_2 \\ \varepsilon_3 \\ \varepsilon_4 \\ \varepsilon_5 \end{bmatrix}$$
(4.4d)

$\mathbf{X}$ is a $N \times p$ matrix, called the **effects matrix** or **model matrix**, having as many columns as there are coefficients in the model and as many rows as there are experiments. $\mathbf{Y}$ is the vector (column matrix) of the experimental responses, β is the vector of the coefficients and $\mathbf{\varepsilon}$ is the vector of the experimental errors.

In the solubility example above, having 10 experiments and with 4 terms in the model, the responses vector $\mathbf{Y}$ and the model matrix $\mathbf{X}$, the coefficients vector β, and the error vector $\mathbf{\varepsilon}$ are:

$$Y = \begin{bmatrix} 6.58 \\ 6.30 \\ 10.18 \\ 9.90 \\ 9.41 \\ 10.03 \\ 14.15 \\ 14.75 \\ 11.70 \\ 11.04 \end{bmatrix} \quad X = \begin{bmatrix} +1 & -1 & -1 & +1 \\ +1 & -1 & -1 & +1 \\ +1 & +1 & -1 & -1 \\ +1 & +1 & -1 & -1 \\ +1 & -1 & +1 & -1 \\ +1 & -1 & +1 & -1 \\ +1 & +1 & +1 & +1 \\ +1 & +1 & +1 & +1 \\ +1 & 0 & 0 & 0 \\ +1 & 0 & 0 & 0 \end{bmatrix} \quad \beta = \begin{bmatrix} \beta_0 \\ \beta_1 \\ \beta_2 \\ \beta_{12} \end{bmatrix} \quad \varepsilon = \begin{bmatrix} \varepsilon_1 \\ \varepsilon_2 \\ \varepsilon_3 \\ \varepsilon_4 \\ \varepsilon_5 \\ \varepsilon_6 \\ \varepsilon_7 \\ \varepsilon_8 \\ \varepsilon_9 \\ \varepsilon_{10} \end{bmatrix} \quad (4.5)$$

The model matrix has been, and will normally continue to be represented as a table, as in table 4.3, below. Here it consists of 10 lines, each corresponding to an experiment in the design, and 4 columns corresponding to the 4 parameters of the model.

Table 4.3 Effects (or Model) Matrix **X** of a Complete 2^2 Factorial Design with Centre Point and Each Experiment Repeated

X_0	X_1	X_2	X_1X_2
+1	-1	-1	+1
+1	-1	-1	+1
+1	+1	-1	-1
+1	+1	-1	-1
+1	-1	+1	-1
+1	-1	+1	-1
+1	+1	+1	+1
+1	+1	+1	+1
+1	0	0	0
+1	0	0	0

The columns of the model matrix for a 2-level factorial design correspond to the linear combinations for calculating its coefficients.

2. Linear regression by the least squares method: estimating the "best" value of the model coefficients

In the model with 2 independent variables and an interaction term, the "true" unknown values of the coefficients are β_0, β_1, β_2, β_{12}. Because experimental result is associated with a random error, ε_i, it is impossible to measure their exact values – we can only estimate these. On replacing the coefficients by their estimators b_0, b_1, b_2, b_{12} we obtain:

$$\hat{y}_i = b_0 + b_1 x_{i1} + b_2 x_{i2} + b_{12} x_{i1} x_{i2}$$

where $\hat{y}_i$ is the response value *calculated by the model* for the point i. Multi-linear least squares regression leads to estimates of the coefficients which minimize the square of the difference between the calculated value and the experimental value (the residual), summed over all of the experiments, called SS_{RESID}.

In *least squares regression* the coefficients b_i, estimates of the true values β_i, are chosen so that the residual sum of squares $SS_{RESID} = \sum(\hat{y}_i - y_i)^2$ is *minimized*.

The estimators are grouped in the vector **B**, equivalent to the vector of the true unknown values β in equation 4.4c. They are calculated by:

$$\mathbf{B} = \begin{bmatrix} b_0 \\ b_1 \\ b_2 \\ b_{12} \end{bmatrix} = (\mathbf{X'X})^{-1}\mathbf{X'Y} \tag{4.6}$$

where $\mathbf{X'}$ is the *transpose matrix* of $\mathbf{X}$, with p rows and N columns and $(\mathbf{X'X})$ is the square $(p \times p)$ *information matrix*, the product of the transpose of $\mathbf{X}$ with $\mathbf{X}$. When inverted it gives $(\mathbf{X'X})^{-1}$, the *dispersion matrix*. $\mathbf{X'X}$ may only be inverted if $\det(\mathbf{X'X}) \neq 0$.

Both the information and the dispersion matrices are of great importance, not only in determining coefficients by multi-linear regression, but also in accessing the quality of an experimental design. They will be referred to quite frequently in the remainder of this chapter and the remainder of the book. In our example the information and dispersion matrices are as follows:

$$(\mathbf{X'X}) = \begin{bmatrix} 10 & 0 & 0 & 0 \\ 0 & 8 & 0 & 0 \\ 0 & 0 & 8 & 0 \\ 0 & 0 & 0 & 8 \end{bmatrix} \qquad (\mathbf{X'X})^{-1} = \begin{bmatrix} \frac{1}{10} & 0 & 0 & 0 \\ 0 & \frac{1}{8} & 0 & 0 \\ 0 & 0 & \frac{1}{8} & 0 \\ 0 & 0 & 0 & \frac{1}{8} \end{bmatrix} \qquad (4.7)$$

If an experimental design has the property of *orthogonality*, this will lead to a *diagonal* information matrix. Inversion of such a matrix is trivial, and can be done immediately without calculation. Most of the designs we have studied up until now have this property but this will not be the case at all for most of the ones we will study from now on. A computer program is needed to invert the information matrices of those designs.

Least squares multi-linear regression is by far the most common method for estimating "best values" of the coefficients, but it is not the only method, and is not always the best method. So-called "robust" regression methods exist and may be useful. These reduce the effect on the regression line of outliers, or apparently aberrant data points. They will not be discussed here, and least squares regression is used in the examples throughout this book.

3. Calculation of the coefficients

Applying equation 4.5 to the design and the response data of table 4.2 we obtain an estimate of each coefficient:

$$b_0 = 10.40 \qquad b_1 = 2.08 \qquad b_2 = 1.92 \qquad b_{12} = 0.28$$

Therefore:

$$\hat{y}_i = 10.4 + 2.08\, x_{i1} + 1.92\, x_{i2} + 0.28\, x_{i1}x_{i2}$$

4. Derivation of the least squares multi-linear regression equation

Equation 4.3 has been presented without proof. The interested reader is advised to consult one of the standard texts on multiple linear regression (1, 2). An outline is presented here. We take the model equation 4.4c:

$$\mathbf{Y} = \mathbf{X}\beta + \varepsilon.$$

In this equation $\mathbf{X}$ (the model matrix) and $\mathbf{Y}$ (the vector of responses values) are known quantities and β and ε are unknown. The vector β can therefore only be estimated, by a vector $\mathbf{b}$, so that the calculated response values are: $\hat{\mathbf{Y}} = \mathbf{Xb}$. Let e_i be the difference between the calculated value $\hat{y}_i$ and the experimental value y_i, and let $\mathbf{e}$ be the corresponding vector of the differences. Thus $\mathbf{e} = \mathbf{Y} - \hat{\mathbf{Y}} = \mathbf{Y} - \mathbf{Xb}$.

The sum of squares of the differences e_i may be written:

$$\sum_{i=1}^{N} e_i^2 = \mathbf{e'e} = (\mathbf{Y} - \mathbf{X}\beta)'(\mathbf{Y} - \mathbf{X}\beta)$$

$$= \mathbf{Y'Y} - \beta\mathbf{X'Y} - \mathbf{Y'X}\beta + \beta'\mathbf{X'X}\beta \qquad (4.8)$$

$$= \mathbf{Y'Y} - 2\beta\mathbf{X'Y} + \beta'\mathbf{X'X}\beta$$

The different terms that are added or subtracted in equation 4.8 are scalar quantities (a vector multiplied by its transpose, or vice versa, is a scalar, see appendix I). $\mathbf{Y'Y}$ is the sum of squares of the responses. The so-called "least-squares" solution to this is the vector of estimators $\mathbf{B}$ which minimizes the sum of squares of the differences. The partial differential of the expression 4.8 with respect to the $\mathbf{b}$ must equal zero for $\mathbf{b} = \mathbf{B}$:

$$\left(\frac{\partial(SS)}{\partial\mathbf{b}}\right)_{b=B} = -2\mathbf{X'Y} + 2\mathbf{X'XB} = 0$$

$$\mathbf{X'XB} = \mathbf{X'Y} \qquad (4.9)$$

Since $\mathbf{X}$ has N rows and p columns, its transpose, $\mathbf{X'}$, will have N columns and p rows (see appendix I). Thus the product of $\mathbf{X}$ with its transpose, $(\mathbf{X'X})$, the information matrix, is a square matrix, and can be inverted, provided $\det(\mathbf{X'X}) \neq 0$, to give the dispersion matrix $(\mathbf{X'X})^{-1}$. Thus, multiplying each side of equation 4.9 by $(\mathbf{X'X})^{-1}$ we obtain the expression 4.6:

$$(\mathbf{X'X})^{-1}\mathbf{X'}\,\mathbf{XB} = (\mathbf{X'X})^{-1}\mathbf{X'Y}$$

$$\mathbf{B} = (\mathbf{X'X})^{-1}\mathbf{X'Y}$$

The structure of the information matrix of a 2-level factorial design, with or without centre-points, is simple, and is easily inverted to give $(\mathbf{X'X})^{-1}$. Inversion for most other designs is by no means trivial, but it can be done rapidly by computer.

We saw the 2^4 full factorial design for a model with all first-order interactions consisted of 16 experiments to determine 11 coefficients. These were estimated from linear combinations of the experimental results. They can also be estimated by least squares regression, with exactly the same results. Therefore least squares regression is an alternative to the method of linear combinations for estimating effects from factorial or Plackett and Burman designs. The calculations are more complex, but this is of little significance as these calculations are invariably carried out using a computer, except in the case of the simplest designs. However, the least squares method is also used to estimate coefficients from the non-standard designs described in chapter 3 (¾ and other irregular fractions, Rechtschaffner, D-optimal) and those for second-order models, where the simpler method cannot be applied.

5. Saturated experimental designs

Here the number of experiments equals the number of coefficients and the model matrix X is square. There is a unique solution and the error cannot be estimated. The calculated responses $\hat{Y}$ are identical to the experimental responses Y. X is a $p \times p$ matrix, and can be inverted to X^{-1}, provided the design is non-singular (see below). We have therefore:

$$B = X^{-1}Y$$

However, the previous method, though apparently more complex, is still valid and the dispersion matrix, $(X'X)^{-1}$, gives the same invaluable information on the quality of the experimental design. Designs with fewer experiments than there are coefficients in the model cannot give model estimates. Certain designs with as many experiments as there are coefficients, or even more, have $\det(X'X)^{-1} = 0$ and these cannot be used to determine the model. They are termed *singular*.

III. ANALYSIS OF VARIANCE (ANOVA)

A. Example of a Single Factor

1. Analysis of variance of the model

We can assume that we have the right model and the correct design. But can we test this and know whether the model and the parameters we are estimating are actually significant? The analysis of variance (ANOVA) is the basis of the statistical analysis that follows the fitting of the model. We have calculated the model coefficients, but we should ask ourselves here whether they are meaningful or whether the deviations of the data points from a constant value are simply due to chance, to random variation of the response, because of measurement errors or variation or drift in uncontrolled factors.

The principle of the method will first be illustrated using the data that were shown in figure 4.1, with the fitted line for the model coefficients (equation 4.1) calculated by least squares regression:

$$\hat{y} = b_0 + b_1 x_1$$

The total sum of squares SS is the sum of the squares of the differences from zero of all the points (see figure 4.2). It is associated with N data. It is usual instead to calculate the sum of squares of the deviations from the mean value of the responses, instead of the sum of squares of deviations from zero, as indicated in figure 4.3. So the *adjusted total sum of squares* SS_{TOTAL} is given by:

$$SS_{TOTAL} = \sum_{i=1}^{N} (y_i - \bar{y})^2 = 985.87$$

and since one parameter, the mean value, is involved in the calculation, we "lose" one degree of freedom. The adjusted total sum of squares is associated with $N-1$ degrees of freedom, where N is the number of data.

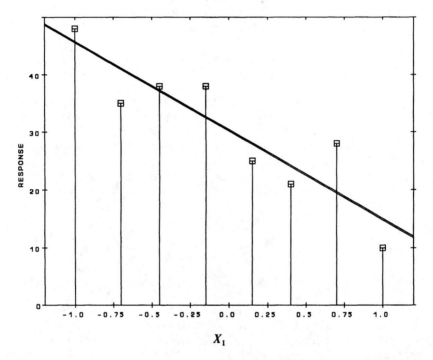

Figure 4.2 Total sum of squares (the sum of the squares of the response values).

This sum of squares of the deviations about the mean is then divided into two parts: SS_{REGR} and SS_{RESID}. We will see later that each of these may be split up its turn.

For SS_{RESID} we take the deviations from the regression line, or residuals, shown in figure 4.1. The *residual sum of squares* SS_{RESID} has already been defined as the sum of squares of the differences between experimental data (y_i) and predicted response values $(\hat{y}_i)$. This sum of squares therefore represents deviations of the experimental data from the model.

$$SS_{RESID} = \Sigma(y_i - \hat{y}_i)^2 = 186.61$$

The remaining sum of squares (usually much larger) represents that part of the data that is explained by the model. This is the *regression sum of squares*, SS_{REGR}, defined by:

$$SS_{REGR} = SS_{TOTAL} - SS_{RESID} = 799.26$$

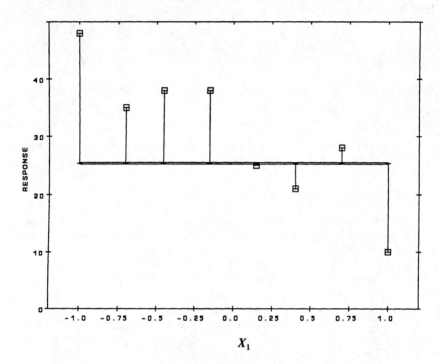

Figure 4.3   Total adjusted sum of squares (about the mean value).

The regression sum of squares SS_{REGR} is associated with $v_1 = p - 1$ degrees of freedom, where p is the number of coefficients in the model, because the constant term has already been subtracted in adjusting the total sum of squares. In this example, where $p = 2$, there is 1 degree of freedom.

The number of degrees of freedom v_2 associated with the residual sum of squares SS_{RESID} is calculated as the number of data points N minus the number of coefficients in the model.

$$v_2 = N - p = 8 - 2 = 6$$

So what use is all of this? We have calculated the coefficients of a model but as noted previously we need to know whether the deviations of the data points from the constant value are meaningful, or simply due to random variation.

2. Significance testing: the F-test

Each of the sums of squares can be divided by the number of degrees of freedom to give a mean square MS,

$$MS_{REGR} = \frac{SS_{REGR}}{p - 1} = 799.26$$

$$MS_{RESID} = \frac{SS_{RESID}}{N - p} = 31.102$$

The **residual mean square** MS_{RESID} can be considered as an estimate of the **variance** and its square root as an estimate of the **standard deviation** of the experimental technique. (A certain degree of caution is in order here – the residuals will, as we have seen, include random variation, but may also be partly a result of shortcomings in the model. For the moment we have no way of telling. When some of the experiments are replicated the dispersion of the results about the mean value allows another estimation of the experimental variance to be obtained, this time without any bias.)

The **regression mean square** MS_{REGR} is the sum of squares per coefficient, explained by the model. It is associated, as is the corresponding sum of squares, with $p - 1$ degrees of freedom. If the dependence of the response on the model parameters were simply a result of chance fluctuations the regression mean square would also be an estimate of the variance. In this case regression and residual means squares would be expected to be similar, and their ratio to be around one.

The ratio F is therefore calculated:

$$F_{v_1, v_2} = \frac{MS_{REGR}}{MS_{RESID}} = 25.69$$

If we are to prove the relationship is statistically significant, then F must exceed a critical value, this critical value being rather greater than 1.

One approach is to calculate the value of $F_{(p,\ N-p-1)}$ and compare it with the critical value in a table. It can be seen in the very partial table of values, table 4.4, given purely for the purpose of illustration, that the critical value decreases quite slowly with the value of $v_1 = p-1$ (except for the first two rows, where they increase slightly) but that if the residual variance is calculated with few degrees of freedom $v_2 = N-p$ the critical value of F can be very high. For statistical calculation at least 2 residual degrees of freedom, and preferably 3 or more, are required to give meaningful results.

For the problem we have been examining, $v_1 = 1$ and $v_2 = 8 - 1 - 1 = 6$. Looking up the critical value of F, for a probability of 95% that the response is not constant but depends on x_1, that is a less than 5% probability (0.05) that the result was due to chance, we obtain in the above table a value of 5.99. The probability that the trend observed is the result of random fluctuation or error is thus less than 5%. It is also greater than the critical value at 1%, 13.74.

We may also look up the critical value at 0.1%, which is 35.51. The calculated value of F is less than this, therefore the probability lies between 99% and 99.9%. The significance is between 0.01 and 0.001 (1% and 0.1%). Alternatively, most computer programs will now give us directly the calculated significance of the value of F, for the given numbers of degrees of freedom.

Table 4.4 Some Critical Values of F at 5% Probability

Degrees of freedom v_2 (residual)	Degrees of freedom v_1 (parameters in model)				
	1	2	3	4	5
1	161.4	199.5	215.7	224.6	230.2
2	18.5	19.0	19.2	19.2	19.3
3	10.13	9.55	9.28	9.12	9.01
4	7.71	6.94	6.59	6.39	6.26
5	6.61	5.79	5.41	5.19	5.05
6	5.99	4.14	4.76	4.53	4.39

3. R^2

R^2 is the proportion of the variance explained by the regression according to the model, and is therefore the ratio of the regression sum of squares to the total sum of squares.

$$R^2 = \frac{(SS_{TOTAL} - SS_{RESIDUAL})}{SS_{TOTAL}}$$

For the above example with only one variable it is equal to the regression coefficient r^2.

B. Analysis of Variance for the Replicated 2^2 Full Factorial Design with Centre Points

1. Analysis of variance of the model

We continue the surfactant mixture (mixed micelles) solubility example introduced in the first part of this chapter. If only 4 experiments are carried out at the factorial points and the model equation contains 3 coefficients plus the constant term, then the design is saturated. The model will fit the data exactly:

$$SS_{REGR} = SS_{TOTAL} \quad \text{and} \quad SS_{RESID} = 0$$

but with no degrees of freedom.

We have seen, in actual fact, that each experiment was duplicated and two experiments were also carried out at the centre. The coefficients in the model:

$$y = \beta_0 + \beta_1 x_1 + \beta_2 x_2 + \beta_{12} x_1 x_2 + \varepsilon$$

were estimated by linear regression. These estimates, b_0, b_1, b_2, b_{12}, are those given in section II.C.3. We may then determine for each datum the difference $y_i - y_m$ between each response and the mean, and hence the total adjusted sum of squares. The response is calculated for each data point x_{i1}, x_{i2}:

$$\hat{y}_i = 10.4 + 2.08 \, x_{i1} + 1.92 \, x_{i2} + 0.28 \, x_{i1} x_{i2}$$

This allows us to calculate the sum of squares explained by the model:

$$SS_{REGR} = SS_{TOTAL} - SS_{RESID} = 67.9 - 3.00 = 64.9$$

The regression sum of squares is associated with 3 degrees of freedom, equal to the number of coefficients in the model (except for the constant term already accounted for in calculating SS_{TOTAL} about the mean). The mean regression sum of squares is obtained by dividing SS_{REGR} by the number of degrees of freedom associated with the regression, $p - 1 = 3$.

$$MS_{REGR} = s_1^2 = SS_{REGR}/3 = 21.63$$

Similarly, the residual sum of squares representing the part of the response not accounted for by the model is associated with $N - p = 6$ degrees of freedom. The mean square value is therefore:

$$MS_{RESID} = s_2^2 = SS_{RESID}/6 = 3.00/6 = 0.50$$

If the model describes the studied response well, then MS_{RESID} is an unbiased estimate of the variance of the experiment σ^2, describing its repeatability. Its square root $s = \sqrt{MS_{RESID}} = 0.70$ is an estimate of the standard deviation of the experimental technique. The determination of the total and residual sums of squares is shown in table 4.5.

Table 4.5 Total and Residual Sums of Squares for the 2^2 Design of Table 4.2

	X_1	X_2	y_i	$\hat{y}_i$	$y_i - y_m$	$(y_i - y_m)^2$	$y_i - \hat{y}_i$	$(y_i - \hat{y}_i)^2$
1	-1	-1	6.58	6.68	-3.82	14.59	-0.10	0.0100
	-1	-1	6.30	6.68	-4.10	16.81	-0.38	0.1444
2	+1	-1	10.18	10.28	-0.22	0.05	-0.10	0.0100
	+1	-1	9.90	10.28	-0.50	0.25	-0.38	0.1444
3	-1	+1	9.41	9.96	-0.99	0.98	-0.55	0.3025
	-1	+1	10.03	9.96	-0.37	0.14	0.07	0.0049
4	+1	+1	14.15	14.69	3.75	14.06	-0.54	0.2916
	+1	+1	14.75	14.69	4.35	18.92	0.06	0.0036
5	0	0	11.70	10.40	1.30	1.69	1.30	1.6900
	0	0	11.04	10.40	0.64	0.41	0.64	0.4096
Σ					$SS_{TOTAL}=$	67.90	$SS_{RESID} =$	3.01

Since certain experiments have been replicated (in this case all of the experiments, but the treatment is the same if only some of the experiments are repeated) the residual sum of squares may be divided further, into two parts: pure error, and lack-of-fit. The *pure error sum of squares* is given by:

$$SS_{ERR} = \sum_{i=1}^{n} \sum_{j=1}^{r_i} (y_{ij} - \bar{y}_i)^2 = 0.667$$

where n is the number of distinct experiments (5 in this case), also referred to as *cells*,

r_i is the number of experiments carried out in the ith cell (number of replications of the ith experiment – in this case 2 for all the cells),

y_{ij} is the result of the jth experiment of the ith cell,

$\bar{y}_i$ is the mean of the experimental results in the ith cell,

and the pure error sum of squares SS_{ERR} is associated with v_3 degrees of freedom:

$$v_3 = \sum_{i=1}^{n} (r_i - 1) = 5$$

As before, we may define an *error mean square*, dividing this by the number of degrees of freedom:

$$MS_{ERR} = s_3 = SS_{ERR}/v_3 = 0.67/5 = 0.133$$

This is an *unbiased* estimate of the experimental variance σ^2. As the number of degrees of freedom increases this becomes a more reliable estimate of the "true" standard deviation σ.

The remainder of the residual sum of squares is the *lack-of-fit sum of squares*:

$$SS_{LOF} = SS_{RESID} - SS_{ERR} = 3.00 - 0.667 = 2.33$$

associated with:

$$v_4 = v_2 - v_3 = 6 - 5 = 1 \text{ degree of freedom.}$$

$$MS_{LOF} = s_4 = SS_{LOF}/v_4 = 2.33/1 = 2.33$$

The various terms for the specific example being treated are summarised in table 4.6 .

Table 4.6   Analysis of Variance of the Regression

	Degrees of freedom	Sum of squares	Mean square	F	Significance
Total	9	67.9	–	–	
Regression	3	64.9	32.131	161.8*	***
Residual	6	3.00	0.50	–	
Lack of fit	1	2.33	2.33	17.45	**
Pure error	5	0.67	0.134		

* Obtained by dividing the regression mean square by the pure error mean square (3, 5 degrees of freedom).

The *significance* in the table is the probability (between 0 and 1) of obtaining a ratio of mean squares greater than F. In table 4.6 and what follows we represent the significance level in the conventional manner:

***	:	< 0.001 (0.1%)
**	:	< 0.01 (1%)
*	:	< 0.05 (5%)
α	:	≥ 0.05 (5%)

The relations between the sums of squares are shown diagrammatically in figure 4.4.

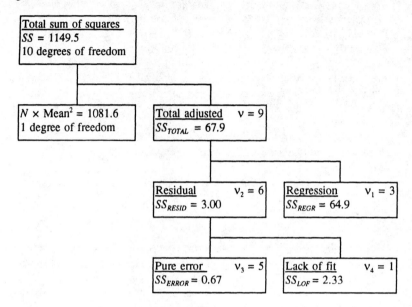

Figure 4.4 ANOVA "tree" showing the relationship of different sums of squares.

Fisher's test ($F = MS_{LOF}/MS_{ERR}$) allows the two estimates of the variance, s_4^2 and s_3^2 to be compared. A ratio much larger than 1 would indicate to us that the estimation s_4^2 is too high and that therefore the model is inadequate, certain necessary terms having been omitted. In Fisher's tables, a value $F_{1,5} = 6.60$ corresponds to a significance level of 0.05 (5%). Two cases may be envisaged:

- The two estimations, s_3^2 and s_4^2, may be considered as significantly different. The mathematical model is thus rejected and it is therefore s_3^2, derived from the error sum of squares which is retained as an estimate of σ^2. Fisher's test may be carried out a second time to compare s_1^2 with our estimate of the experimental variance s_3^2: $F = MS_{REGR}/MS_{ERR} = 161.8$. With this value the significance level of the Fisher test $F_{3,5}$ is much better than 0.001 (0.1%), and the regression is therefore highly significant (symbolised by ***). Therefore although there are statistically significant deviations from the mathematical model, the model itself is still highly significant and explains the major part of the deviations of the responses from their mean value $\bar{y}$.
- The estimations s_3^2 and s_4^2 are *not* considered to be significantly different.

The mathematical model may be accepted and a more precise estimate of σ^2 obtained by combining our two estimations. In this case it is s_2^2 derived from the residual sum of squares, which is retained as an estimation of σ^2. Fisher's test may be carried out again, this time to compare s_1^2 with s_2^2: $F = MS_{REGR}/MS_{RESID} = 32.13/0.5 = 62.26$. The significance level is still lower than 0.001 and the regression is therefore highly significant.

It is clear that it is the former case that applies here, as $F = 17.45$. Note that in the absence of repeated experiments the significance of the regression is calculated by means of the test $F = MS_{REGR}/MS_{RESID}$ but the validity of the model may not be tested.

As we continue with the example we will use the following values for the estimated experimental variance and standard deviation:

$$s^2 = 0.134 \qquad\qquad \text{: estimation of } \sigma^2 \text{ with } \nu = 5 \text{ degrees of freedom}$$

$$s = \sqrt{0.134} = 0.36 \qquad \text{: standard deviation (experimental error)}$$

This is an elementary example where the results of linear regression and analysis of variance could have been obtained by apparently simpler methods. Nevertheless, we shall see that this approach will prove necessary for more complex models and further development of the analysis of variance is the basis of statistical tests for analysis and testing of models against the data.

In particular, we shall see that the regression sum of squares can also be subdivided, each part being associated with certain terms in the model – a sum of squares for the linear model, another for the quadratic and interaction terms. The statistical significance of the different terms in the model can then be assessed.

3. R^2 and R^2_{adj}

R is the *multiple regression coefficient*, also called the coefficient of determination. It is the square of the coefficient which is always calculated and used.

The proportion of the variance explained by the regression R^2 can be calculated as in section III.A.3 as the ratio of the regression sum of squares to the total adjusted sum of squares.

$$R^2 = \frac{SS_{REGR}}{SS_{TOTAL}} = \frac{66.90}{67.90} = 0.955$$

As R^2 approaches 1, the better is the fit and the better apparently is the model. However there are limitations to its use in multiple linear regression, especially in comparing models with different number of coefficients, fitted to the same data set. A saturated model will inevitably give a perfect fit and a model with almost as many coefficients as there are data is quite likely to give a high value for R^2. The following correction for the number of degrees of freedom takes this into account by replacing the sums of squares by the equivalent mean squares *MS*.

$$R^2_{adj} = \frac{(MS_{TOTAL} - MS_{RESID})}{MS_{TOTAL}}$$

$$= 1 - \frac{(1 - R^2)(N - 1)}{(N - p)}$$

Its value is rather less than R^2. For this example we have:

$$R^2_{adj} = \frac{(7.544 - 0.500)}{7.544} = 0.934$$

C. Statistical Significance of the Coefficients

1. Variance

When we have an estimate of the experimental variance (either from repeating experiments, as here, or by having carried out more distinct experiments than there are coefficients in the model, or from having done some preliminary experiments), the statistical significance of the individual coefficients may be calculated.

In the dispersion matrix $(\mathbf{X'X})^{-1}$ let c^{ij} be the element of the ith row and the jth column. The c^{ii}, the element of the ith row and the ith column, are thus the diagonal terms of the matrix. The *variance of the ith coefficient* of the model is given by:

$$\sigma_{b_i}^2 = c^{ii}\sigma^2$$

and replacing σ^2 by the estimation s^2 we have:

$$s_{b_i}^2 = c^{ii}s^2$$

For example, the variance of b_0 can be obtained by:

$$s_{b_0}^2 = c^{11}s^2 = \frac{1}{10} \times 0.134 = 0.0134$$

that is, the standard deviation for the constant term is $\sqrt{0.0134} = 0.116$. Column 3 of table 4.7 lists the standard deviations of the coefficients for the solubility example. We will call these standard deviations the *standard errors* of the coefficients.

2. Statistical significance

The significance of each coefficient may then be calculated by means of Student's t variable with $v = 5$ degrees of freedom for the experimental variance. This is defined as $t = b_i/s_{b_i}$ (see column 4 of table 4.7). Comparison with the values of $t_{v,\alpha/2}$ in the tables of Student's t allow us to determine the significance level α, given in column 5 of the table. We will represent this significance level in the conventional manner in what follows (as we have done for Fisher's test):

***	:	$\alpha < 0.001$ (0.1%)
**	:	$\alpha < 0.01$ (1%)
*	:	$\alpha < 0.05$ (5%)
α	:	$\alpha \geq 0.05$ (5%)

It can be seen in table 4.7 that the constant term and the coefficients of the main effects are highly significant. On the other hand the first-order interaction is negligible.

3. Confidence limits

At a significance level α, the individual **confidence interval** of a coefficient is estimated by:

$$b_i - t_{v,\alpha/2} \times s_{b_i} \leq \beta_i \leq b_i + t_{v,\alpha/2} \times s_{b_i}$$

where v is the number of degrees of freedom. This confidence interval, $t_{v,\alpha/2} \times s_{b_i}$ is shown in the final column of table 4.7, for a two-way significance of 0.025 and $v = 5$ ($t_{5,\,0.025} = 2.57$). The probability of the coefficient being within these limits is 95%.

Table 4.7 Standard Deviations, Significance and Confidence Limits of the Coefficient Estimations

	Coefficient	Standard deviation	t	Significance	Confidence limits $t_{v,\alpha/2} \times s_{b_i}$
b_0	10.40	0.116	89.8	***	0.30
b_1	2.08	0.129	16.12	***	0.33
b_2	1.92	0.129	14.88	***	0.33
b_{12}	0.28	0.129	2.17	< 10%	0.33

D. Replication of a Factorial Design

1. Replication of runs within a factorial design

Replicated experiments within a design allow for estimates of the repeatability. For quantitative factors, it is usually the centre points which are added and replicated. The design may then be analysed by ANOVA instead of using the saturated design methods. Further examples are given in chapters 5 and 9.

However since pharmaceutical processes are usually quite reproducible it is fairly unusual to replicate whole designs to improve the precision of the experimental measurements, though there are a few examples in the pharmaceutical literature (3, 4). Exceptions are experiments involving animals or carried out with biological material where the inherent variability is often high (5). This run-to-run variability is a combination of the variation between animals (standard deviation σ_b), and that within the same animal and that due to other experimental variation (combined standard deviation σ_s).

Formulations of the kind described above have been used quite widely in solubilising lipophilic drugs and vitamins, for oral, parenteral, and transdermal drug delivery. We consider the 2^2 design plus centre point tested in an animal model for the pharmacokinetic profile or pharmacological effect of the solubilised active substance. If each of the formulation was tested in two animals, with no testing of different formations on the same animal, the analysis would be identical to that of the solubility experiment above. The major part of the variance σ^2 consists of that *between animals* σ_b^2 and *within the same animal* σ_s^2, where $\sigma^2 = \sigma_b^2 + \sigma_s^2$. It is this variance σ^2 which is associated with the pure error SS_{ERR} and 5 degrees of freedom in the analysis. This is unlikely to be enough. Methods that can be used to improve precision, and allow more powerful significance testing are increased numbers of replications, crossover studies, and blocking.

2. Replication of factorial designs and size of animal experiments

Where a single animal is used for a single run the experimental variation is the total variation between animals (σ). This can sometimes lead to the groups of animals in individual combinations of factor levels being larger than necessary, as the experimenter feels that he requires sufficient units for each treatment to be able to reach some sort of conclusion, calculating a mean and standard deviation for each treatment. Festing, analysing toxicological studies (6-8), has argued that this leads to unnecessarily large experiments and that many designs are over-replicated. What counts is not the number of degrees of freedom per treatment, but the total number of animals and the overall number of degrees of freedom in the design. There are three main considerations.

- Replication improves precision. The standard deviation for a given distinct experimental result is divided by a factor $\sqrt{n}$ when the experiment is replicated n times.
- Analysis of variance requires a sufficient number of degrees of freedom. For pharmacological, toxicological and pharmacokinetic experiments carried out

in order to identify statistically significant effects, at least 10 degrees of freedom are normally needed. There is little to be gained by increasing the number above 20 (9). For the non-biological experiments described in the remaining chapters, we will propose designs with rather less than 10 degrees of freedom.

• For ethical reasons as well as economic, the number of animals used should be minimized, within the constraints of the first two points.

3. Replication in blocks

We saw in chapter 3 how a factorial experiment could be carried out in a number of blocks if it is too large to be carried out at one time in uniform conditions. Thus, the results of changes in conditions are included in a block effect and do not distort the calculated effects. Here, the whole experimental design may be replicated as many times as is required, each replication being a block, and there may also be one or more replications within a block. Further blocks may be added later, until sufficient precision is obtained.

Thus, if the 5 formulations of the experimental design are being tested in rats, and 10 rats can be treated at a time, then the block consists of the duplicated design. At least one, and perhaps two, further blocks will be necessary. The form of the mathematical model is identical to that of the preceding case, except that a block effect, μ_i, with mean value zero over all blocks is introduced in the model for the block i. The error term ε includes all other variation, including the variation *between different* animals. With two blocks there will be 12 degrees of freedom as before. This would normally be enough provided the experimental precision is adequate. However, since the repeatability includes intersubject variation, the precision of the estimation of the formulation effects will be $\frac{1}{4}\sqrt{\sigma_s^2 + \sigma_b^2}$ and further replications may be necessary here.

4. Crossover designs

It is sometimes possible to administer several treatments to the same animal, allowing variation between animals to be separated from run to run variation in the same animal. In a complete crossover design each animal is administered all treatments. Each animal is treated as a block and the size of the design has to be determined right from the start. Consider for example that the pharmacokinetic profiles of the 5 formulations of these are to be tested on 4 dogs. The total number of experiments is $5 \times 4 = 20$, each animal being tested with all formulations. There are 4 parameters in the synergistic model to which we add a term μ_i referring to the animal being tested, with mean value zero over all animals. Thus:

$$y = \beta_0 + \mu_i + \beta_1 x_1 + \beta_2 x_2 + \beta_{12} x_1 x_2 + \varepsilon$$

Table 4.8 shows simulated data for such a design, and the results of the calculations.

Table 4.8 Effect of Formulation on the Pharmacological Response to Drug at Constant Dose in 4 Subjects

(a) Experimental Design and Simulated Data for Crossover Study

No.	X_1 X_2	Concentration of bile salt (M)	Lecithin-cholate molar ratio	Response in subjects A - D (%)			
				A	B	C	D
1	-1 -1	0.075	0.6 : 1	15	70	34	44
2	+1 -1	0.125	0.6 : 1	45	115	67	60
3	-1 +1	0.075	1.4 : 1	25	80	42	26
4	+1 +1	0.125	1.4 : 1	46	90	50	120
5	0 0	0.100	1.0 : 1	39	74	80	26

(b) Results of Multilinear Regression

Coefficient estimates	$b_0 = 57.4$ $b_1 = 16.1$ $b_2 = 1.8$ $b_{12} = 0.6$
Block (subject) effects	A: -23.4 B: 28.4 C: -2.8 D: -2.2

(c) Analysis of Variance of the Regression

	Degrees of freedom	Sum of squares	Mean square	F	Significance
Total	19	16314	–	–	
Regression	3	4186	1395	3.45	*
Blocks	3	6834	2278	5.60	**
Residual	13	5295	407		

The formulation has a significant effect on the response. The only factor influencing the response is the concentration of bile salt. However we also see that there is almost as much residual variation (residual sum of squares) as there is variation explained by the model (regression sum of squares). The systematic differences between subjects (block sum of squares) are of similar importance. Note that the block effects sum to zero (compare with the presence-absence model for qualitative variables in chapter 2).

If the block effects had been neglected we would have had to add the block sum of squares to the residual sum of squares. It is easy to show that under these circumstances the formulation would not have been be found to have a significant effect.

5. Blocked crossover designs

If the design is too large for all animals to be administered with all treatments, then it may be carried out in blocks. Take for example a formulation to be tested according to a 2^3 design. This may be partitioned on the **123** column (see chapter 3) into two 2^{3-1} blocks and each block of 4 experiments carried out as a crossover, using 3 animals for each, with 12 experimental runs in each block. The model will be the synergistic model of chapter 3, with an extra term μ_i for the animal used. There are 24 experiments and the 23 degrees of freedom in the analysis of variance are partitioned as follows:

Main effects:	d.f. = 6	Animals 3×2 -1	d.f. = 5
Interactions:	d.f. = 3	Error:	d.f. = 10

The block effect is confounded with the animal effects. Again, the error refers only to differences within the same animal and so it is much reduced by the crossover design.

Even when not using whole animals the pharmaceutical scientist will sometimes have to take biological variation into account in the testing of different formulations in animal models for bioavailability, or tolerance, or that of transdermal formulation screening using human skin samples. We have referred here to the two level factorial designs, but similar considerations apply to the optimization designs in the remaining chapters.

IV. QUALITY OF A DESIGN

The methods and statistical criteria we have just studied allow the model coefficients to be estimated, the fitness of the model to be assessed, and its acceptance or rejection. They also allow estimation of the experimental variance s^2, and from this the variances and intervals of confidence of the coefficient estimates can be calculated. It is important, however, to test the quality of the design *before* the experiment, to allow, where necessary, the comparison of different strategies and to justify the choice of one plan or design rather than another (10). There are many objective criteria for doing this. We will list a few of the most important without theoretical justification, but describing their practical significance. Their use will be illustrated in later examples.

A. Statistical Criteria for Design Quality

1. Orthogonality

Multiplication of the dispersion matrix by σ^2 gives the *variance-covariance matrix*. Its diagonal terms correspond to the variances of the estimators of the coefficients. A design is *orthogonal* if its dispersion matrix $(\mathbf{X'X})^{-1}$ is diagonal, the non-diagonal terms being equal to zero. The estimations of the coefficients are then independent of each other. The non-diagonal terms of the matrix correspond to the covariances of the estimators, which are therefore zero. Thus the orthogonality of the matrix may be verified before carrying out any experiments, as it depends only on the model matrix **X**, and not on the responses vector **Y**.

The effect of non-orthogonality is illustrated as follows. Consider a response y depending on 2 variables X_1 and X_2. We propose the first-order model:

$$y = \beta_0 + \beta_1 x_1 + \beta_2 x_2 + \varepsilon$$

to be determined by the 2^2 full factorial design in the domain -1, +1 of both variables, X_1, X_2, with each experiment carried out twice. We imagine two possible situations.

(i) It is possible to carry out experiments over the whole of the domain. We may therefore do design A, the factorial design of figure 4.5a. In the figure are the *mean* values, used for simulation. Consider that these are the "true" values of the response, with zero random error. They correspond to the following "true" values of the coefficients, which will be used in the simulation:

$$\beta_0 = 10.5 \qquad \beta_1 = 2.1 \qquad \beta_2 = 1.9$$

(ii) Two regions of the domain are inaccessible, because they do not give an interpretable response: the region about the point $\{x_1 = -1; x_2 = -1\}$ and that about $\{x_1 = +1; x_2 = +1\}$. We propose eliminating these two (duplicated) experiments from the plan and replacing them by the 4 (unduplicated) experiments shown in figure 4.5b (design B). Mean "true" response values are shown corresponding to the same values for the coefficients as before.

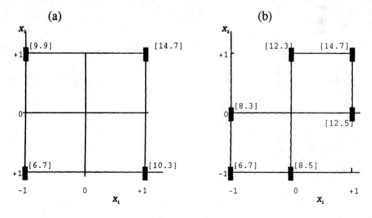

Figure 4.5 (a) Orthogonal 2^2 design. (b) Truncated design.

We first analyse the dispersion matrices of the two designs. The matrix for design A is orthogonal, with non-diagonal elements zero and diagonal elements equal to ⅛. The matrix for design B is not orthogonal; in particular the terms $c^{12} = c^{21}$ are non-zero.

$$(\mathbf{X'X})^{-1} = \begin{bmatrix} 0.125 & 0 & 0 \\ 0 & 0.3 & -0.2 \\ 0 & -0.2 & 0.3 \end{bmatrix}$$

The non-diagonal terms represent the dependence of the estimates of one parameter on the estimate of another. For example, if we were to carry out the experiments of each design many times we could obtain a different estimate for b_1 and b_2, on each occasion. These estimates would be distributed about a mean value with a standard deviation obtained in the way we have already described.

In figures 4.6 and 4.7 are plotted simulations for 100 replications of the two designs. The responses were varied randomly (according to a gaussian distribution) about their "true" values, given in figures 4.5a and 4.5b. Each point represents the pair of estimates b_1, b_2, for one simulated repetition of the design.

Now for *design A* (figure 4.6) we see that the cloud of points is roughly circular, indicating that b_1 and b_2 are uncorrelated.

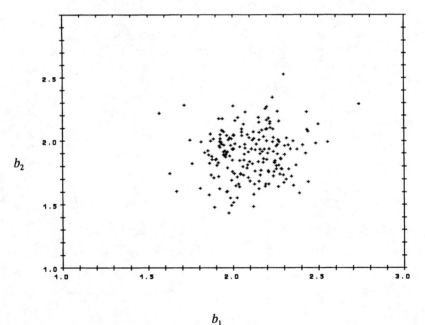

Figure 4.6 Simulation of estimations of b_1 and b_2, for the 2^2 design of figure 4.5a, showing no correlation between the estimations.

For *design B* (figure 4.7) the "cloud" is elliptical in form and oriented so that there is a tendency for high values of b_1 to be associated with low values of b_2, and vice versa. Thus, the estimates of the two coefficients are correlated.

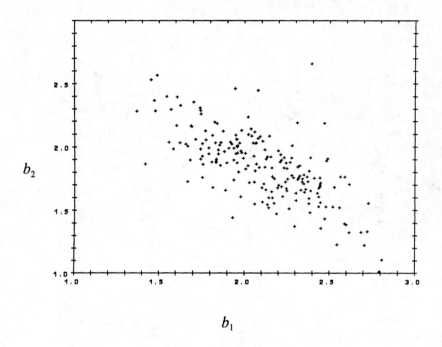

Figure 4.7   Simulation of estimations of b_1 and b_2 with the non-orthogonal design of figure 4.5b, showing negative correlation of the estimations.

This correlation is inherent in the design and can be deduced by analysis of the dispersion matrix $(\mathbf{X'X})^{-1}$. The corresponding elements in the dispersion matrix, c^{23}, which is at the intersection of the column corresponding to X_1 with row X_2, and c^{32}, the intersection of row X_1 with column X_2, are therefore finite and negative, showing a negative correlation between the two coefficients. This same correlation can be seen in the arrangement of the points in figure 4.5b; the values of X_1 are inversely correlated with those of X_2.

The *diagonal terms*, c^{11}, c^{22}, c^{33}, of the two matrices also differ slightly. They are slightly higher for matrix B than for matrix A. The better precision of design A is a consequence of its slightly more extensive experimental domain.

Figure 4.8 shows a similar design, this time for 3 quantitative variables, still in the domain $\{-1, +1\}$ but for which points $\{x_1 = -1 ; x_2 = -1\}$ and $\{x_1 = +1 ; x_2 = +1\}$ are inaccessible for all X_3. No points are duplicated in this design of 12 experiments. It appears that there is co-linearity between X_1 and X_2 as before, but experiments. It appears that there is co-linearity between X_1 and X_2 as before, but X_3 is uncorrelated with either of the other two variables in the design. Examination of the dispersion matrix for the design confirms this.

The c^{23} and c^{32} are finite and negative, but c^{24}, c^{42}, c^{34}, and c^{43} are all equal to zero. The coefficient estimates b_1 and b_2 are negatively correlated, as in the previous example, but there is no co-linearity between b_3 and either b_1 or b_2.

$$(\mathbf{X'X})^{-1} = \begin{bmatrix} c^{11} & c^{12} & c^{13} & c^{14} \\ c^{21} & c^{22} & c^{23} & c^{24} \\ c^{31} & c^{32} & c^{33} & c^{34} \\ c^{41} & c^{42} & c^{43} & c^{44} \end{bmatrix} = \begin{bmatrix} 0.083 & 0 & 0 & 0 \\ 0 & 0.167 & -0.083 & 0 \\ 0 & -0.083 & 0.167 & 0 \\ 0 & 0 & 0 & 0.083 \end{bmatrix}$$

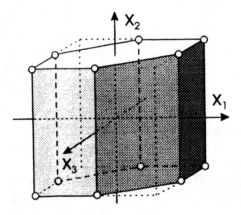

Figure 4.8 Non-orthogonal design for 3 variables.

2. Variance inflation factors (VIF)

It is not always possible to obtain an orthogonal experimental design. For example, although most of the designs for screening and factor studies are orthogonal, the D-optimal design of chapter 3 is not - neither are the Rechtschaffner designs, nor are the great majority of designs for second-order models. It is then quite difficult to measure the deviations from orthogonality; each non-diagonal term in the dispersion matrix needs to be inspected.

Variance inflation factors (3, 4, 5, 6) provide a simpler, overall measure of the non-independence of the estimators, that is, the co-linearity of the variables of the model.

Let the postulated polynomial model $y = f(x_1,..., x_k)$ be written as a sum of monomes:

$$y = \beta_0 + \beta_1 z_1 + \beta_2 z_2 + ... + \beta_{p-1} z_{p-1} + \varepsilon$$

The monome z_k could represent, for example, x_i, or x_i^2 or $x_i x_j$.

The inflation factor for the coefficient b_j ($j = 1, 2, ..., p-1$) is :

$$\text{VIF}_j = f_j = c^{jj} \sum_{i=1}^{N} [z_{ij} - \bar{z}_j]$$

An alternative and equivalent definition is the following. If there is no co-linearity, the values for Z_i over the N experiments will be independent of the values of the other coded variables Z_j ($i \neq j$). Thus if we were to carry out multiple linear regression of Z_i on all the other $p - 2$ terms in the model, Z_j, and there were no co-linearity, we would obtain a multiple regression coefficient of zero. (It is important to note that the regression is on the variables, *not* on the experimental results y_i.) If the settings of the variable being tested were totally dependent on a combination of the settings of the other variables (perfect co-linearity) the multiple regression coefficient for that variable would be 1. For a good model, the value would be small or zero. If R_i is the multiple regression coefficient of the regression of Z_i on all other terms in the model, the inflation factor of the ith regression coefficient is:

$$\text{VIF}_i = (1 - R_i^2)^{-1}$$

The values of the inflation factors demonstrate the independence or else the correlation between the estimates of the coefficients. It can be shown that all VIF ≥ 1. In an orthogonal design all VIF are equal to 1, and *vice versa*. A design is normally considered acceptable if all VIF are less than 4, but obviously the closer the value to 1 the better is the design. (Some authorities accept values of less than 10.) We use this quality criterion mainly for non-standard matrices, D-optimal designs in particular. No VIF is defined for the constant term in the model.

The design of figure 4.5 has a VIF of 1.8 for both main effects. That of figure 4.8 has a VIF of 1.33 for b_1 and b_2, but the VIF = 1 for b_3.

3. Determinant of the dispersion matrix: D-optimality

$(X'X)^{-1}$ being square, its determinant can be calculated. This determinant, written $|(X'X)^{-1}|$, is proportional to the precision of the estimators of the coefficients. Thus, minimizing its value corresponds to increasing the precision of the coefficients in the model.

This property may be demonstrated most easily when the design is orthogonal and $(X'X)^{-1}$ is diagonal. Its determinant is then equal to the product of the diagonal terms:

$$|(X'X)^{-1}| = \prod c^{ii}$$

To minimize this determinant, by the choice of an appropriate experimental design, is to minimize the product of the coefficients of variance. Thus, the product of the standard errors of the coefficients is also made as small as possible. This gives a D-optimal design, which is defined as one where the determinant of the associated dispersion matrix is minimal. Although the design's construction is described quite fully in chapter 8, the criterion of D-optimality is valid for all designs. Note again that the determinant may be calculated *before* carrying out the experiment, as only the model matrix X enters into the relationship 6.8, not the results vector Y.

B. Applications to Various Designs

1. Plackett-Burman and two-level factorial designs

We consider one of these designs, 2^k full-factorial, fractional-factorial or Plackett-Burman design, of N experiments. The postulated model of p coefficients consists of a constant term, main effects and possibly first-order interactions. All diagonal coefficients (coefficients of variance) in the dispersion matrix are equal to $1/N$ and all non-diagonal terms are zero.

$$c^{ii} = 1/N \qquad \text{and} \qquad c^{ij} = 0 \qquad \text{for } i \neq j$$

The product of the experimental variance σ^2 with each of the diagonal terms, c^{ii}, of the $(X'X)^{-1}$ matrix, is the variance σ_{bi}^2 of each of the coefficient estimates. This variance is thus σ^2/N and the standard error is $\sigma/\sqrt{N}$ for *all* coefficients. All *non-diagonal terms* being equal to zero, the designs are orthogonal and all effects are estimated independently of one another.

The *VIF* equal 1 for all coefficients, confirming the orthogonality of the design. The **determinant** of the dispersion matrix is N^{-p}, which is the smallest possible value for N experiments in a cubic domain of limits ± 1. 2-level factorial and Plackett-Burman designs allow the best and most precise estimates of coefficients.

2. Rechtschaffner designs

These saturated designs allow all main effects and first-order interactions to be estimated and were described in chapter 3, section V.B. Examination of the $(X'X)^{-1}$ matrix gives the standard deviations of the estimations of the coefficients, and the VIF. For designs of 6 and 7 factors the VIF values are greater than 1, thus indicating some slight co-linearity, but nonetheless they may be considered to be acceptable. They are higher for 8 factors (VIF = 1.75) and it was concluded that a D-optimal design of 37 experiments would be more satisfactory. Rechtschaffner designs are therefore not orthogonal, though the departure from orthogonality is relatively slight for 7 factors or less.

3. ¾ factorial designs

These consist of 3 fractional factorial blocks of a factorial design. For the construction of the 12 experiment design (¾ of 2^4), see chapter 3, section V.A. For the 24 experiment design, see chapter 3, section III.E.4.

The coefficients are all estimated with the same precision: the standard error of estimation is $\sigma/\sqrt{\frac{2}{3}N}$, $\frac{2}{3}N$ being the number of experiments in 2 of the blocks. They are not completely orthogonal, with variance inflation factors between 1.33 and 1.5. Their properties are summarized in table 4.8, along with those of the other designs discussed above.

Table 4.8 Summary of Quality Criteria for Various Designs

Design	Number of experiments	Orthogonal	Standard error of coefficient estimates	VIF
Plackett-Burman	$N = 4q$	yes	$\sigma/\sqrt{N}$	1.00
2-level factorial	$N = 2^k$	yes	$\sigma/\sqrt{N}$	1.00
Rechtschaffner				
6 factors	22	no	0.2285σ	1.14
7 factors	29	no	0.2243σ	1.42
8 factors	37	no	0.2245σ	1.75
¾ factorial	12	no	$\sigma/\sqrt{8}$	1.0 - 1.5
	24	no	$\sigma/\sqrt{16}$	1.13 - 1.5
	48	no	$\sigma/\sqrt{32}$	1.13 - 1.5

References

1. N. Draper and H. Smith, Applied Regression, 2nd edition, J. Wiley and Sons, N. Y., 1981.
2. D. C. Montgomery and F. A. Peck, Introduction to Linear Regression Analysis, J. Wiley and Sons, N. Y., 1992.
3. N.-O. Lindberg, C. Tufvesson, and L. Olbjer, Extrusion of an effervescent granulation with a twin screw extruder, Baker Perkins MPF50D, *Proc. 6th Int. Conf. Pharm. Tech. (APGI)*, 187-209 (1996).
4. P. Kleinebudde and H. Lindner, Experiments with an instrumented twin-screw

extruder using a single-step granulation process, *Int. J. Pharm.* **94**, 49-58 (1993).

5. L. Jian L and A. Li Wan Po, Effect of xylometazoline and antazoline on ciliary beat frequency, *Int. J. Pharm.* **86**,59-67 (1992).

6. M. F. W. Festing, The scope for improving the design of laboratory animal experiments, *Lab. Anim.* **26**, 256-267 (1992).

7. M. F. W. Festing, Reduction of animal use: experimental design and quality of experiments, *Lab. Anim.* **28**, 212-221 (1994).

8. M. F. W. Festing and D. P. Lovell, Reducing the use of laboratory animals in toxicological research and testing by better experimental design, *J. Roy. Stat. Soc. Ser. B* **58**, 127-140 (1996).

9. R. Mead, The Design of Experiments, Cambridge University Press, Cambridge, 1988.

10. M. Sergent, D. Mathieu, R. Phan-Tan-Luu, and G. Drava, Correct and incorrect use of multilinear regression, *Chemom. Intell. Lab. Syst.*, **27**, 153-162 (1995).

11. R. D. Snee, Some aspects of non-orthogonal data analysis. Part I: Developing prediction equations, *J. Qual. Techn.*, **5**, 67-79 (1973).

12. G. J. Hahn, W. Q. Meeker and P. J. Feder, The evaluation and comparison of experimental designs for fitting regression equations, *J. Qual. Techn.*, **8**, 140-157 (1976).

13. B. Chatterjee and B. Price, Regression Analysis by Example, John Wiley and Sons, N. Y., 1977.

14. D. W. Marquardt, Generalized inverses, ridge regression, biased linear estimation, and non-linear estimation, *Technometrics*, **12**, 592-612 (1970).

Further Reading

•   S. N. Deming and S. L. Morgan, Experimental Design: A Chemometric Approach, Elsevier, Amsterdam, 1987.

•   D. C. Montgomery, Design and Analysis of Experiments, 2nd edition, John Wiley and Sons, N.Y., 1984

5

RESPONSE SURFACE METHODOLOGY

Predicting Responses Within the Experimental Region

I. Introduction
    Objectives of response surface methodology (RSM)
    Recognising the situation requiring RSM
    The place of response surface modelling in a project
II. Mathematical models used in RSM
    Prediction by first and second degree models: simplified approach
    Some statistical considerations in RSM
    General mathematical model for RSM
III. The choice of strategy for RSM
    Postulated mathematical model
    Defining the experimental domain
    RSM: a step in experimental research
IV. Experimental designs for first degree models
    Equiradial designs for 2 factors
    Equiradial designs for more than 2 factors – the simplex design
V. Strategies for a spherical domain and second degree model
    Composite experimental designs
    Uniform shell (Doehlert) designs
    Hybrid and related designs
    An equiradial design for 2 factors – the regular pentagon
    Box-Behnken designs
VI. Strategies for a cubic shaped domain
    Standard designs with 3 levels
    Non-standard designs
    Comparison of designs in the cubic domain
VII. Strategies on rejection of the second degree model
    Cubic models are little used in RSM
    Strategies

I. INTRODUCTION

A. Objectives of Response Surface Methodology (RSM)

The methods and examples described in the first part of this book enable us to identify important and significant factors and also to know something of how they interact - how the effects of one factor will be influenced by changes in the levels of another. When we develop a formulation or process we would want to understand as fully as possible the effects of modifying them, over the whole of the experimental domain.

We would also want to use our knowledge of the response's dependence on the input variables to *predict* this response over the whole of the domain, and possibly also at its periphery. We have said *response*, but it is evident that we would usually mean *responses*, because we will usually be measuring a number of properties. Some of these are of vital importance as regards the product, others being less significant. The prediction may be carried out if there is a known mathematical model, generally empirical, for each response, which adequately represents changes in the response within the zone of interest. By "adequately" we mean that the value calculated with the mathematical model is sufficiently close to the value we would obtain if we were to do the experiment.

The purpose of response surface methodology (RSM) is to obtain such a model.

B. Recognising the situation requiring RSM

This is relatively frequent, especially in industrial research and development. The main uses of predictive models are:
- Optimizing the process for formulation, i.e. maximizing or minimizing one or more of the responses, keeping the remainder within a satisfactory range.
- Carrying out simulations with the model equation.
- Obtaining a process or product with properties (responses) within a fixed range of values.
- Understanding the process better, thus assisting development, scale-up, and transfer of formulations and processes.
- Being capable of knowing at any time the optimum manufacturing conditions to obtain a product with a particular set of properties, "à la carte".
- Plotting the responses.

An additional objective might be to find conditions where the result of the process (the product) is insensitive to process variation – the process being therefore robust.

Response surface methodology has certain specific characteristics which distinguish it from screening and factor studies. In particular the experimenter will often think of the experimental domain, or region of interest, in terms of a centre of interest and the region that surrounds it. That is why the domain is most often considered spherical, even when defined by the permitted ranges of each factor. The notion of a centre of interest corresponds to the idea that if one could only do a single experiment, it is there at that point that the experiment would be carried out. Also, up until now we have tried to identify the effects of *individual* variables. RSM comes at the stage where we know, roughly, whether a variable is significant or not. Rather than estimating the effect of each variable directly we want to fit the coefficients in the model equation to the response and map the response over the whole of the experimental domain in the form of a surface. We are concerned with how good the *model* is, but on the whole we are rather less interested in interpreting the individual coefficients of the model.

This mapping of the response is performed by describing it in terms of a model equation. This equation describes the response in terms of a function of variables which are normally *quantitative* and *continuous*. This function can then be visualised using contour plots or three-dimensional diagrams. Evidently the technique is only useful where the response is sufficiently reproducible, and its dependence on the process or formulation variables can be described by a mathematical model. It cannot be used, at least not directly, for discontinuous responses. A change of state within the experimental domain, for example, would preclude its use.

The model must be predictive over the whole of the domain and should therefore be validated. Rather more validation work is required here than for the screening or factor study designs, should the resources in time and materials be available.

This chapter is exclusively concerned with demonstrating an optimum strategy for obtaining the "best" possible model. Here for the most part we will assume that a single response, the most important one, is being treated. This might be the yield of pellets, the dissolution rate of a controlled release tablet, the degree of degradation, or the solubility of an active substance. A large part of the following chapter will be devoted to showing how the models obtained by RSM can be used for optimization, both in the case of single responses and of the much more complex problem of treating *multiple response data*. Parts of the chapters describing the mixture problem discuss response surface methodology and optimization where the constituents of the mixture cannot be varied independently.

C. The place of response surface modelling in a project

The models most often used to describe the response are first-, second-, and, very occasionally, third-order polynomials. The number of coefficients in a polynomial increases very rapidly with the number of variables and the degree of the model and the number of experiments needed increases at least as rapidly. Also a given model will describe the phenomenon that we are trying to model better over a restricted experimental domain than over a wide one. These three considerations imply certain conditions for using response surface modelling.

- The strategy is often sequential. The response over the domain is analysed according to increasingly complex models. A first-order model is initially postulated, and one would only go on to a higher order model, with the further experimentation needed, if the simple model were found to be inadequate for describing the phenomenon.

- The number of factors studied is usually between 2 and 5 and in any case should not exceed 7. Thus RSM is often preceded by one or more screening and factor-effect studies, enabling the most important factors to be selected and isolated. These also allow adjustment and modification of the experimental domain, recentring the design, if necessary, in a more interesting region.

The centre of interest introduced in paragraph B corresponds either to conditions already tested in practice (usual manufacturing conditions) or to the result of an optimization using one of the direct methods described in chapter 6 (simplex, steepest ascent, optimum path).

II. MATHEMATICAL MODELS USED IN RSM

The first-order and synergistic models have already been discussed in chapters 2 and 3 and the experimental designs used in RSM with the first-order model are for the most part the same as those used in screening. Other designs for the linear model are introduced in section IV. This chapter will be concerned mainly with the second-order, quadratic model. The use of third-order and higher models is rare.

One important restriction in both the first-order and synergistic models is that any maximum or minimum value is always on the edge of the experimental domain. This is not always realistic. There are likely to be cases where the maximum or minimum value of a response is to be found nearer the centre of the region of interest. Having identified the critical variable it is often necessary to carry out a more detailed study and here the first-order or synergistic model used up to now is augmented by quadratic terms.

To demonstrate this we use the simple example that was introduced in chapter 4, that of solubility in a mixed surfactant system. The treatment is in two stages, the first being a intuitive rather than mathematical demonstration of testing for lack of fit and curvature of a response surface. Then, in section II.B, we will carry out a more detailed, statistical analysis of the same process, showing how prediction confidence limits are calculated and the use of ANOVA in validating a model.

A. Prediction by First and Second Degree Models: Simplified Approach

1. Summary of the problem: solubility in a surfactant mixture

The effect on the solubility of an experimental drug of varying the concentration of bile salt and the lecithin:bile salt ratio was described in chapter 4. The domain is defined by its *centre of interest*: bile salt concentration = 0.1 M, molar lecithin-cholate molar ratio = 1:1. The limits in bile salt concentration are about ± 0.025 M and those of the lecithin/bile-salt ratio are about ± 0.6. The design space, initially defined in table 4.1, is shown in figure 5.1a.

2. Mathematical model

Bearing in mind the relatively narrow domain we might assume as a first approximation that the solubility can be described adequately by a first-order, purely additive, model:

$$y = \beta_0 + \beta_1 x_1 + \beta_2 x_2 + \varepsilon$$

We saw in chapter 3 that a complete factorial design 2^2 is suitable for determining this model. The *experimental design, plan* and *experimental responses* for this design are given in table 5.1.

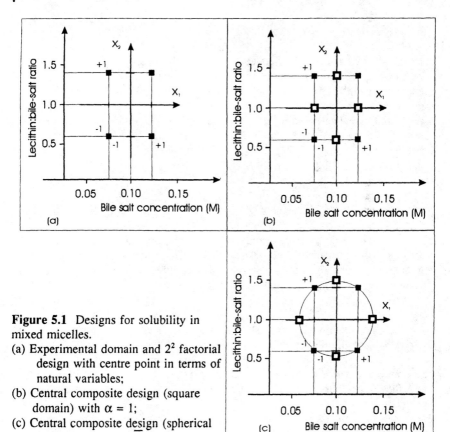

Figure 5.1 Designs for solubility in mixed micelles.
(a) Experimental domain and 2^2 factorial design with centre point in terms of natural variables;
(b) Central composite design (square domain) with $\alpha = 1$;
(c) Central composite design (spherical domain) with $\alpha = \sqrt{2}$.

Table 5.1 2^2 Full Factorial Design for Solubility of a Drug in Mixed Micelles

No.	X_1	X_2	Concentration of bile salt (M)	Lecithin-cholate molar ratio	Solubility (mg/mL)		Difference (mg/mL)
					experimental	calculated	
1	-1	-1	0.075	0.6 : 1	6.44	6.16	0.28
2	+1	-1	0.125	0.6 : 1	10.04	10.33	-0.28
3	-1	+1	0.075	1.4 : 1	9.72	10.00	-0.28
4	+1	+1	0.125	1.4 : 1	14.45	14.17	0.28

3. Estimation of the linear coefficients and validation of the model

We can obtain estimates of the coefficients β_0, β_1, β_2, using linear combinations (contrasts) or by multi-linear regression:

$$b_0 = 10.16 \qquad b_1 = 2.08 \qquad b_2 = 1.92$$

As we indicated earlier, this model is intended to be used to predict solubility within the experimental domain, that is to calculate the solubility in solvents whose compositions differ from those of the 4 experiments in the design. We need therefore to be assured of the model's validity. We will examine the statistical aspects of this validation later in the chapter but our approach here will be more intuitive. We note first of all that the mathematical model, with the estimates of the coefficients:

$$\hat{y} = b_0 + b_1 x_1 + b_2 x_2$$

gives calculated values very similar to the experimental values, with differences (residuals) of only ± 0.28 (table 5.1).

Our first step in model validation is to try to estimate higher order terms than those of the model we are testing. One supplementary term, the first order interaction between X_1 and X_2, may be calculated directly from the design points:

$$b_{12} = 0.28$$

Its value is small compared with the other effects. This leads us to suppose that this interaction is negligible, or at least has little influence.

A second method of verifying the predictive capabilities of the model is to carry out one or more additional experiments (not repeats of those already carried out) called *test-points*, and to compare the experimental results for these points with the predictions of the model. The fact that the predicted and experimental values are close to one another will not *prove* the validity of the model in the remainder of the domain, but it will increase our confidence in its being satisfactory. On the other hand, a measured value very different from the predicted one will lead to doubts as to the model's validity. The centre point is a good choice for testing, as it is the furthest possible from the design points. The predicted value at the centre point, according to the model, is the constant term b_0 = 10.16, which can be compared with the measured value.

Therefore, the solubility in a medium containing 0.1 M bile salt (X_1 = 0) and an equimolar ratio of lecithin (X_2 = 0) was determined in duplicate (experiments 5 and 5′ as described in chapter 4) at the same time as the measurements at the factorial points. (Note that experiments at the test points should be done, if possible, at the same time as the other experiments, all of them in a random order.) The experimentally measured solubilities were 11.70 and 11.04 mg mL^{-1}, a mean value of 11.37. This is a difference of 1.21 mg mL^{-1} with respect to the model calculation of 10.16, which appears large in comparison with the differences observed previously. We therefore believe that the response surface may not be an inclined plane, but a curved surface. We have detected this curvature in the centre of the domain, and we therefore require a more complex mathematical model. This conclusion for the moment is entirely subjective as we have not yet considered any statistical tests. We will demonstrate later on (section II.B) how it is possible to test if this difference is significative and we will show that in such a case it may be attributed to the existance of squared terms in the model. For the moment we will limit ourselves to the conclusion that a more complex mathematical model is necessary.

4. Quadratic model and complementary design

To allow for possible curvature in the response surface, we add squared terms in x_1 and x_2. We now have to estimate the 6 terms of the quadratic model:

$$y = \beta_0 + \beta_1 x_1 + \beta_2 x_2 + \beta_{11} x_1^2 + \beta_{22} x_2^2 + \beta_{12} x_1 x_2 + \varepsilon$$

Up to now, we have the results of only 5 experiments carried out under distinct conditions. At least one other experiment is needed, but there is no satisfactory way of adding only one experiment to the design, in terms of precision of estimate and orthogonality.

The 2^2 design with centre point can be extended by adding experiments along each of the axes at values $\pm \alpha$ of the other coded variable ($X_1 = \pm \alpha$, $X_2 = 0$ and $X_1 = 0$, $X_2 = \pm \alpha$). These are called *axial points*. The result is the *central composite design* for 2 factors. If α is set equal to ± 1, as in figure 5.1b, the design is also a full factorial design at 3 levels (3^2), quite often used for studies on 2 factors.

The design can be further modified if we are able to go outside the original square experimental domain. We define the domain as a circle passing through all the factorial points and move out the axial points, in this case so they lie on the circle. The value of α is thus increased to $\pm\sqrt{2}$. The resulting design is shown in figure 5.1c. The interest of the design lies in its excellent properties, which we will describe later in some detail, and also in the fact that it can be carried out in two stages, first the factorial design and centre point(s), with the "star" design (axial points plus more centre points) to follow.

Although all the design points (other than the centre point) here lie on the circumference of the circle limiting the experimental domain, this is not the case for all such composite designs. In particular, for 3 and 5 or more variables and a spherical (or more accurately for more than 3 variables, hyperspherical) domain, the star points usually lie just inside the boundary.

It is essential to this sequential approach that the experimenter is reasonably confident that he can extend the domain from ±1 to ±1.414. If he takes what he thinks are reasonable lower and upper limits for each variable, he must do his first 4 experiments *well* within those limits. If an experiment is to be done in more than one stage the subsequent steps must be anticipated not only in defining the kind of design to begin with, but also in fixing the boundaries. In the case of the present example, it is difficult to increase the lecithin:cholate molar ratio beyond about 1.6 or the solution becomes too viscous for easy manipulation. If a value greater than 1.4 had been selected in the factorial experiment for level +1 it would not have been possible to have set $\alpha = \sqrt{2}$ in the central composite design.

The results of the complementary design ("star" or axial design), with the new coded variables, are given in table 5.2. The runs were duplicated, but only the mean values (right hand column) will be considered at this stage.

Table 5.2 Solubility in Surfactant System: Complementary Points to a 2^2 Factorial to Give a Central Composite Design

N°	X_1	X_2	concentration of bile salt (M)	lecithin-cholate molar ratio	solubility (mg mL^{-1}) expt. values		mean
6	-1.414	0	0.065	1:1	7.80	7.63	7.72
7	+1.414	0	0.135	1:1	14.50	14.09	14.30
8	0	-1.414	0.100	0.43:1	7.70	7.43	7.57
9	0	+1.414	0.100	1.57:1	12.35	12.75	12.55
10	0	0	0.100	1:1	11.79	10.90	11.35

5. Calculation of the coefficients of the quadratic model

The effects are not calculated directly, but by multilinear regression.

$$b_0 = 11.36 \qquad b_{11} = -0.270$$
$$b_1 = 2.207 \qquad b_{22} = -0.746$$
$$b_2 = 1.845 \qquad b_{12} = 0.282$$

We see that b_0 is close to the experimentally measured value at the centre of the domain in both the first and second parts of the design. The differences between measured values and those calculated by the quadratic model are between -0.39 and +0.36. The experimental values obtained for the centre points in the both series of experiments agree with one another. It seems that there is no block effect.

A contour-plot of the solubility prediction is shown in figure 5.2. This would enable a formulation to be selected, in combination with the analysis of other relevant parameters (viscosity, stability, cost). It shows the solubility increasing with the bile salt concentration and also with increasing lecithin:bile salt ratio (the increase being greater at lower values of the ratio).

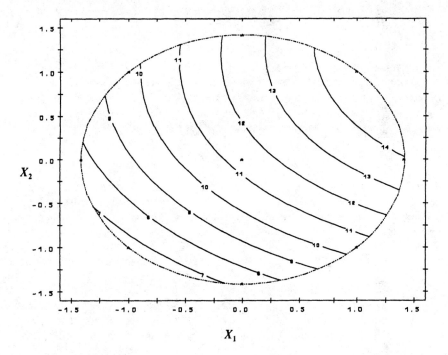

Figure 5.2 Calculated solubility in mixed micelle solution.

B. Some Statistical Considerations in RSM

To estimate a model we need to carry out as many experiments as there are coefficients in the model. More experiments than this may be done, either replicating some of the experiments or from the addition of test points, or simply because of the nature of the experimental design. Certain statistical treatments then become possible. We will illustrate these using the example presented in section II.A, first treating the data of the factorial design plus centre point, then taking the full central composite design.

The previous treatment was based on mean data, but in fact each solubility experiment was duplicated. The full results for the factorial design and axial points, plus centre points in each case were given in tables 4.2 and 5.2 respectively.

1. Prediction confidence interval

The variance of the calculated response

Here we continue analysing the first design (2^2 factorial) using the first-order model to predict the response elsewhere in the domain as we have already indicated. Using the estimated values of the coefficients, we can calculate a response, in this case the solubility, at any point A in the domain by:

$$\hat{y}_A = b_0 + b_1 x_{1A} + b_2 x_{2A} \tag{5.1}$$

The estimators of the coefficients (b_0, b_1 and b_2) are random variables obtained by linear combinations of experimental values, themselves random variables. There is uncertainty in the calculated response $\hat{y}_A$ at point A in the domain, defined by its variance, var($\hat{y}_A$). This in turn depends on the experimental variance σ^2, which is assumed constant, and the variance function d_A at that point:

$$\text{var}(\hat{y}_A) = d_A.\sigma^2 \qquad \text{where} \quad d_A = \mathbf{x}'_A(\mathbf{X'X})^{-1}\mathbf{x}_A \tag{5.2}$$

d_A depends on the coordinates of A (experimental conditions in the form of coded variables) and on the experimental design, in terms of the dispersion matrix $(\mathbf{X'X})^{-1}$. The matrix notation introduced in chapter 4 is used; $\mathbf{x}_A$ is the column matrix, or vector, describing the point A and $\mathbf{x}'_A$ is its transpose, the row vector $(1, x_{1A}, x_{2A})$. We will neglect systematic errors.

The main point to emerge from equation 5.2 is that that the uncertainty of prediction of the solubility at point A depends on three things:

- The variance of the experimental response σ^2, which itself depends on the quality of measurement in the individual experiments (that is, the experimental method, the quality of the equipment, and the skill of the operator).
- The experimental design, in terms of its dispersion matrix $(\mathbf{X'X})^{-1}$.
- The coordinates $\mathbf{x}_A$ of the point A in factor space.

Equation 5.2 allows us to calculate the precision of prediction within the domain provided that we have an estimate of the experimental variance. Even if we have no such estimate, the relative precision may be known, as the variance function d_A can be estimated over the experimental region.

Confidence interval for the true value

Let the true (unknown) value of the response, whose measured value at point A is y_A, be written as η_A. Then η_A is the expectation of y_A:

$$\eta_A = E[y_A]$$

If we have an estimate s^2 of the variance σ^2, with v degrees of freedom, we may then calculate a confidence interval for η_A about $\hat{y}_A$ at point A, for a given statistical significance α. This interval is $\pm\, t_{\alpha/2,v}(d_A)^{1/2}\, s$ and so:

η_A lies within the range: $\hat{y}_A \pm t_{\alpha/2,v}(d_A)^{1/2}\, s$

This assumes that the mathematical model is of the correct form!

Confidence interval for prediction of future measurements
It is likely that the experimenter will wish to carry out experiments in order to verify the model predictions. The measured value y_A will not necessarily be within the above limits, the range (again for a probability $1 - \alpha$) being somewhat wider:

y_A lies within the range: $\hat{y}_A \pm t_{\alpha/2,v}(1 + d_A)^{1/2}\, s$

If several measurements are made we can calculate a confidence interval for their mean. Let $\bar{y}_A$ be the mean of m future measurements at point A.

$\bar{y}_A$ lies within the range: $\hat{y}_A \pm t_{\alpha/2,v}(1/m + d_A)^{1/2}\, s$

and approaches the confidence limits for the true value as m increases. Equations 5.2 and the expressions for the confidence intervals are general and may be applied to any experimental design. We apply them below to the 2^2 design.

2. Example of the application of confidence intervals

Take the centre point 5 as test point A. The run was carried out twice, so $m = 2$. We have already seen that the estimated value at the centre $\hat{y}_A = 10.16$.

Calculations involving equation 5.2 are usually done by a computer program and the experimenter need not concern himself with details of the calculation. Here the matrix algebra is particularly simple and it can be verified using the rules given in appendix I that $d_A = 1/8$. An estimate $s = \sqrt{0.134} = 0.366$ with 5 degrees of freedom was obtained for the variance (by ANOVA on the 5 pairs of replicated experiments, as explained in chapter 4). Taking $\alpha = 5\%$, and referring to a table of Student's t, we obtain $t_{0.05/2,v} = 2.571$. Thus the predicted solubility $\hat{y}_A$ at the centre point A is 10.161, and the confidence limits for the true solubility η_A are given by:

$$\eta_A = \hat{y}_A \pm 2.571 \times 0.354 \times 0.366 = 10.161 \pm 0.333 \text{ mg/mL}$$

or

$$9.828 \le \eta_A \le 10.494$$

We may likewise calculate the confidence interval for predicted value for solubility at the centre point being the 95% confidence limits for the mean of two measurements, $\bar{y}_s$:

$$\bar{y}_s = \hat{y}_A + t_{\alpha/2,\nu}(\tfrac{1}{2} + d_A)^{\frac{1}{2}} \, s = 10.161 \pm 0.744 \text{ mg/mL}$$

or

$$9.42 \leq \bar{y}_s \leq 10.90$$

This is to be compared with the experimental solubility value for the mean of the two experimental points at the centre, which was 11.37 mg/mL. It lies just outside the confidence interval for the predicted mean of two experimental points (figure 5.3), indicating that the first-order model is to be rejected.

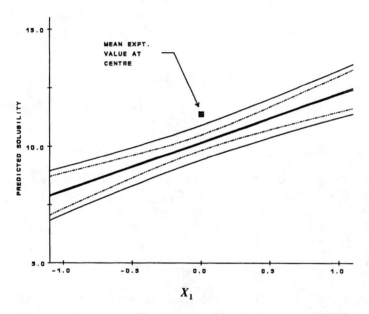

Figure 5.3 Predicted solubility η for a lecithin-bile salt ratio = 1 (double line). compared with the mean experimental entre-point value (filled square). The single lines are with 95% confidence limits of the predictions of the model for the mean of two data and the dotted lines are 95% confidence limits for η.

3. Curvature of the response surface

The mean value of the response at the centre points is another estimate of the constant term in the model, which we will call b'_0. It may be compared with the estimate b_0 from the 4 duplicated points of the factorial experiment. Thus we find:

$$b_0 - b'_0 = 11.37 - 10.16 = 1.21$$

We have already supposed that this difference, which is the difference between calculated and measured values at the centre point, indicates a curvature in the response surface. We will now demonstrate this. We add two terms, $\beta_{11}x_1^2$ and $\beta_{22}x_2^2$, to the model, representing the quadratic effects of the factors X_1 and X_2, and also add columns for X_1^2 and X_2^2 to the model matrix as shown in table 5.3.

Table 5.3 2^2 Design: the Model Matrix for Added Quadratic Terms

No.	X_0	X_1	X_2	X_1^2	X_2^2
1	+1	-1	-1	+1	+1
2	+1	+1	-1	+1	+1
3	+1	-1	+1	+1	+1
4	+1	+1	+1	+1	+1
5	+1	0	0	0	0

The columns X_0, X_1^2 and X_2^2 can be seen to be identical for the factorial experiments. Following the same reasoning as in chapter 3, we conclude that the estimator b_0 for the constant term, obtained from the factorial points, is biased by any quadratic effects that exist. On the other hand the estimate b'_0 obtained only from the centre point experiments is *unbiased*. Whatever the polynomial model, the values at the centre of the domain are direct measurements of β_0. So the difference between the estimates, $b_0 - b'_0$, (which we can write as b_{11+22} using the same notation as in chapter 3) is a measure of the curvature $\beta_{11} + \beta_{22}$. The standard deviation of b_0 is $s/\sqrt{8}$ (as it is the mean of 8 data of the factorial design) and that of b'_0 is $s/\sqrt{2}$. (being the mean of 2 centre points). We define a function t as:

$$ t = \frac{(b_0 - b'_0)}{s\sqrt{\tfrac{1}{2} + \tfrac{1}{8}}} = \frac{11.37 - 10.16}{0.366 \times \sqrt{0.625}} = \frac{1.21}{0.289} = 4.18 $$

In a table of Student's t we find the critical value for 95% probability and 5 degrees of freedom to be 2.57. It therefore appears that the coefficient b_{11+22} is statistically significant and the model is inadequate.

4. ANOVA of the regression on the 2^2 design with 2 centre points

An alternative approach is to include all the points in the regression and its analysis of variance. There are 10 experimental data (see table 4.2) to estimate 3 coefficients. The standard analysis of variance of the regression (first order model) is shown in table 5.4. The figures differ slightly from those of table 4.6 as the model for the regression here does not include the interaction term $b_{12}x_1x_2$.

The 5 replications allow the variance to be estimated with 5 degrees of freedom (s_3^2). The 5 distinct experiments themselves give 5 more degrees of freedom partitioned as follows: 1 degree for the mean value and 2 for the regression, leaving 2 for another estimation s_4^2 of the experimental variance. If this second variance is significantly greater than s_3^2 it will be because of lack of fit of the linear model. The comparison of these two variances by an F test shows that they are significantly different.

Table 5.4 ANOVA of the Regression: 2^2 Design Plus Centre Points

	Degrees of freedom	Sum of squares	Mean square	F	Probability >F
Total	9	67.90	-	-	
Regression	2	64.26	32.131	239.8	< 0.0001
Residual	7	3.64	0.520	-	
Lack of fit	2	2.97	1.486	11.11	0.0145
Pure error	5	0.67	0.134		

We conclude that there are significant differences between the predictions of the first order model and the experimental data. The significance of the regression is determined by dividing the mean square of the regression, 32.131, by the pure error mean square ($s_3^2 = 0.134$). The resulting value, F = 239.8, shows the regression itself to be very significant. The model, on the other hand, is inadequate and may be rejected.

5. ANOVA of the regression on the complete central composite design

The individual results for the duplicated star points and the extra centre point were listed in table 5.2. These, when added to the factorial design of table 4.2, make up the duplicated central composite design. Analysis of variance for regression may be carried out on the complete data set of the composite design for the second degree model (20 data). The results are summarised in table 5.5.

In the ANOVA table, the regression sum of squares has been divided into the part associated with the first-order terms (with 2 degrees of freedom) and that which is explained by the second-order terms (with 3 degrees of freedom).

From the residual sum of squares $SS_{RESID} = 2.15$, with 14 degrees of freedom, a residual mean square can be calculated ($MS_{RESID} = 2.15/14 = 0.154$). It is this value which is used to calculate F-ratios for assessing the statistical significance of the first-order and second-order terms. (Note that the F-ratio for the first order model is obtained by dividing the mean square for the regression by the residual mean square calculated for the first order model, not by the value in the table, 0.154, which refers to the quadratic model.)

The quadratic terms are evidently significant at better than 99.9%. Lack of fit is not statistically significant. The second order model is therefore accepted for predicting the response in the experimental region.

A value $s^2 = 0.154$, calculated with 14 degrees of freedom, is taken for the experimental variance. This corresponds to a standard deviation of 0.39 mg/mL.

Table 5.5 ANOVA of the Regression: Central Composite Design

	Degrees of freedom	Sum of squares	Mean square	F	Probability >F
Total	19	139.93	-	-	
Regression	5	137.78	27.56	179.2	***
Linear	2	132.06	66.03	429.4	***
Quadratic	3	5.72	1.91	12.40	***
Residual	14	2.15	0.154	-	
Lack of fit	3	0.87	0.291	2.50	0.114
Pure error	11	1.28	0.116		
$R^2 = 0.985$	$R_{adj}^2 = 0.979$		$s = \sqrt{0.154} = 0.39$		

In addition to separating the regression sums of squares associated with the linear and quadratic terms (and also cubic terms in the case of a third-order model), the sums of squares for individual coefficients may be isolated in the same way and significant interaction and quadratic terms may be identified. Table 5.6 lists the values and the significances of the coefficients of the second degree model.

Table 5.6 Coefficients, Standard Deviations and Statistical Significances

Coefficient	Estimate	Standard deviation	Statistical significance
b_0	11.36	0.20	***
b_1	2.21	0.10	***
b_2	1.84	0.10	***
b_{11}	-0.27	0.13	0.057
b_{22}	-0.75	0.13	***
b_{12}	0.28	0.14	0.061

Thus, non-significant terms might be eliminated and the model simplified. This is rarely necessary in response surface modelling, where the fit of the total model is important, but the individual coefficients are not in general analysed.

6. Isovariance curves

Equation 5.2 is general whatever the model. We can write it down as:

$$\text{var}(\hat{y}_A) = \mathbf{d}_A.\sigma^2$$

where $\mathbf{d}_A = \mathbf{x}'_A \, (\mathbf{X'X})^{-1}\mathbf{x}_A$, which, for a given experimental design and model, depends only on the position of the point A. It does not depend at all on the experimental results. Note that the vector $\mathbf{x}'_A$ is different from the case described previously, as it includes second degree terms:

$$\mathbf{x}'_A = (1, x_{1A}, x_{2A}, x_{1A}^2, x_{2A}^2, x_{1A}x_{2A})$$

The experimental variance being assumed constant within the whole of the experimental domain, the function $\mathbf{d}_A$, which we will call the **prediction variance function** demonstrates well the quality of the experimental design in terms of its ability to predict. The function may be represented in terms of curves joining points with the same prediction variance function – otherwise called contours of isovariance. Figure 5.4 gives the isovariance curves for the central composite design that we have been discussing.

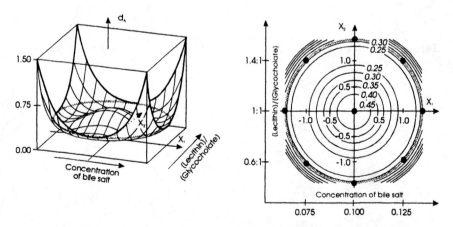

Figure 5.4 Variance function for 2 factor central composite design.

The variance function within the domain varies between 0.23 and 0.48. The concentric curves show that the confidence interval of the prediction is constant at a given distance from the centre. The variance function decreases (and thus the precision of the prediction improves) as we move from the centre of the domain up to a distance of about 1 unit, then it increases quite rapidly up to the edge of the domain. It can be shown that d_A becomes greater than 1 just outside the domain boundary – that is, the variance of prediction outside the domain (extrapolation) is greater than the experimental variance. To obtain the variance of prediction, multiply it by the estimation of σ^2.

The variance function for the point i in the design d_i is often known as the *leverage* of that point.

C. General Mathematical Model for RSM

Response surface modelling is based on the following premise: any function whether known or unknown may be expressed in the neighbourhood of a given point by a Taylor series, provided that the function may be differentiated an infinite number of times.

We stated above that the experimental domain is often thought of as being a centre of interest (point) around which extends the region of interest (experimental domain). We have also pointed out that the domain is relatively limited in extent, certainly with respect to the domains of other types of studies, factor studies, and especially screening and direct optimization by simplex (see chapter 6). We have also to suppose that the response is continuous, without steep variations, peaks or breaks. It is imagined as a curved surface, whether peak, valley or otherwise, with moderate slope or change of slope. There is an absence of sharp peaks, "wells", and "cliffs", as RSM will not work under such conditions.

The Taylor series for the response function is a polynomial of a given degree. For the model most commonly used, which is the second degree equation, it is:

$y = \beta_0$	constant term
$+ \beta_1 x_1 + \ldots + \beta_k x_k$	first order terms
$+ \beta_{11} x_1^2 + \ldots + \beta_{kk} x_k^2$	square terms
$+ \beta_{12} x_1 x_2 + \ldots + \beta_{k-1,k} x_{k-1}\ x_k$	rectangular terms
$+ \varepsilon$	error

The total number of coefficients p in a polynomial model of degree d and k variables is given by:

$$p = (k+d)!/(k!d!)$$

where $k!$ represents factorial k (see table 5.7).

The number of experiments in the design must be at least equal to the number of coefficients, and frequently much greater. A model with more than 30 coefficients would not often be a realistic option. This virtually excludes the use of polynomials of order greater than 3 and means that the third-order polynomials themselves are very rarely employed.

Table 5.7 Number of Coefficients in a Polynomial

Order	N° of factors k					
d	2	3	4	5	6	7
1	3	4	5	6	7	8
2	6	10	15	21	28	36
3	10	20	35	56	84	120

III. THE CHOICE OF STRATEGY FOR RSM

A. Postulated Mathematical Model

We need to be able to judge both the qualities and the defects of designs. However their advantages and deficiencies are only such for a given mathematical model. A design could be excellent for determining one model and far less good for another. Therefore, the model is the first criterion for choosing a design. We ask, for example, whether the phenomenon is purely additive (first-order model), or whether there is a curvature in the response surface (quadratic or second-order model), or yet again if there is an inflexion (third-order or cubic model). Very often the experimenter will not know which, if any, of these models is correct. It is his responsibility to choose, from among the several which are possible, the approach that he believes is the most likely or the most sure, depending on his knowledge of the problem. We may envisage two possibilities.

- The experimenter does not know which model to use. He therefore decides on a sequential strategy, by which he may begin with a first-order model, go on to a quadratic model, and possibly even go on from there to a third-order model. There are relatively few experiments in the first stage and the design is completed subsequently in one or more steps. The experimental designs used at each stage are not necessarily the best possible (optimal, in terms of efficiency, precision of the coefficient estimates, "value for money"), but they are adapted to the experimenter's knowledge, or possibly lack of knowledge, of the phenomenon.

- Alternatively, he may consider that one particular model is probable. For example, if taking into account the dispersion of the experimental results he knows that there is little chance of a curvature being detected within the experimental domain, he will select a first-order model. On the other hand, if a preliminary study of one of the factors shows a maximum of the response within the domain a quadratic model will be postulated. Here, the chosen design may be one not easily modified in order to treat a more complex model. (There is, however, always the possibility of recovery from an error in the initial choice by using the special techniques described in chapter 8: D-optimal designs adding runs by using an exchange algorithm. Nevertheless, these may be difficult to set up and are to be avoided if possible.)

B. Defining the Experimental Domain

1. Shape of the domain

The shape is most often considered *spherical* (as in the example just treated). Designs for spherical domains consist of points on one or more concentric spheres or hyperspheres. The experimental conditions are limited by the boundaries for each factor but the experimental designs do not include any experiments with combinations of extreme values of more than one factor. Such combinations of extreme conditions (at the corners of a cube) are therefore considered to be outside the experimental domain.

There are situations where, for technical, physico-chemical, or economic reasons, the total region that must be considered is defined by the limits of each factor and it is thus *cubic* in shape. An experimental design with a cubic rather than a spherical structure is then preferred (see section VI).

The domain may also be *mixed*, or cylindrical, that is spherical in some factors and cubic in others. See chapter 8 for the construction of designs in this type of domain.

2. Number of levels

Also, some factors may take a large number of levels of any value without complicating the experiment. These could be the percentage of an additive, such as water. Other factors, even if quantitative, continuous, and controllable, might not be so easy to adjust. The designs in this chapter have very different properties with respect to this particular criterion.

3. Open and closed experimental domains

Lastly, we consider whether the domain is *open* or *closed*. Is it possible that after completion of the study the experimenter will want to continue by moving the centre of interest to a neighbouring area, or even to a zone quite separate from that of the initial experiment? In optimization it is quite common to find that the optimum region appears to be on the edge or even outside the zone initially chosen. The conclusions are therefore extrapolations, of poor precision, unreliable, and require further experimentation. Some of the strategies proposed in this chapter allow such extrapolations, and in certain favourable cases experiments carried out in the first stage may be re-used, and incorporated into the new design.

C. RSM: A Step in Experimental Research

We have already pointed out several times (and we shall do so again) that there is little point in insisting *only* on the best experimental design at a given stage. One should rather try to look at each stage globally, taking into account what has already been done, and considering what are the possibilities for the next stages in the experimentation.

Once the stage has been reached where RSM is applicable, we have the advantage of experiments carried out already. Ideally, these will have been done according to the experimental designs already studied in this book. The choice of a design for RSM may be guided by our wish to re-use experimental data already available. Thus one might classify experimental designs for response surface studies into three categories:

- Those constructed incorporating a factorial design (where the preceding phase was a factor or possibly a screening study).
- Those which include a "simplex" (equilateral triangle, regular tetrahedron, etc.) in their basic structure, often used where the centre and zone of interest are adjusted or shifted about in the factor space during the study (see chapter 6 for the use of these designs in optimization).
- Those which have no basic sub-structure, and where one cannot incorporate previous experiments, even partially.

Particular constraints that may affect the choice of design are:

- The impossibility of carrying out all experiments at the same time because of lack of raw material or other reasons, implying the possible existence of block effects.
- The possibility of carrying out only a very limited number of experiments, for reasons of cost or time.
- The necessity or possibility of postulating an incomplete model (absence of certain terms). An example will be treated in chapter 8.

IV. EXPERIMENTAL DESIGNS FOR FIRST DEGREE MODELS

The first degree model consists of a constant, plus first-order terms (also called main effects). We will not devote a great deal of space to this problem as we have already looked at some of the designs used. The same screening and factorial designs of chapters 2 and 3 may also be used here to determine the model. The essential difference is that here the phenomenon is continuous and it is really thought to be linear within the domain. This domain in an RSM study is in general much less extensive than that of a screening study. In addition to factorial and Plackett-Burman designs, two other kinds of designs are frequently used in the case of the first degree model and merit discussion.

A. Equiradial Designs for 2 Factors

These consist of N points on a circle about the centre of interest in the form of a regular polygon. Taking into account the limited number of parameters ($p = 3$) in the model, only 3 to 6 sided polygons are used. The design consists of all experiments where:

$$x_1 = \sin 2i\pi/N \quad \text{and} \quad x_2 = \cos 2i\pi/N$$

with i taking all integral values between 0 and $N - 1$.

The designs may be rotated by any angle and still retain the same properties. Certain orientations more commonly used than others are shown in figure 5.5 and table 5.8. We shall see later that the pentagonal and hexagonal design, with addition of centre points, may also be used with a second-order model (see sections V.B and D).

We have chosen arbitrarily to construct the design inside a circle of radius 1. Any other radius could have been chosen. The quality of a design depends on the relative positions of the experiments one to another.

Table 5.8 Equiradial Designs for 2 Factors

No.	Triangle		Square		Pentagon		Hexagon	
	X_1	X_2	X_1	X_2	X_1	X_2	X_1	X_2
1	0.000	1.000	-0.707	-0.707	1.000	0.000	1.000	0.000
2	-0.866	-0.500	0.707	-0.707	0.309	0.951	0.500	0.866
3	0.866	-0.500	-0.707	0.707	-0.809	0.588	-0.500	0.866
4			0.707	0.707	-0.809	-0.588	-1.000	0.000
5					0.309	-0.951	-0.500	-0.866
6							0.500	-0.866

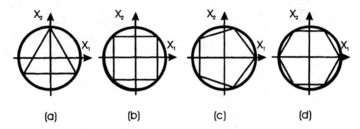

Figure 5.5 Equiradial designs for 2 factors. (a) Equilateral triangle, (b) square, (c) regular pentagon, and (d) regular hexagon.

In tables 5.3 and 5.8 we see that factors may be set at a number of different levels, which leads us to suppose that they may vary continuously. Coded variables, written as X_i, may be converted to natural ones (the real experimental conditions), written as U_i. This is by means of an equation such as:

$$U_i = U_i^{(0)} + X_i \times \Delta U_i$$

$U_i^{(0)}$ is the value of the natural variable at the centre of interest and ΔU_i is the change in the factor corresponding to a variation of 1 unit in the corresponding coded variable X_i. Taking, for example, the values for experiment 1 in table 5.3:

$$U_1^{(0)} = 0.100 \text{ M}; \qquad \Delta U_1 = 0.025 \text{ M}; \qquad X_1 = -1.414$$

$$U_1 = 0.100 - 1.414 \times 0.025 = 0.065 \text{ M corresponds to } X_1$$

B. Equiradial Designs for More than 2 Factors – The Simplex Design

Amongst all the possibilities, only two kinds of designs are used in this context (first-order model, $k \geq 3$). These are complete or fractional factorial designs at 2 levels with resolution III or better and the simplex. Factorial designs were covered in chapters 2 and 3 and will not be discussed further here. The second design, the simplex, is the most economical that exists, as there are only as many experiments as there are coefficients in the first-order model. For 2 factors it is an equilateral triangle, and for 3 factors it is a regular tetrahedron. Analogous designs may be constructed for any number of factors.

Any orientation of the simplex is satisfactory. The general design is presented in table 5.9 in an orientation equivalent to that of figure 5.5a.

Table 5.9 Simplex Designs of 2-8 Factors for RSM.

No.	X_1	X_2	X_3	X_4	X_5	X_6	X_7	X_8
1	-0.500	-0.289	-0.2041	-0.1581	-0.1291	-0.1091	-0.0945	-0.0833
2	0.500	-0.289	-0.2041	-0.1581	-0.1291	-0.1091	-0.0945	-0.0833
3	0	0.577	-0.2041	-0.1581	-0.1291	-0.1091	-0.0945	-0.0833
4	0	0	0.6124	-0.1581	-0.1291	-0.1091	-0.0945	-0.0833
5	0	0	0	0.6325	-0.1291	-0.1091	-0.0945	-0.0833
6	0	0	0	0	0.6455	-0.1091	-0.0945	-0.0833
7	0	0	0	0	0	0.6547	-0.0945	-0.0833
8	0	0	0	0	0	0	0.6614	-0.0833
9	0	0	0	0	0	0	0	0.6667
R		0.577	0.6124	0.6325	0.6455	0.6547	0.6614	0.6667

In each column, X_j, the negative elements are equal to $-1/\sqrt{2j(j+1)}$ and the positive elements to $1/\sqrt{2j(j+1)}$. The experiments are inscribed in a circle of radius $R = k/\sqrt{2k(k+1)}$ and all the coordinates must be divided by this value to normalise the simplex to a sphere of radius 1. We see an example of this in the coordinates of the triangle in table 5.8.

It is illustrated by a simplex of 4 factors (regular tetrahedron). The design, in table 5.9, is enclosed by a dotted line. Dividing each element by the radius $R = 0.6325$ we obtain:

	X_1	X_2	X_3	X_4
1	-0.778	-0.449	-0.318	-0.250
2	0.778	-0.449	-0.318	-0.250
3	0	0.898	-0.318	-0.250
4	0	0	0.953	-0.250
5	0	0	0	1.000

Similarly for a simplex with 5 factors (6 runs) we take the design enclosed by a double dotted line. After division by $R = 0.6455$, this gives:

	X_1	X_2	X_3	X_4	X_5
1	-0.775	-0.448	-0.377	-0.245	-0.200
2	0.775	-0.448	-0.377	-0.245	-0.200
3	0	0.894	-0.377	-0.245	-0.200
4	0	0	0.949	-0.245	-0.200
5	0	0	0	0.980	-0.200
6	0	0	0	0	1.000

V. STRATEGIES FOR A SPHERICAL DOMAIN AND SECOND DEGREE MODEL

A. Composite Experimental Designs

1. General method of construction

The central composite design, as an examination of the literature will show, is the one most used in response surface analysis. For this reason, and also because its construction follows on naturally from that of the factorial designs covered previously, we will examine it first. However, even if the evidence of the pharmaceutical literature is overwhelmingly in favour of its employment, we do not recommend it to the reader with quite the same insistence. There are other methods, covered later in the chapter, which may be superior in quite frequent circumstances, and which certainly merit equal consideration.

As the name indicates, the composite design for k factors consists of several groups of experiments:

- N_F experiments of a factorial (complete 2^k or fractional 2^{k-r}) design.
- $2k$ experiments of a "star" or *axial* design (2 experiments on the axis associated with each factor at a value $x_i = \pm \alpha$, with all other variables set at zero).
- N_0 centre points: the number chosen according to the desired properties of the design.

The central composite design for 2 factors has already been discussed in section II. We now look in detail at a slightly more complex example of 3 factors.

2. Use of the central composite design in the formulation of an oral solution

Introduction

In section II.B.2 we treated the 2^3 factorial design using data published by Senderak and coworkers. In their paper (2) they described the formulation of a solution of a slightly soluble drug aided by factorial experimental design followed by RSM. They *first* tested the effects of 4 factors on the cloud point and turbidity of the resulting solutions over a fairly wide range of the factor variables, using a complete 2^4 design. The factors were the percentages of surfactant (polysorbate 80), propylene glycol, ethanol, and of invert sugar medium. The remaining components were either constant (at low levels), except for water which was a "slack variable". Experiments were also carried out at the centre of the domain. The experimental domain and the results of this study are summarised in table 5.10 for the cloud point and turbidity. The reader is referred to the original paper for full details of the factorial design and its experimental data.

Table 5.10 Study of the Effects of 4 Factors on Cloud Point and Turbidity of a Liquid Formulation: Experimental Domain and Coefficient Estimations from 2^4 Design [from reference (2), with permission]

Component	Limits %	$\log_{10}$(turbidity)		Cloud point °C	
		Parameter estimate	Signif.	Parameter estimate	Signif.
Polysorbate 80	2 - 4	6.54	**	- 0.495	0.16
Propylene glycol	5 - 25	7.01	**	-0.415	**
Invert sugar	25 - 65	-15.59	**	0.125	**
Alcohol	1 - 5	1.46	0.10	-0.014	0.87

It was concluded that:
- The first three factors listed above had an influence on the turbidity or cloud point within the limits studied, whereas the remaining factor, concentration of alcohol, could be omitted from subsequent studies.
- Although main effects and interactions could be estimated, the analysis of variance showed considerable and significant lack of fit to the model because of curvature. The model could not be used for predicting the response over the domain.

If the model is totally inadequate, there are normally two things that can be done. The first is to increase its complexity, adding quadratic terms for example. Additional experiments are then carried out. (An advantage of the central composite design is that it is easy to make the corresponding additions to the factorial design. We saw this for 2 factors; it is equally the case for 3, 4 or 5.)

The second solution is to reduce the size of the experimental domain. Usually the smaller the domain, the less the total curvature, and therefore there exists a simpler satisfactory model within the reduced domain. Third- or higher-order designs are rarely used (except for mixture models) and for the quadratic model to be sufficient the extent of the experimental region must normally be restricted.

In this case, both actions were taken. A more restricted experimental domain was selected, eliminating regions which, according to the 2^4 factor study, would result in a high turbidity. Alcohol concentration was omitted as a variable (it was fixed at 2%) and a 3 factor second-order model was postulated.

$$y = \beta_0 + \beta_1 x_1 + \beta_2 x_2 + \beta_3 x_3 + \beta_{11} x_1^2 + \beta_{22} x_2^2 + \beta_{33} x_3^2 + \beta_{12} x_1 x_2 + \beta_{13} x_1 x_3 + \beta_{23} x_2 x_3 + \varepsilon$$

The experiments were carried out according to a central composite design. Certain of these experiments, the factorial part of the design, have already been discussed in chapter 3, section II.B.2, where they were used to illustrate the 2^3 full factorial design.

The central composite design for 3 factors
The design is shown in table 5.11 and figure 5.6. It consists of a 2^3 factorial design plus 6 "star" or axial points, plus points at the centre of the domain. The two parts of the experiment may be carried out separately, provided enough experiments are done at the centre of the domain as part of each block.

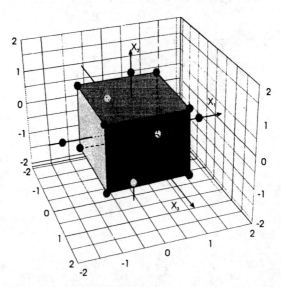

Figure 5.6 Central composite design for 3 factors.

Table 5.11 Central Composite Design for the Formulation of an Oral Solution [data from reference (2), with permission]

No.	X_1	X_2	X_3		Polysorbate 80 (%)	Propylene glycol (%)	Sucrose invert medium (mL)	Turbidity (ppm)	Cloud point (°C)
1	-1	-1	-1	Factorial design	3.7	17	49	3.1	75.7
2	+1	-1	-1		4.3	17	49	2.8	73.5
3	-1	+1	-1		3.7	23	49	3.9	80.5
4	+1	+1	-1		4.3	23	49	3.1	83.7
5	-1	-1	+1		3.7	17	61	6.0	62.0
6	+1	-1	+1		4.3	17	61	3.4	69.5
7	-1	+1	+1		3.7	23	61	3.5	69.9
8	+1	+1	+1		4.3	23	61	1.8	70.8
9	-1.68	0	0	Star design	3.5	20	55	4.9	73.4
10	+1.68	0	0		4.5	20	55	3.3	74.6
11	0	-1.68	0		4.0	15	55	4.5	81.9
12	0	+1.68	0		4.0	25	55	5.1	69.5
13	0	0	-1.68		4.0	20	45	3.3	80.7
14	0	0	+1.68		4.0	20	65	3.2	59.9
15	0	0	0	Centre points	4.0	20	55	2.3	67.7
16	0	0	0		4.0	20	55	3.8	74.9
17	0	0	0		4.0	20	55	2.9	72.7
18	0	0	0		4.0	20	55	2.4	74.8
19	0	0	0		4.0	20	55	3.5	70.4
20	0	0	0		4.0	20	55	3.2	72.8

For the star design, one can imagine starting at the centre point then moving along each axis (in each direction, positive and negative) a set distance $\alpha = \pm 1.68$. The general rules for the choice of α will be given later, in section V.A.4.

Experimentation plan: results and linear regression

The plan is shown in table 5.11, with the measured responses for the turbidity and cloud point. (Although the experiments are listed in the standard order, here as in the rest of the book, they should be performed in random order.) Only the turbidity data will be discussed here. Some of these data, those for the factorial part of the design, have been shown already and have been analysed in chapter 3. We have noted that this kind of design is advantageous in that it is possible to carry out factorial and star designs in separate stages. Although the experiment could have been done in stages in this way, this did not in fact happen, and the central composite design was carried out in one step.

The model coefficients were then estimated by least squares linear regression (table 5.12). Compare the values with those obtained with the factorial points, in chapter 3, section II.B.4. We see that only the interaction terms are unmodified; the axial data are not used in calculating these.

Table 5.12 Estimate of Coefficients for the Second-Order Model (Turbidity)

	Coefficient	Estimate	Standard error	Significance
b_0	Constant	3.039	0.282	***
b_1	polysorbate 80	-0.593	0.187	**
b_2	propylene glycol	-0.146	0.187	0.455
b_3	sucrose invert medium	0.120	0.187	0.538
b_{11}		0.243	0.182	0.212
b_{22}		0.491	0.182	*
b_{33}		-0.058	0.182	0.760
b_{12}		0.050	0.245	0.842
b_{13}		-0.400	0.245	0.133
b_{23}		-0.650	0.245	*

Testing the significance of the quadratic model by ANOVA

Analysis of variance is carried out as described in chapter 4. As before, the (adjusted) total sum of squares SS_{TOTAL} is calculated as the sum of the squares of the differences between the experimental data and the data mean. It can be separated into:

- The sum of squares explained by the second-order model, SS_{REGR}.
- The residual sum of squares, SS_{RESID}, which represents the random experimental variation in the system, plus any lack of fit of the model.

There were 6 replicated experiments, all at the centre of the domain. The differences between the results can only be due to experimental variation. (We assume that the variation is random and that it is constant throughout the experimental domain.) Therefore, it will be possible to partition the residual sum of squares, distinguishing between the part due to random variation and that which is a result of inadequacy of the model.

The pure error sum of squares SS_{ERR} is therefore the sum of squares of the differences between the response for each replication and their mean value:

$$SS_{ERR} = \sum_{i=1}^{N_0} (y_{0i} - \bar{y}_0)^2$$

where y_{0i} are the individual responses of the N_0 experiments at the centre and $\bar{y}_0$ is their mean value. SS_{ERR} is associated with 5 degrees of freedom (one less than the number of replicated experiments) and the residual sum of squares is the sum of the pure error and the lack of fit sums of squares, $SS_{RESID} = SS_{ERR} + SS_{LOF}$. The various sums of squares are detailed in table 5.13 for the turbidity.

The ratio $F_{LOF} = MS_{LOF}/MS_{ERR} = 1.69$ is not significantly high, showing that there is no significant lack-of-fit, and the mean squares for the lack-of-fit and the pure error are comparable. Thus, the residual mean square MS_{RESID} can be used as our estimate for the experimental variance. Taking its square root, the experimental standard deviation is estimated as 0.69, with 10 degrees of freedom.

Fisher's test may then be carried out to compare the regression mean square MS_{REGR} with MS_{RESID}. The ratio, $F = 3.29$, shows the regression to be significant, with only 4% probability that the observed deviations from a constant value are the result of random error. Calculation of R^2 shows that the regression model accounts for 75% of the total variance. The estimated model may therefore be used to trace the response surface for the turbidity. Evidently, if the process were more highly reproducible and the pure error were lower, the regression would be more significant and there would be a greater possibility of finding statistically significant lack-of-fit.

Table 5.13 ANOVA for the Response Turbidity

	Degrees of freedom	Sum of squares	Mean square	F	Sign.
Total (adj)	$N - 1 =$ 19	$SS_{TOTAL} =$ 19.00			
Regression	$p - 1 =$ 9	$SS_{REGR} =$ 14.20	$MS_{REGR} =$ 1.578	3.29	*
Residual	$N - p =$ 10	$SS_{RESID} =$ 4.80	$MS_{RESID} =$ 0.480		
Lack-of-fit	$n-p-N_0+1=$ 5	$SS_{LOF} =$ 3.01	$MS_{LOF} =$ 0.603	1.69	0.29
Pure error	$N_0 - 1 =$ 5	$SS_{ERR} =$ 1.79	$MS_{ERR} =$ 0.358		

Analysis of the response surface

We take first the full quadratic model and calculate the lines of equal response. As there are 3 variables, the factor space is in 3 dimensions, and to represent this on a 2-dimensional paper surface or computer screen, we need to take slices with one of the variables held constant. Here the polysorbate concentration, X_1, was chosen as the fixed variable. Taking $x_1 = -1$, 0, +1 (real values 3.7, 4.0, 4.3% polysorbate), lines of equal turbidity are plotted in the X_2, X_3 plane (propylene glycol, sucrose invert medium), as shown in figure 5.7.

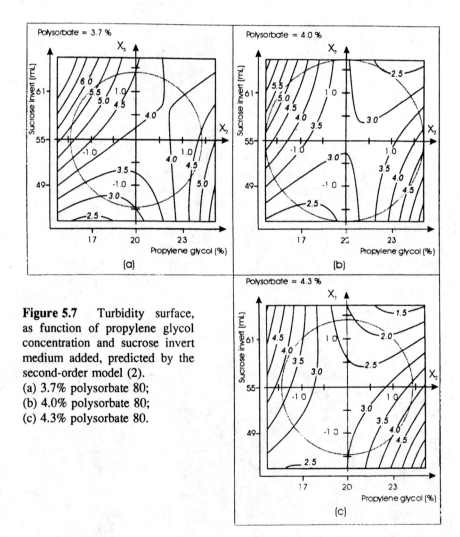

Figure 5.7 Turbidity surface, as function of propylene glycol concentration and sucrose invert medium added, predicted by the second-order model (2).
(a) 3.7% polysorbate 80;
(b) 4.0% polysorbate 80;
(c) 4.3% polysorbate 80.

Note the boundary of the domain. Since the outline of the printed graph is rectangular and the domain is spherical with radius $\sqrt{3}$, it is useful, even essential, to be able to visualise this on the contourplot.

We verify at this stage that no physically impossible (or for that matter no highly improbable) values of the response are predicted in any part of the domain. This can happen, for it is unlikely that any polynomial model would describe the system perfectly and the quadratic model is a very simple polynomial. A negative turbidity would be wrong! If we were calculating the response surface of a yield, or the percentage of drug dissolved in a given time, then the whole of the calculated response surface should lie between 0 and 100%. Quite often it does not, and we then need to decide what to do with those negative yields and unduly productive predictions. An appropriate transformation can sometimes remove the problem, otherwise one can decide to live with it, considering all negative values as zero and cutting off all high values at 100%. The plotting program can then be told to start at 5% and finish at 95%!

3. ANOVA of the regression with partition of the regression sum of squares

The regression sum of squares is the total sum of squares explained by the model. The model consists of first-order terms, square terms and first-order interactions and the regression sum of squares can be divided among these three sets of terms as follows. For simplicity we will consider all the second-order terms together. We fit the first-order model:

$$y(x_1, x_2, x_3) = \beta_0 + \beta_1 x_1 + \beta_2 x_2 + \beta_3 x_3 + \varepsilon$$

to the data, rather than fitting the second-order equation. In this way we obtain identical estimated values for the first-order coefficients (though the estimate for the constant term is not quite the same). The first-order regression sum of squares $SS_{REGR(1)}$ and the corresponding residual sum of squares $SS_{RESID(1)}$ may be calculated in exactly the same way that SS_{REGR} and SS_{RESID} were calculated previously:

$$SS_{REGR(1)} = 5.28 \qquad SS_{RESID(1)} = 13.72$$

Of the total adjusted sum of squares, 28% is accounted for by the first-order terms alone, and 75% by *all* the terms of the second-order model, first-order, square, and interaction. The residual sum of squares of the first order model includes the sum of squares for the second-order terms. So the difference between the residual sums of squares for the two models is the second-order regression sum of squares $SS_{REGR(2)}$:

$$SS_{REGR(2)} = SS_{REGR} - SS_{REGR(1)} \qquad = 14.16 - 5.28 = 8.92$$

The full ANOVA table is therefore that of table 5.14.

The mean square is obtained by dividing each sum of squares by the corresponding degrees of freedom. The F statistics are got by dividing each mean square by the residual mean square (provided the lack of fit test is not statistically significant).

From the calculated probabilities of each value of the F statistic we conclude that the second-order terms in the model are "almost" statistically significant and improve the fit of the model to the experimental data.

Table 5.14 ANOVA for the Turbidity: Separation of First- and Second-Order Sums of Squares

	Degrees of freedom	Sum of squares	Mean square	F	Sign.
Total (adj)	$N - 1 =$ 19	$SS_{TOTAL} = 19.00$			
Regression	$p - 1 =$ 9	$SS_{REGR} = 14.16$	$MS_{REGR} =$ 1.573	3.25	*
1st order	3	$SS_{REGR(1)} =$ 5.28	$MS_{REGR(1)} =$ 1.76	2.05	0.147
2nd order	6	$SS_{REGR(2)} =$ 8.92	$MS_{REGR(2)} =$ 1.49	3.10	0.055
Residual	$N - p =$ 10	$SS_{RESID} =$ 4.84	$MS_{RESID} =$ 0.484		
Lack of fit	$N-p-N_0+1=$ 5	$SS_{LOF} =$ 3.05	$MS_{LOF} =$ 0.610	1.70	0.291
Pure error	$N_0 - 1 =$ 5	$SS_{ERR} =$ 1.79	$MS_{ERR} =$ 0.358		

Using a similar approach, it is possible to calculate the effect of omitting the interaction terms or the square terms in the model. In fact, these calculations can be taken still further. We can work out the contribution of each parameter in the model to the total sum of squares and, using the F tables, determine the statistical significance of each parameter. It is a common practice to simplify the model accordingly, taking only the parameters that are statistically significant. Doing this we find that the model can be simplified to:

$$y(x_1, x_2, x_3) = \beta_0 + \beta_1 x_1 + \beta_2 x_2 + \beta_3 x_3 + \beta_{22} x_2^2 + \beta_{23} x_2 x_3 + \varepsilon$$

This simplification technique is of doubtful interest for response surface modelling, and for optimization. For here we are not really interested in the values of any of the coefficients in the model, but only in the overall effect of the coefficients on the response surface. There is no reason to suppose that the simplified model is "better" than the original model. It may apparently have more residual degrees of freedom and the regression may be more "significant". But it will not be a better model for prediction of the response over the domain and for finding the optimum.

It is especially dangerous to eliminate "non-significant" terms from the model when treating saturated or nearly saturated designs. It is possible to obtain as a result apparently highly significant regressions and small error estimates which do not have a great deal of meaning, especially where the original model has few degrees of freedom.

4. Properties

Rotatability

It is evident that the quality of a design matrix for RSM is a question of the quality of prediction of the response within the domain. It is described by the variance function. We have already seen that:

$$d_A = x_A'(X'X)^{-1}x_A$$

at point A in the factor space. Thus, d_A depends only on the information matrix (which in turn depends only on the design and model equation) and on the position of point A in the factor space, defined by the experimental coordinates x_A. The relative variance (and thus the standard deviation) of prediction of the responses may be calculated over all of the domain.

For the composite design there is a value of α for which the variance of prediction of the response depends only on the distance from the centre and not on the direction in which one moves from the centre. The isovariance lines or surfaces are spherical. This property is known as *rotatability*.

Once this property is respected, no direction is favoured above another in the prediction. To obtain rotatability in a composite design, set the axial distance to:

$$\alpha = \sqrt[4]{N_F}$$

where N_F is the number of experiments in the factorial part of the design. So for 4, 8, 16, and 32 experiments, α is respectively 1.414, 1.682, 2, 2.378.

We examine the effect of the choice of α on the isovariance curves of a 2 factor central composite design with 3 centre points, shown in figure 5.8. The rotatability of the design with $\alpha = 1.414$ can clearly be seen in figure 5.8c.
In all 3 cases the variance function increases rapidly at the limits of the experimental zone, showing that extrapolation outside the domain is most unreliable. However with $\alpha = \sqrt[4]{4} = 1.414$ the variance function equals 1 at the edge of the domain (a circle of radius $\sqrt{2} = 1.414$), whereas with $\alpha = 1$ the variance function is very large at the domain's edge, on the X_1 and X_2 axes.

Resulting values of α for the spherical domain are given in table 5.16. Some them appear high (up to 2.378 for 6 factors) and might seem to require experiments a long way out along the axes. We should note however that the "factorial" experiments themselves are situated $\sqrt{k}$ units from the centre of interest. For 6 factors, this is equal to 2.449. In fact, the star points are always close to the sphere defining the edge of the domain, either on that sphere or just inside it.

If a composite design is carried out sequentially (see below), it is very important to take this into account and accordingly reduce the step size for the factorial design so that the star experiments may be carried out and are not in an inaccessible part of the factor space, excluded by technological or economic factors. If this does happen and we need to compromise on rotatability, the loss of rotatability may be analysed graphically, as in figure 5.8.

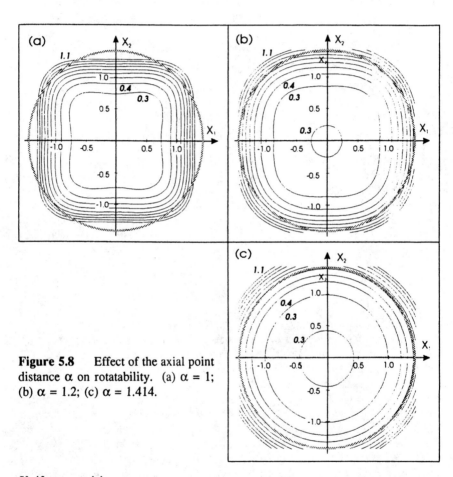

Figure 5.8   Effect of the axial point distance α on rotatability. (a) $\alpha = 1$; (b) $\alpha = 1.2$; (c) $\alpha = 1.414$.

Uniform precision
It is clear that if we want uniform precision, that is a variance function that remains constant within the domain, we first of all require a rotatable design. Thus α must be set to $\sqrt[4]{N_F}$ as described above, so that d_A varies only with the distance from the centre. This variation in precision with distance may be modulated, and "evened out" by adjusting N_0, the number of experiments at the centre. We illustrate this with the 2 factor composite design.

A completely constant variance of prediction is not possible. Mathematically it can be shown that the closest to uniform precision can be obtained for $N_0 = 5$. However, if we compare the isovariance curves (figure 5.9) for $N_0 = 5$ with those for $N_0 = 3$, we see that the difference may not be great enough to justify carrying out the extra experiments if these are very costly. (Note that the variance function may be transformed to a relative standard deviation by taking the square root.) On the other hand a single experiment at the centre appears to be insufficient.

In table 5.16, we give theoretical values of N_0 giving uniform precision along with what we consider to be the *minimum* number of experiments necessary to give an adequate quality in prediction.

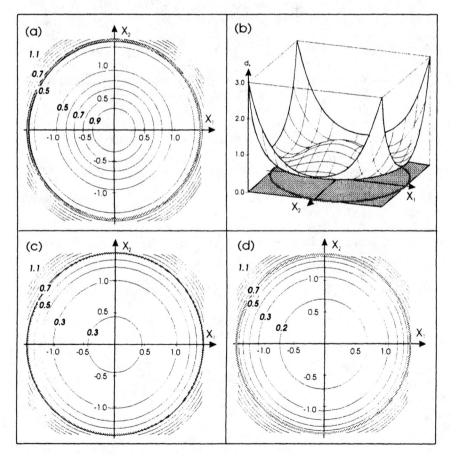

Figure 5.9 Uniform precision: effect of the number of centre points on the variance function for a 2 factor composite design. (a-b) $N_0 = 1$; (c) $N_0 = 3$; (d) $N_0 = 5$.

Use of a fractional factorial in a composite design
We saw in the example treated above that adding star points to the factorial design did not change the estimates of the interaction terms. The axial points were not used to estimate the rectangular coefficients b_{ij}, which thus depended only on the factorial design. If we are to determine the full second-order design, including all the b_{ij} terms, the factorial design should preferably be complete or of resolution R_V, or better. Therefore, full factorial designs should be used for 2, 3, or 4 factors. In the case of 5 or 6 factors, the fractional designs 2^{5-1} and 2^{6-1} offer an economical alternative (table 5.16). See references 3-4 for examples of the use of a 5 factor composite design.

Small composite designs
There is an alternative, rather more economical, composite design for 3 factors, based on a (resolution III) 2^{3-1} design, a star design and centre points (5). It allows determination of the second-order model, but its properties are inferior to those of the full design. In particular it is not rotatable. This design may be useful as it allows sequential experimentation with a small total number of experiments (about 12-14). Thus the exploratory 2^{3-1} design may be completed by the axial points. It has been used for the optimization of a solid dispersion (6).

A similar design for 4 factors exists, but the 2^{4-1} design on which it is based must be of resolution III. It is unlikely such a design would be used for an initial factor study. If, on the other hand, the resolution IV 2^{4-1} design is augmented by axial and centre points, the interaction terms remain aliased among themselves. This design is not useful, because if an economical design is sought the hybrid design (see below) is better.

The $\frac{3}{4} \times 2^4$ design, of chapter 3, section V.A, table 3.31, may be augmented by axial points at $\alpha = 1.68$ and centre points, to give a small design of quite good quality.

See reference 5 for a discussion of so-called "small" composite designs.

Sequentiality and blocking
One of the more important properties of composite designs is the possibility of carrying out the experiments in two or more blocks. This is already illustrated in the solubility example, where a 2^2 design plus centre points was carried out first and then augmented by a star design, also with centre points.

As a general rule the designs may be split up into:
- A fractional factorial design with centre points carried out to determine the main effects – that is, a first-degree model.
- A second factorial matrix (also with centre points), carried out to complete the factorial design (complete or fractional of resolution R_V) and determine the interactions.
- The square terms, estimated last of all by adding the axial (star) points.

Each block should contain the same number of experiments at the centre. We see possible partitions listed in table 5.15, stages 1 and 2 only, stage 3 in each case being the axial design, allowing estimation of the b_{ii}.

Table 5.15 Partitions of Composite Designs (Excluding Stage 3: Axial Design)

k	Stage 1	Estimated effects	Stage 2	Estimated effects
2	2^2	b_0, b_1, b_2, b_{12}	—	—
3	2^{3-1} $I \equiv 123$	b_0, b_i	2^{3-1} $I \equiv -123$	b_0, b_i, b_{ij}
4	2^{4-1} $I \equiv 1234$	b_0, b_i	2^{4-1} $I \equiv -1234$	b_0, b_i, b_{ij}
5	2^{5-1} $I \equiv 12345$	b_0, b_i, b_{ij}	—	—
6	2^{6-1} $I \equiv 123456$	b_0, b_i, b_{ij}	—	—
6	2^{6-2} $I \equiv 123 \equiv 3456 \equiv$ 12456	b_0, b_i	2^{6-2} $I \equiv -123 \equiv -3456 \equiv$ 12456	b_0, b_i, b_{ij}

5. Summary of recommended central composite designs

The factorial design should be at least of resolution V (and in any case must not be R_{IV}). Full factorial designs are usually employed for 3 and 4 factors (but see the section on small designs, above), whereas fractional factorial designs are used for 5 and 6 factors. The composite design for 7 factors contains more than 80 experiments, for a model of 36 coefficients. Not being considered of practical interest here it is not given in the table. The values of α and N_0 are determined by the properties looked for in the design, in particular the property of uniform precision. The optimum numbers of experiments (according to this and other criteria) are often considered too high. We have therefore suggested minimum numbers. Below these values the uniformity of precision of the predictions is greatly compromised (table 5.16).

Table 5.16 Summary of the Most Commonly used Central Composite Designs for 2-6 Factors

Number of factors: k	2	3	4	5	6
Factorial design	2^2	2^3	2^4	2^{5-1} $I \equiv 12345$	2^{6-1} $I \equiv 123456$
N_F no. factorial design points	4	8	16	16	32
α axial points spacing	1.414	1.682	2	2	2.378
N_A no. axial design points	4	6	8	10	12
N_0 optimum n° of centre points for uniform precision	5	6	7	5	9
N_0 minimum n° of centre points	2-3	3	3-4	5	5

B. Uniform Shell (Doehlert) Designs

The uniform shell design (7) is a (rhombic) lattice of uniform density in normalized factor space. The design itself is normally just one part of the lattice – a single point and its nearest neighbours. It may be extended in any direction. This includes the possibility of adding additional factors without any adverse effects on the quality of the design.

We consider these designs to be highly recommendable as tools for pharmaceutical development. Although we will see that they may be inferior to the central composite designs according to some of our criteria, they are still of excellent quality and they have a number of interesting additional properties that make them particularly suitable for solving certain problems.

1. The uniform shell design for 2 factors

This is a regular hexagon with a centre point, thus requiring a minimum of 7 experiments. The mathematical model used to treat it may be first-order, synergistic or quadratic. We will assume a quadratic model and try to judge the quality of the design in the usual way. The design is highly efficient, the R-efficiency being 86%. As shown in figure 5.10, it is also rotatable, nearly orthogonal and has uniform variance of prediction [unlike the designs for more than 2 factors (see section V.B.6)]. If we wish to do a minimum of experiments for 2 variables only, and to analyse the results according to a quadratic model, then this hexagonal design of only 7 experiments is to be recommended. It would be better to do more than one experiment at the centre point (see table 5.24).

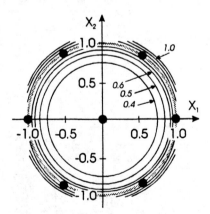

Figure 5.10 Variance function for 2 factor Doehlert design with 3 centre points.

We will now look at the practical use of the Doehlert design, demonstrating its various properties of sequentiality.

2. Expanding a simplex to obtain a Doehlert design

Two factors

To illustrate this possibility we use some data for the properties of tablets produced at different compression forces and percentage of disintegrant (8). The full design was a hexagon with 3 centre points, the experiments being indicated by squares in figure 5.11. Assume initially a first-order model for the response y.

The most economical design possible is a *simplex*. In normalized factor space this is an equilateral triangle, shown enclosed by thick lines in figure 5.11. We suppose, for the sake of illustration, that these experiments at the vertices of the simplex had been carried out first of all, that is, a simplex in a more restricted domain. The results of the tablet hardness are indicated for the three experiments. In figure 5.11, we see that, starting from each point of the simplex, we can construct a different hexagon (shown by a dotted line, a thin continuous line, and a thicker line). We can choose between these designs by taking the "best" point of the simplex as the centre of the new domain. Assume that we wish to maximize the hardness. We take the point with hardness of 10.6 kg as the new centre, and construct a new design in that direction. Squares joined by thicker lines show the actual design.

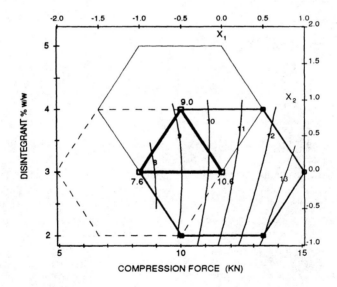

Figure 5.11 Expansion of a simplex to a Doehlert design for 2 factors and contour plot of tablet hardnesses [adapted from reference (8), with permission].

Expanding the simplex to a hexagonal design thus requires carrying out 4 extra experiments (plus extra centre points). The results, from reference 8, may be analysed according to both the first and the second order models. It is inevitable that the quadratic model will give an apparently better fit to the data, but the improvement appears to be both statistically and pharmaceutically significant. Therefore we use the second-order model and plot the response curves for the yield in the experimental domain.

We have illustrated the expansion of a simplex here on the basis of just one response – hardness. It is probable that the experimenter would wish to consider other key response variables, such as disintegration and dissolution times. The data of the original paper showed disintegration times and dissolution times increasing with the hardness. Some form of compromise needs to be found in optimizing the formulation rather than relying on a single response. The various possible methods are discussed in chapter 6.

Three or more factors

A simplex for 3 factors may be expanded in the same way, though the problem becomes more complex. It is possible to construct a simplex in any dimensional factor space (see section IV.B) and this is the most economical design for a first-order model. For 3 variables, the simplex is a regular tetrahedron and its coordinates are (usually):

No.	X_1	X_2	X_3
1	0	0	0
2	1	0	0
3	0.5	0.866	0
4	0.5	0.289	0.816

The uniform shell design for 3 factors consists of 13 experiments and includes these 4 points. It has the form of a cuboctahedron, obtained by joining together the mid-points of the edges of a cube. The coordinates of the remaining 7 experiments are found from these 4 by subtraction. So here also the stepwise approach is possible. We can go from a first order model, requiring very few experiments to a quadratic model, covering a larger domain and with a certain degree of choice in the direction of the expansion of the domain.

3. Sequential approach using the Doehlert design: wet pelletization in a high-shear mixer

Defining the problem: experimental domain

We illustrate expansion and displacement of the domain using another example in process optimization, the formation of pellets by granulation in a high shear mixer, described by Vojnovic and coworkers (9). The object of this study was to map the effects of 3 factors, impeller speed (X_1), amount of granulation liquid (X_2), and the spheronization (or pelletization) time (X_3), on two responses, the yield of pellets between 800 and 1250 µm and the geometric mean diameter by mass of the particles. The experimental domain, its choice based on the results of smaller scale experiments, is shown in table 5.17.

Table 5.17 Experimental Domain: Wet Pelletization in a High-Shear Mixer

Factor	Associated variable	Lower level (coded -1)	Upper level (coded +1)
Impeller speed (rpm)	X_1	250	400
Amount of binding solution (%)	X_2	60	70
Pelletization time (min)	X_3	5	20

First step: study of two factors
To begin with, we keep the pelletization time, X_3, constant at what is thought to be an acceptable "mean" value $x_{3,0} = 12.5$ min and determine the effects of varying X_1 and X_2. The design and experimental results are shown in table 5.18.

The quadratic model parameters can be estimated by multi-linear regression and the model analysed by ANOVA. The adjusted model is:

$$\hat{y} = 1594 - 90.8x_1 + 120.3x_2 - 20.0x_1^2 - 389.2x_2^2 - 222.8x_1x_2$$

and the contour lines of the response surface are given in figure 5.12.

Table 5.18 Doehlert Design (2 Factors) for Wet Pelletization in a High-Shear Mixer: Design A – Experimental Plan – Data for Mean Size [data from reference (9), with permission]

No.	X_1	X_2	X_3	Impeller speed (rpm)	Binding solution (%)	Pelletization time (min)	Mean size (µm)
1	0	0	0	325	65.0	12.5	1594
2	1	0	0	400	65.0	12.5	1398
3	0.5	0.866	0	362	69.3	12.5	1352
4	-0.5	0.866	0	288	69.3	12.5	1466
5	-1	0	0	250	65.0	12.5	1750
6	-0.5	-0.866	0	288	60.7	12.5	1064
7	0.5	-0.866	0	362	60.7	12.5	1337

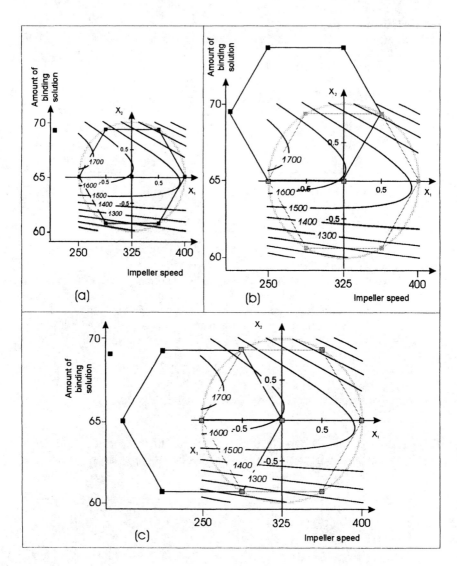

Figure 5.12 Response surface for the mean particle size (2 factor design) with two possible displacements of the domain.

Displacing the experimental domain
We sometimes find that the initial choice of experimental domain was not the best possible one. In the above example it is evident that the design is not centred on the optimum and we may wish either to carry out further experiments at different values of X_1 and X_2 or construct a new design. The uniform shell design gives us this possibility of extending or displacing the design quite cheaply. New hexagonal designs may be constructed, as shown previously in any of the 6 directions, overlapping with the old design and requiring only 3 extra experiments in the case of the 2 factor design. Two possible displacements are shown in figure 5.12.

Expanding the domain by adding another factor
We recall that there were three factors to be studied but the third factor, pelletization time X_3, was held constant at a mean level, thought to be approximately optimum. The hexagonal design can be extended to three dimensions, by adding 6 experiments, in the form of two equilateral triangles, at a low and a high pelletization time, at $x_3 = \pm 0.816$ (figure 5.13). The additional experiments and corresponding experimental results are shown in table 5.19.

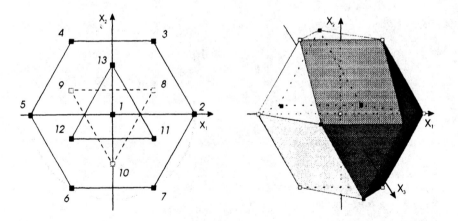

Figure 5.13 Expansion of the Doehlert design by addition of a 3rd variable factor (added experiments are shown by filled squares in the 3-D diagram on the right-hand side).

Table 5.19 Supplementary Experiments for the Addition of a Third Factor [data from reference (9), with permission]

	X_1	X_2	X_3	Impeller speed (rpm)	Binding solution (%)	Pelletization time (min)	Mean size (µm)
8	0.5	0.289	0.816	362	66.4	18.6	1533
9	-0.5	0.289	0.816	288	66.4	18.6	1617
10	0	-0.577	0.816	325	62.1	18.6	1332
11	0.5	-0.289	-0.816	362	63.6	6.4	1409
12	-0.5	-0.289	-0.816	288	63.6	6.4	1560
13	0	0.577	-0.816	325	68.9	6.4	1431

The experimental design for 2 factors is a sub-set of the design for 3 factors, and the design for 3 factors is a sub-set of the design for 4 factors and so on. This is demonstrated later in the chapter, in table 5.22, which gives Doehlert designs for up to 5 factors.

The complete quadratic model determined by multilinear regression for the mean particle size is:

$$\hat{y} = 1594 - 97.4\,x_1 + 117.6\,x_2 + 16.7\,x_3 - 20.0\,x_1^2 - 379.2\,x_2^2 - 70.9\,x_3^2$$
$$- 222.8\,x_1 x_2 + 119.6\,x_1 x_3 + 82.8\,x_2 x_3$$

The coefficients of x_1 and x_2 are little changed from the two-factor model, and thus the new surfaces in the $\{X_1, X_2\}$ plane, are very close to those of figure 5.12.

Response surfaces in the $\{X_1, X_3\}$ are plotted at three levels of X_2 in figures 5.14a-c. It should be noted on these diagrams that the circle in these figures represents the intersection of a sphere of radius 1 (the experimental domain) with the $X_1 X_3$ plane, and it is therefore considerably smaller at $x_2 = \pm 0.5$ than at $x_2 = 0$.

These figures 5.14a to 5.14c show considerable changess in the dependence on impeller speed and pelletization time (X_1, X_3) as the percent binder solution is varied between $x_2 = \pm 0.5$. Second-order coefficients were found to be influential so the quadratic model was maintained and not simplified further. In particular, as can be seen in the isoresponse curves, the amount of liquid, and its interaction with the impeller speed had large effects on the particle size. The effect of the pelletization time was less significant.

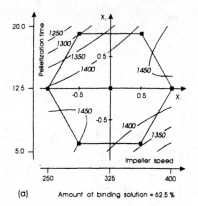

(a)   Amount of binding solution = 62.5 %

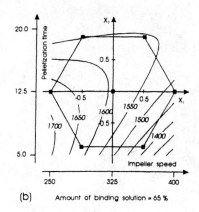

(b)   Amount of binding solution = 65 %

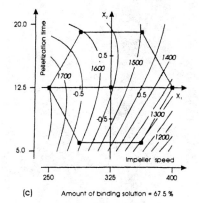

(c)   Amount of binding solution = 67.5 %

Figure 5.14 Mean particle size as a function of impeller speed and pelletization time at different percentages of binder solution.

Displacing the experimental domain with 3 factors

The 3 factor design can also be displaced in the same way as for 2 factors. The domain is translated by drawing a second design alongside the initial design. Six of the 13 original experimental points can be re-used. The method was used by Vojnovic and coworkers in the above study to examine the possibility of obtaining pellets of a particle size smaller than 1000 μm.

For most of the experiments and over most of the domain the particle size was rather greater than this. However at one experimental point (experiment 6) the response, both measured and calculated, approached 1000 μm and so the domain could be recentred at that point. This point was chosen to be the centre of a new Doehlert design B, defined in terms of new coded variables X_i. This design required only 7 additional experiments, the remaining 6 runs having already been performed as part of the design A (figure 5.15a).

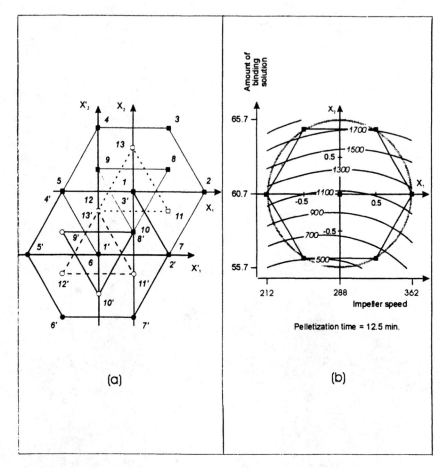

Figure 5.15 (a) Displacement of the domain. (b) Isoresponse curves in the new domain (mean particle size as function of impeller speed and binding solution with 12.5 min. pelletization time).

The new design is exactly the same as the old in terms of the coded variables. We however modify the correspondence between natural and coded variables. Experiment 6 ($x_1 = -0.5$; $x_2 = -0.866$; $x_3 = 0$) is the new centre. Thus to obtain values of the natural variables that correspond to the coded variables in the new design we carry out the following transformations:

- For X'_1 (mixer speed) the new centre point is at -0.5 with respect to the old set of coded variables, and at 0 with respect to the new set. Values of the natural variables are obtained by subtracting 0.5 × 75 = 37.5 rpm (equivalent to 0.5 in terms of coded variables), from each of the 13 values in design A.

- Similarly, for X'_2 (binder percentage), with a centre point -0.866 in terms of the old axis X_2, we subtract 0.866 × 5 = 4.33% from each value of the natural variable.
- X'_3 remains unchanged from X_3, in going from design A to design B.

The new domain and the resulting new design are given in tables 5.20 and 5.21.

Table 5.20 Wet Pelletization in High-Shear Mixer: Domain for Displaced Design B

Factor		Associated variable	Lower level (coded -1)	Upper level (coded +1)
Impeller speed	rpm	X'_1	212.5	362.5
Amount of binding solution	%	X'_2	55.7	65.7
Pelletization time	min	X'_3	5	20

Table 5.21 Design B (Displaced Design): Plan and Experimental Results [data from reference (9), with permission]

No	*	X'_1	X'_2	X'_3	Impeller speed (rpm)	Binding solution (%)	Pelletization time (min)	Mean size (µm)
1'	6	0	0	0	287.5	60.7	12.5	1064
2'	7	1	0	0	362.5	60.7	12.5	1337
3'	1	0.5	0.866	0	324.5	65.0	12.5	1594
4'	5	-0.5	0.866	0	250.5	65.0	12.5	1750
5'	-	-1	0	0	212.5	60.7	12.5	1124
6'	-	-0.5	-0.866	0	250.5	56.4	12.5	620
7'	-	0.5	-0.866	0	324.5	56.4	12.5	642
8'	10	0.5	0.289	0.816	324.5	62.1	18.6	1332
9'	-	-0.5	0.289	0.816	250.5	62.1	18.6	1512
10'	-	0	-0.577	0.816	287.5	57.8	18.6	649
11'	-	0.5	-0.289	-0.816	362.5	59.3	6.4	966
12'	-	-0.5	-0.289	-0.816	250.5	59.3	6.4	977
13'	12	0	0.577	-0.816	287.5	64.6	6.4	1560

* Corresponding experiment in the A design.

The quadratic model is fitted to the data, allowing the isoresponse curves to be drawn for the new displaced domain in figure 5.15b. This enables conditions to be identified for manufacturing beads of the correct particle size. This mean size and the yield of pellets between 800 and 1250 μm appear to be particularly sensitive to the added volume of liquid in this part of the domain, but not to be greatly affected by impeller speed or processing (pelletization) time. The results confirm the conditions for obtaining pellets of 1000 μm diameter, but also indicate conditions for manufacturing smaller sized particles.

4. Doehlert designs for 3 to 5 factors: numbers of levels

The standard uniform shell designs are listed in table 5.22 for 3, 4, and 5 factors.

An important property of the Doehlert design lies in the number of levels that each variable takes. In the normal form of the design the number of levels of X_1, X_2, X_3 is 5, 7, 3. In fact, no matter how many variables we have in the design, the minimum number of levels will be 3, then 5, 7, ... 7. The levels are evenly spaced. Thus in the case of 5 variables the numbers of levels are 5, 7, 7, 7, 3. So if there are difficulties in adjustment of the levels of a given factor, then it would be as well to set it to the variable that has only 3 levels in the design.

5. Choice of the number of levels: example of dissolution testing

The dissolution profile of some coated pellets was investigated as a function of pH, buffer concentration and stirring speed. Using a pharmacopoeial dissolution apparatus of 7 stirred vessels, 7 experiments may be carried out in one run. However all 7 of these experiments must be at the same stirring rate, as it is not possible to stir the individual flasks independently of one another.

The variable X_3 takes 3 levels, so it was chosen to correspond to the stirring rate. The normal rate for the dissolution testing of the formulation was 75 rpm, and the extremes to be tested were 75 ± 25 rpm. The coded levels ± 0.866 correspond therefore to 50 and 100 rpm. (It would also have been possible to set the levels ±1 to 50 and 100 rpm, in which case the levels coded ± 0.866 would have corresponded to 53 and 97 rpm.) If each experiment were carried out twice, these experiments could be done in two dissolution runs. We assume here that the run to run reproducibility is sufficiently good. There remain 7 experiments at the level zero of stirring speed (75 rpm). These experiments also were carried out in duplicate. This allowed the whole experimental design to be carried out using only 4 dissolution runs as shown in table 5.23.

Examination of this design in the light of table 5.22 shows that the property of minimum number of levels for X_3 is lost as soon as the number of factors is increased. As soon as another factor is added the factor that previously had 3 levels acquires 4 more and it is the new variable that takes 3 levels.

Table 5.22 Uniform Shell (Doehlert) Designs for 2 to 5 Factors

k	X_1	X_2	X_3	X_4	X_5
2	0	0	0	0	0
	1	0	0	0	0
	0.5	0.866	0	0	0
	-0.5	-0.866	0	0	0
	0.5	-0.866	0	0	0
	-0.5	0.866	0	0	0
	-1	0	0	0	0
3	0.5	0.289	0.816	0	0
	-0.5	0.289	0.816	0	0
	0	-0.577	0.816	0	0
	0.5	-0.289	-0.816	0	0
	-0.5	-0.289	-0.816	0	0
	0	0.577	-0.816	0	0
4	0.5	0.289	0.204	0.791	0
	-0.5	-0.289	-0.204	-0.791	0
	0.5	-0.289	-0.204	-0.791	0
	0	0.577	-0.204	-0.791	0
	0	0	0.612	-0.791	0
	-0.5	0.289	0.204	0.791	0
	0	-0.577	0.204	0.791	0
	0	0	-0.612	0.791	0
5	0.5	0.289	0.204	0.158	0.775
	-0.5	-0.289	-0.204	-0.158	-0.775
	0.5	-0.289	-0.204	-0.158	-0.775
	0	0.577	-0.204	-0.158	-0.775
	0	0	0.612	-0.158	-0.775
	0	0	0	0.632	-0.775
	-0.5	0.289	0.204	0.158	0.775
	0	-0.577	0.204	0.158	0.775
	0	0	-0.612	0.158	0.775
	0	0	0	-0.632	0.775

Table 5.23 Uniform Shell Design for Characterising the Dissolution Profile of Coated Pellets, as Function of pH, Buffer Concentration and the Stirring Speed

No.	X_1	X_2	X_3	No of experiments	Stirring speed (rpm)
1	0	0	0	2	75
2	1	0	0	2	75
3	0.5	0.866	0	2	75
4	-0.5	-0.866	0	2	75
5	0.5	-0.866	0	2	75
6	-0.5	0.866	0	2	75
7	-1	0	0	2	75
8	0.5	0.289	+ 0.816	2	50
9	-0.5	0.289	+ 0.816	2	50
10	0	-0.577	+ 0.816	2	50
11	0.5	-0.289	- 0.816	2	100
12	-0.5	-0.289	- 0.816	2	100
13	0	0.577	- 0.816	2	100

6. Properties of the Doehlert designs

These are summarised in table 5.24. In the case of the recommended number of centre points, where two figures are given, the first lower figure may be considered acceptable but the second is to be preferred. It can be seen that the R-efficiencies are comparable to those of the composite designs. In most cases, slightly fewer experiments are needed. However, their main advantage, as we have seen, is the possibility of extending or displacing the design in various directions.

Table 5.24 Properties of Doehlert Designs

No. of factors	No. of centre points	Total number of experiments	R_{eff}	VIF (max)	d_A min	d_A max
2	2-3	9	66%	1.12	0.4	0.8
3	3	16	63%	1.20	0.3	1.0
4	3-4	24	63%	1.40	0.3	0.9
5	5	35	60%	1.46	0.3	0.9
6	6	49	57%	1.55	0.2	0.9

Except for the 2-factor design (see V.B.1) they are not rotatable, but their deviation from this property is not excessive. Nor is the precision uniform, as shown by the maximum and minimum values of the variance function d_A. However these upper and lower limits of the function remain acceptable.

C. Hybrid and Related Designs

We include under this heading a number of designs of different origins. The most important characteristic they have in common is that they are the most economical in terms of the number of experiments. The redundancy of most of the designs we have looked at so far increases with the number of variables. Hybrid designs and the others discussed here are intended to have an R-efficiency as near to 100% as possible. They are therefore saturated or near saturated.

The hybrid designs proper consist of:

• a factorial design for all variables but one, and the remaining variables held at a constant level;
• a star design for the same variables, the remaining variable being held at a different constant level;
• 2 or more axial points;
• one or more centre points may be added, and are needed in certain designs.

The similarity to the central composite designs is clear. These also consist of a factorial and an axial design, but which involve all of the variables and are both centred at the same point. Hybrid designs exist for 3, 4 and 6 variables. They are designed for a second-order model, for the investigation of a fixed spherical zone of interest, most of the experiments being on the surface of a sphere. They are not suitable for a sequential approach. All experiments should be done in one block, in a random order.

The initial designs of this type were proposed by Roquemore (10), who derived designs for 3, 4 and 6 factors. These were modified by Franquart (11) and Peissik (12) using the methods of exchange algorithms described in a chapter 8.

A hybrid design for 3 factors is shown in table 5.25 and figure 5.16. Others, for 4 and 6 factors, are listed in appendix III. They are useful for optimization, where resources are insufficient for a large number of experiments, that we are sure of being in the optimum region and that the response surface is not too complex. As well as being highly efficient, they have the advantage of maximum precision of estimation of the coefficients, and they are also rotatable, as measured by Khuri's index (13). Like the small central composite designs (above) and the minimal designs for cuboid design spaces described later, they suffer from the inevitable disadvantage that there are no degrees of freedom for testing model fit, and so they are sensitive to an erroneous datum. The 3 factor design is improved by the addition of a second centre point.

Table 5.25 A Hybrid Design for 3 Factors: 311A Matrix (12)

No.	X_1	X_2	X_3	
1	0	0	-1	axial points
2	0	0	1	
3	-0.632	-0.632	0.447	factorial design
4	0.632	-0.632	0.447	
5	-0.632	0.632	0.447	
6	0.632	0.632	0.447	
7	-0.894	0	-0.447	star design
8	0.894	0	-0.447	
9	0	-0.894	-0.447	
10	0	0.894	-0.447	
11	0	0	0	centre point(s)

A saturated design for 5 variables (22 experiments), derived by Franquart (11), is given in appendix III. It is not related structurally to the hybrid designs but may be used in similar circumstances. Its properties are satisfactory but it does have the major disadvantage of there being as many distinct levels for each factor as there are experiments.

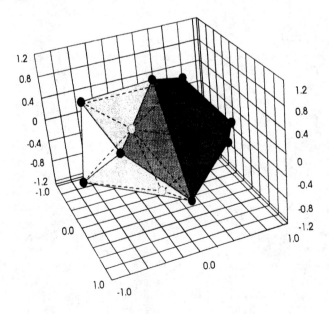

Figure 5.16 Hybrid design for 3 factors.

D. An Equiradial Design for 2 Factors – The Regular Pentagon

Equiradial designs are those whose points (other than the centre point) are regularly spaced on the surface of one or more concentric spheres. Although many useful designs come into that category (2 level factorial designs based on the cube, composite designs) these are described elsewhere. Two factor designs of 4 to 7 points used for the first-order model were shown in section IV.A. Of these the regular pentagon and the regular hexagon may also be used, on adding a centre point, with great efficiency, for determining a second-order model. The hexagon is the Doehlert design for 2 factors.

The pentagonal design with 6 points is saturated. It is therefore an excellent design for RSM in a circular domain, when a minimum of experimentation is required. Being saturated, it has the disadvantage that the model cannot be tested for lack of fit. The design was shown in table 5.8 and in figure 5.5, and a D-optimal design partly based on it is discussed in detail in chapter 8.

E. Box-Behnken Designs

These are designs for a spherical domain whose most interesting property is that each factor takes only 3 levels (14). Each combination of the extreme values of two of the variables is tested, the remaining variables taking a coded level of zero. They are therefore a subset of the 3 level full factorial design. The design for 3 variables is shown in table 5.26 and figure 5.17. The points are each on the centres of the edges of a cube, at a distance of $\sqrt{2}$ from the centre-point. Its geometry is the same as that of the 3 factor Doehlert design, but in a different orientation. An example of its use in fluidized-bed coating is described by Turkoglu and Sahr (15).

Similarly, the 4 variable Box-Behnken design has all possible combinations of factorial designs in 2 of the factors with the remaining 2 factors at zero. There are 6 combinations. The design consists of 24 experiments plus the centre points, the same number as for the central composite design. Hileman *et al.* describe its use in extrusion-spheronization (16).

In general, Box-Behnken designs of k factors are composed of $k(k-1)/2$ factorial designs 2^2, taking each pair of factors in turn, and keeping the other factors at the level coded zero. The number of experiments increases quite rapidly with increasing number of factors.

Most of these designs (but not the $k = 3$) may be partitioned into 2 or 3 orthogonal blocks (as discussed above for composite designs). Added to these advantages we have the fact that inflation factors are normally low (less than about 1.3), provided sufficient centre points are added. The estimators are thus nearly independent, and the matrix nearly orthogonal. However, their R-efficiencies are often low, which may make them an expensive design to carry out. Their properties are summarised in table 5.27.

Table 5.26 Box-Behnken Design for 3 Factors

No.	X_1	X_2	X_3
1	-1	-1	0
2	+1	-1	0
3	-1	+1	0
4	+1	+1	0
5	-1	0	-1
6	+1	0	-1
7	-1	0	+1
8	+1	0	+1
9	0	-1	-1
10	0	-1	+1
11	0	+1	-1
12	0	+1	+1
13	0	0	0

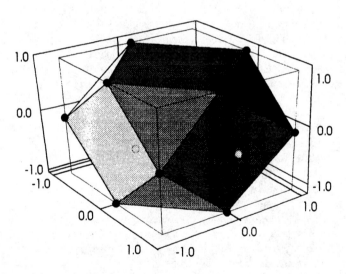

Figure 5.17 Box-Behnken design for 3 factors.

Table 5.27 Some Characteristics of Box-Behnken Designs

k	N	N_0	R_{eff}	Blocks	VIF_{max}
3	15	3	67%	1	1.01
4	27	3	56%	3	1.25
5	42	2	50%	2	1.96
	46	6	46%		1.20
6	50	2	56%	2	1.68
	54	6	52%		1.30

The experimental domain is spherical, radius $\sqrt{2}$, whatever the number of factors, but with the experimental points arranged in a (cubic) pattern with only 3 levels for each factor. Thus there may be an advantage in using the Box-Behnken design when a spherical domain is required for prediction, but where there are limitations to the numbers of levels to which certain factors can be set.

VI. STRATEGIES FOR A CUBIC SHAPED DOMAIN

Up to now we have considered the experimental domain as being spherical. This is the most common situation and corresponds to the circumstances in which the RSM method is normally applied and for which it was developed. However not all problems can be treated satisfactorily using matrices designed for a spherical domain. We consider two restrictions, so-called technological constraints, which the experimenter is likely to encounter from time to time.

The first is that the factors may have *individual limits*, related to the feasibility of the experiments, to economic imperatives, etc. For example, in tableting, the level of stearate may have to be between 0.5% and 2%. The experimental domain is no longer defined by a fixed radius from a centre of interest, but by the impossibility of going above or below certain values of the factors. Such constraints result in a cubic domain; the corners of the domain, even though they are combinations of the extreme values, belong to the zone of interest, just as much as do the experiments that were carried out nearer to the centre of the domain.

The second restriction is that of limitations in the *number* of levels, or of difficulties in setting a factor to certain levels. We have already mentioned this potential problem when discussing the Doehlert experimental design. If the model order in a certain factor is d, then this factor must take at least $d + 1$ levels. Thus for a second-order model at least 3 levels are required, normally equidistant. If it becomes necessary to minimize the number of levels for more than 1 factor, this may mean choosing a design with all factors taking 3 levels. The possible designs are all sub-sets of experiments from the 3^k factorial designs, and the domain is again cubic.

There are many publications describing how these problems may be resolved, and a large number of designs have been proposed. We will first describe designs with a specific general structure. Many others are constructed by special methods. We have therefore given a general list of designs, both the standard ones, and those special designs which seem to us to be the most useful in terms of their structures and statistical properties.

A. Standard Designs with 3 Levels

1. Factorial designs 3^k

The 3^2 full factorial design consists of 9 experiments. Its use in development and optimization has been described quite frequently in the pharmaceutical literature (see for example references 17-19). An example is reported by Wehrlé et al. (17) who used it to study the scale-up of a wet granulation process. The experimental domain and design are shown in figure 5.18 with the plotted surface for the disintegration times. The design consists of all possible combinations of 3 levels of each factor, coded as -1, 0, and +1.

It is identical to the 2-factor central composite design. The matrix is "almost orthogonal" if no extra centre-points are added. It is not rotatable, nor is its precision uniform in the experimental domain.

If a third variable is added, the number of combinations of the 3 levels increases to 27. The design becomes quite large, and as only 10 coefficients are estimated, the use of a 3^3 full factorial design becomes expensive, though not prohibitively so (20). The efficiency R_{eff} is 37%. With a fourth variable the efficiency decreases still further to 18.5%. For this reason the 3^k full factorial designs are almost totally unused for more than 3 variables. Nor do there exist many fractional factorial 3^{k-p} designs. The only one likely to be useful is the 3^{4-1} design of 27 experiments for 4 factors (see appendix III).

As stated above, the other 3 level designs may be thought of as sub-sets of the full factorial 3^k designs.

2. Central composite designs ($\alpha = 1$)

We gave the general method for constructing these designs in section V.A5. Box and Wilson (1) showed that α, the distance of the axial points from the centre, must take certain specific values if one is to obtain a rotatable design. If we do not require, or are prepared to relinquish the property of isovariance by rotation we may set the axial distance $\alpha = 1$. The resulting designs, with all points on the surface of a cube, are still of excellent quality according to most criteria. The 2 factor composite design with $\alpha = 1$ is identical to the 3^2 design (figures 5.1b and 5.18).

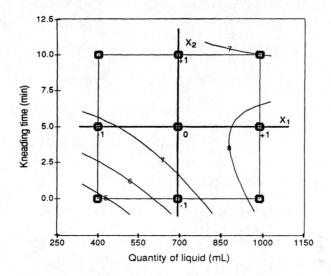

Figure 5.18 Use of the 3^2 design to study a granulation process, with response surface of disintegration time in minutes [adapted from reference (17), with permission].

3. Box-Behnken designs

The Box-Behnken designs discussed in the previous section are subsets of the 3^k full factorial design, and may therefore be set up in the cubic domain. In general their use is not recommended, as predictions are poor for combinations of extreme values of more than 2 factors with very high values of d_{max}. For these reasons, and also because they require a large number of experiments, none of the Box-Behnken designs appear in table 5.30, which lists the best and most efficient designs for the cubic domain.

B. Non-Standard Designs

Various workers have suggested designs requiring fewer experiments than the central composite. Some of these designs are restricted to 3 levels (-1, 0, and +1), the experiments being thus selected from the 3-level factorial. Examples are those described by Mitchell and Bayne (23), Lucas (24), Welch (25), Notz (26), and Hoke (27). Box and Draper (21) and Dubova and Federov (22) have proposed saturated designs with optimized non-standard levels. Those which are saturated or near-saturated designs may be used in similar circumstances to the hybrid designs in the spherical domain, and some of the non-standard unsaturated designs may also be used instead of the central composite or 3 level factorials.

Although the designs have low variance inflation factors, they are not rotatable, often showing considerable deviations from the property. Also as a general rule the saturated or almost saturated designs show considerable differences in the variance of prediction over the domain. The maximum prediction variance function in the (hyper-)cube can be high ($d_{max} > 1.8$).

The most useful ones are compared with the standard designs in the next section and set out in full in appendix III.

C. Comparison of Designs in the Cubic Domain

Peissik (12) has analysed the designs for 2 to 6 factors, enabling them to be selected according to the criteria of (a) minimum number of experiments and (b) minimum number of levels. The results are summarised in table 5.28.

Table 5.28 Summary of minimum designs for a cubic domain

	3 levels						> 3 levels					
	Minimum N			Any N			Minimum N			Any N		
k	Design	N	R_{eff}	Design	N	R_{eff}	Design	N	R_{eff}	Design	N	R_{eff}
2	MB207	7	78%	BW209	9	66%	BD206	6	100%	BD208	8	75%
3	L311	11	91%	BW315	15	66%	BD310	10	100%			
4	WH415	15*	100%	MB421	21	71%	DF415*	15	100%			
5	NO521	21	100%	MB524	24	88%						
				MB530	30	70%						
				BW527	27	78%						
6	HD628	28*	100%	HD634	34	82%						

BW:	Box and Wilson (1)	BD:	Box and Draper	(21)
MB:	Mitchell and Bayne (23)	DF:	Dubova and Federov	(22)
L:	Lucas (24)			
WH:	Welch (25)			
NO:	Notz (26)			
HD6:	Hoke (27) D6 designs			

* d_{max} is especially high for the following designs:

WH415	$d_{max} = 2.00$
DF415	$d_{max} = 2.03$
HD628	$d_{max} = 2.45$

In each case we selected the design or designs with the best combination of mathematical characteristics: a design closest to orthogonality, best precision of estimation of the coefficients, lowest variance inflation factors (VIF less than 2). Where we have given a design with factors set at *more* than 3 levels this indicates that its properties are better than those of the design with 3 levels only and that it should be preferred unless the restriction in the number of levels is of primary importance. If no design of more than 3 levels is indicated ($k = 5$, 6) this signifies either that the design does not exist or that the 3-level design is of better quality.

To identify the designs we have used Peissik's notation (see the bottom of table 5.28). The full notation she uses consists of a group of letters indicating the type of design, followed by a single digit signifying the number of factors, and then two digits giving the number of experiments. This notation is used in appendix III, where the non-standard matrices are written out in full. The Box-Wilson designs (BW209, BW315, BW527) refer to central composite designs with the axial points at $\alpha \pm 1$ (1), each having 1 centre point.

VII. STRATEGIES ON REJECTION OF THE SECOND DEGREE MODEL

A. Cubic Models are Little Used in RSM

If the second-order model is found to be insufficient after carrying out experiments according to one of the second-order designs described in this chapter (whether for a spherical or cubic domain), we could postulate a third-order model. In practice, cubic models are extremely seldom used because of:

(a) the number of experiments required to estimate the large number of coefficients (17 coefficients for 3 factors, 35 for 4 factors);
(b) the inferior predictive properties of designs for these models, leading occasionally to misleading predictions when interpolating, and above all when extrapolating;
(c) the lack of suitable designs in the literature – partly a consequence of (a) and (b).

Mixture models and designs are an exception to this general rule and third order models are used quite frequently in the case of mixtures. These however are usually reduced cubic models (see chapter 9).

B. Strategies

On rejecting the model, we propose three possible approaches.

1. Contraction of the experimental domain

If the second-order model is insufficient, this may be because the domain is too extensive. It is then probable that the model would be sufficient if the domain were much reduced. Examination of the shape of the calculated curve and of the residuals will most likely indicate whether this is the case. Even if the fit is poor the results will probably be sufficient for selecting a part of the domain, or even a point that is more interesting than the rest.

A new design is then set up in the contracted domain. This may or may not re-use some of the experiments already carried out.

2. Addition of extra terms to the model

The second solution is also to examine the residuals. These residuals, or our general knowledge of the phenomenon, or the fact that certain effects are much greater than others may suggest that a limited number of third-order terms be included. We could decide to add all possible third-degree terms in X_1 for example, that is X_1^3, $X_1^2X_2$, $X_1^2X_3$, $X_1X_2X_3$. The number of extra terms is not excessive. Extra experiments will be required for the design to have the properties required for determining the model. The resolution of such a special model generally requires construction or augmentation of a design by means of an exchange algorithm, which is the subject of chapter 8, D-optimal designs.

If the design has factors at more than 3 levels, it might often appear theoretically possible to introduce one or more higher order terms, without additional experiments. (This is a similar situation to that of the factorial design with centre points.) The temptation should be resisted and the design analysed only according to the model for which is was devised. The RSM designs, composite, Doehlert, and most other designs that we have treated in this chapter may allow the model to be shown as inadequate, but they are insufficient for analysis with a more complex model.

3. Use of an "inadequate" design and model

Even though there may be significant lack of fit between the results of the design and the model, the experimenter may decide that the lack-of-fit root mean square, although statistically significant, is within the range of error that he is prepared to tolerate, and the results are precise enough to optimize the process or formulation, at least for that stage of the process.

The statistical analysis serves as a guide for the experimenter to make logical and informed decisions, but it does not dictate those decisions.

References

1. G. E. P. Box and K. B. Wilson, On the experimental attainment of optimum conditions, *J. Roy. Statist. Soc., Ser. B*, **13**, 1-45 (1951).
2. E. Senderak, H. Bonsignore and D. Mungan, Response surface methodology as an approach to optimization of an oral solution, *Drug. Dev. Ind. Pharm.*, **19**, 405-424 (1993).
3. J. B. Schwartz, J. R. Flamholz and R.H. Press, Computer optimization of

pharmaceutical formulations I: general procedure, *J. Pharm. Sci.,* **62**, 1165 (1973).

4. A. D. McLeod, F. C. Lam, P. K. Gupta, and C. T. Hung, Optimized synthesis of polyglutaraldehyde nanoparticles using central composite design, *J. Pharm. Sci.,* **77**, 704-710 (1988).

5. N. R. Draper and D. K. J. Lin, Small response-surface designs, *Technometrics,* **32**, 187-194 (1992).

6. K. Takayama, H. Imaizumi, N. Nambu, and T. Nagai, Mathematical optimization of formulation of indomethacin/polyvinylpolypyrrolidone/methyl cellulose solid dispersions by the sequential unconstrained minimization technique, *Chem. Pharm. Bull.,* **33**, 292-300 (1985).

7. D. H. Doehlert, Uniform shell designs, *Applied Statistics,* **19**, 231-239 (1970).

8. S. Dawoodbhai, E. R. Suryanarayan, C. R. Woodruff, and C. T. Rhodes, Optimization of tablet formulations containing talc, *Drug. Dev. Ind. Pharm.,* **17**, 1343-1371 (1991).

9. D. Vojnovic, P. Rupena, M. Moneghini, F. Rubessa, S. Coslovich, R. Phan-Tan-Luu, and M. Sergent, Experimental research methodology applied to wet pelletization in a high-shear mixer, Part 1, *S.T.P. Pharma Sciences,* **3** (2) 130-135 (1993).

10. K. G. Roquemore, Hybrid designs for quadratic response surfaces, *Technometrics,* **18**(4), 419-424 (1976).

11. P. Franquart, Thesis: "Multicriterial optimization and methodology of experimental research", University of Aix-Marseilles III, France (1992).

12. A. Peissik, Thesis: "Properties and characteristics of experimental design matrices for second degree polynomial models", University of Aix-Marseilles III, France (1995).

13. A. I. Khuri, A measure of rotatability, *Technometrics,* **30**, 95-104 (1988).

14. G. E. P. Box and D. W. Behnken, Some new three level designs for the study of quantitative variables, *Technometrics,* **2**(4), 455-475 (1960).

15 M. Turkoglu and A. Sakr, Mathematical modelling and optimization of a rotary fluidized-bed coating process, *Int. J. Pharm.,* **88**, 75-87 (1992).

16. G. A. Hileman, S. R. Goskonda, A. J. Spalitto, S. M. Upadrasha, Response surface optimization of high dose pellets by extrusion and spheronization, *Int. J. Pharm.,***100**, 71-79 (1993).

17. P. Wehrlé, Ph. Nobelis, A. Cuiné, and A. Stamm, Response surface methodology: an interesting tool for process optimization and validation: example of wet granulation in a high shear mixer, *Drug Dev. Ind. Pharm.,* **19**, 1637-1653 (1993).

18. D. M. Lipps and A. M. Sakr, Characterization of wet granulation process parameters using response surface methodology. 1. Top-spray fluidized bed, *J. Pharm. Sci.,* **83**, 937-947 (1994).

19. P. S. Adusumilli and S. M. Bolton, Evaluation of cetacean citrate complexes as matrices for controlled release formulations using a 3^2 full factorial design, *Drug Dev. Ind. Pharm.,* **17**, 1931-1945 (1991).

20. P. Merku, O. Antikainen, and J. Yliruusi, Use of 3^3 factorial design and multilinear stepwise regression analysis in studying the fluidized bed

granulation process. Part II, *Eur. J. Pharm. Biopharm.*, **39**, 112-116 (1993).

21. G. E. P. Box and N. R. Draper, On minimum-point second order designs, *Technometrics*, **16**, 613-616 (1960).

22. I. S. Dubova and V. V. Federov, Tables of optimum designs II (Saturated D-optimal designs on a cube), Preprint n° 40, issued by Interfaculty Laboratory of Statistical Methods, Moscow University.

23. T. J. Mitchell and C. K. Bayne, D-optimal fractions of three-level factorial designs, *Technometrics*, **20**, 369-383 (1978).

24. J. M. Lucas, Which response surface is best? *Technometrics*, **18**, 411-417 (1976).

25. W. J. Welch, Computer-aided design of experiments for response surface estimation, *Technometrics*, **18**, 217-224 (1984).

26. W. Notz, Minimal point second order designs, *J. Stat Plann. Infer.*, **6**, 47-58 (1982).

27. A. T. Hoke, Economical second-order designs based on irregular fractions of 3^n factorials, *Technometrics*, **16**, 375-423 (1974).

Further reading

• R. H. Myers and D. C. Montgomery, Response Surface Methodology, Wiley Interscience, N. Y., 1995

• G. E. P. Box, W. G. Hunter, and J. S. Hunter, Statistics for Experimenters, Wiley, N. Y., 1978, chapters 14 and 15.

6

OPTIMIZATION

Pharmaceutical Process Optimization and Validation

I. INTRODUCTION

Using the methods of chapter 5 we can determine a (polynomial) model and use it to predict the value of a response over the whole of the experimental region. In addition to characterizing the process or system in general, we may use this mapping to find the best processing conditions and the best formulation.

A. The Shape of the Response Surface

The polynomial model estimated for the measured response is almost invariably
first- or second-order. The response surface for the first-order model is an inclined
plane. The second-order model is the more usual and will be discussed here. The
number of variable factors is almost always restricted to less than 6 or 7. If there
are only 2 factors in the model, then it is easy to see the shape of the response
surface by inspection. When there are 3 or more factors this is far more difficult.
However, there are only a limited number of shapes that are possible for the
calculated response surface. Second-order polynomials are *conic functions*. For 2
factors, they have the form of ellipsoids or hyperbolas, as shown in figure 6.1.a-b.

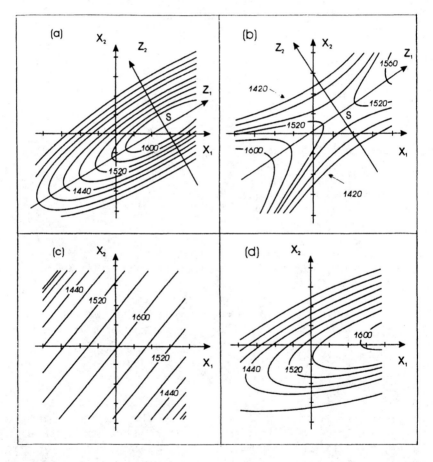

Figure 6.1 Conics: (a) maximum; (b) saddle point; (c) ridge; and (d) rising
ridge.

The functions may be analysed by *canonical analysis*, a method of transforming the model equation so that the form of the response surface can be identified immediately (1, 2). Each function has a *stationary point* S, where the slope of the response surface with respect to each of the variables is zero. This point could be a *maximum*, as for the ellipsoid of figure 6.1a. Any change in the factor variables, in whatever direction, gives a decrease in the predicted response. If it were a *minimum*, the form of the response surface would also be an ellipsoid. Point S may also be a *saddle point*, shaped rather like a mountain pass (figure 6.1b), or alternatively the surface may also sometimes be very "elongated", in the form of a ridge (figure 6.1c).

Principal axes Z_i, also shown in figures 6.1a-d, may be defined. These are perpendicular to one another, their orgin is the stationary point, and there are as many principal axes as there are variables. We see in the case of the saddle point of 6.1b that changes in the experimental conditions from point S result in either an increasing response along the Z_1 axis or a decrease along the Z_2 axis. The equation of the response surface can thus be transformed to its *canonical form*, where instead of being expressed in terms of the coded variables X_1, X_2,...,X_k, the calculated response is shown as a function of the principal axes of the conic Z_1, Z_2,..., Z_k. The equation then takes a very simple form:

$$\hat{y} = \hat{y}_s + \lambda_1 z_1^2 + \lambda_2 z_2^2 + \lambda_3 z_3^2 \dots + \lambda_k z_k^2 \tag{6.1}$$

where $\hat{y}_s$ is the calculated response at the stationary point. Equation 6.1 gives the nature of the response surface. If all the λ_i are positive then the stationary point is a *minimum* and if all are negative then it is a *maximum*. However as is more usual, if *some* of the λ_i are positive and the rest negative then the stationary point is a saddle point.

If, for example, λ_1 is positive, then any displacement along the principal axis Z_1 away from the stationary point S leads to an increased response. The larger the value of λ_1, the greater is the increase in response. The surface is thus steepest along the Z_i axis that corresponds to the largest absolute value of λ_i. If one of the $\lambda_i \approx 0$, then moving along its principal axis will have no influence on the response and the surface is a stationary ridge (or valley). Canonical analysis thus tells us the position of the stationary point, and whether it exists within the experimental domain or outside it, and also the shape of the reponse surface.

Transformation to the canonical form is usually carried out by computer. Equation 6.1 may be used directly to optimize a single response (3). It is a purely numerical method, relatively difficult to master, and it is now little used by the experimenter. It is also possible, although more difficult, to apply the canonical analysis method to multiple responses. The technique has been mentioned at this point, not as an optimization method to be undertaken by the non-specialist, but to demonstrate the basic forms of the second order response surface. All the examples cited in this chapter show one or other of these different shapes.

B. The Optimization Process

1. Optimization of a Single Response

If there is only a single measured response, optimization may be defined very simply as the determination of the experimental conditions which lead to the best value of that response. This is often a maximum or minimum. If the optimum is neither of these but a *target value*, defined by an upper or lower limit, the response which is maximized here will be a *satisfaction index*, increasing as the response approaches the target value. We may define either a single point or an optimum zone within the experimental region.

The experimenter should be reasonably sure that there is no better point or zone (always bearing in mind the experimental precision) within the experimental region, or its neighbourhood. It should be unnecessary to continue experimentation to improve the response, unless it is to change the experimental region, add additional factors or modify the objectives.

An optimization may also be halted when the results appear satisfactory, without having reached the absolute optimum. We might call this an *improvement* rather than an optimization, and it may open up the way to further modification and improvement. Screening and factor studies frequently result in better results for processes and formulations, and factor studies especially indicate trends which show how a factor might be varied to favour a given response. This improvement, although useful in itself, is not the same as optimization. We have already indicated that such designs on their own, Plackett-Burman, factorial, etc., are not usually adequate for the kind of optimization we are looking for here, and we should use rather the methods described in chapter 5.

The implicit approach of chapter 5 was to optimize the process or formulation by examining the response surface directly. But other methods are both useful and necessary when there are many factors (desirability) or when the optimum is outside the experimental region (steepest ascent and optimum path). Also there are direct optimization methods available (sequential simplex) which do not involve mapping the response surface at all.

2. Multiple responses

In practice, a product or manufacturing process is rarely described by a single response variable, but by a number of responses. It is unlikely that optimizing one of them will lead to the best values of the others, unless the responses are so correlated with one another that the operation will be equivalent, essentially, to the optimization of a single response. We need to redefine what we mean by "optimization". Since it is impossible to obtain best values for all responses, optimization becomes a question of finding experimental conditions where the different responses are most satisfactory, overall. And so for a maximum, minimum, or target value of a response, we substitute the more vague, intuitive notion of a compromise or generally satisfactory region. Certain responses can oppose one another; changes in a factor that improve one response may have a negative effect on another. There is a certain degree of subjectivity in weighing up their relative importance. Nonetheless the idea can be quantified; we shall see this under the heading of *desirability*.

If the "satisfaction criteria" are modified, the optimization must also be re-examined. These criteria must be clearly and rigorously defined, to justify the final optimized choice.

3. Choosing an optimization method

We discuss the most important of the numerous optimization methods, indicating where each technique may be applied, emphasising its place in the context of the experimental strategy. The choice of method should be dependent on the previous steps and probably on our ideas about how the project is likely to continue. The various methods are listed in table 6.1, with a very brief and approximate summary of the circumstances in which they are used.

Table 6.1   Choosing an Optimization Method

Circumstances for use	Method
Mathematical model of any order Normally no more than 4 factors* Multiple responses	GRAPHICAL ANALYSIS
Mathematical model of any order From 2 up to 5-6 factors Multiple responses	DESIRABILITY FUNCTION
First-order model Optimum outside domain Single response	STEEPEST ASCENT
Second-order model Optimum outside domain Single response	OPTIMUM PATH
No mathematical model Direct optimization Single or multiple responses	SEQUENTIAL SIMPLEX
Industrial situation Little variation possible	EVOP

* When used alone. More factors may be investigated when used with the optimum path or desirability methods.

The criteria are:

- Are there one or several responses to optimize?

- Is there a known mathematical model which describes the response adequately within the domain?
- Do we expect to continue the optimization *outside* the experimental region that has been studied up to this point?
- Is it necessary to map the response surfaces about the optimum? Can the experimental conditions be changed slightly without unacceptable variation in the responses? This is the *ruggedness* of the optimum.

II. GRAPHICAL OPTIMIZATION OF MULTIPLE RESPONSES

A. Response Surface Analysis

It is usually relatively simple to find the optimum conditions for a single response that does not depend on more than 4 factors, once the coefficients of the model equations have been estimated – provided, of course, that the model is correct. Real problems are usually more complex. In the case of extrusion-spheronization it is not only the yield of pellets that is important, but their shape (how near to spherical), friability, smoothness, and their ease of production. The optimum is a combination of all of these.

One possible approach is to select the most important response, the one that should be optimized – such as the yield of pellets. For the remaining responses we can choose acceptable upper and lower limits. Response surfaces are plotted with only these limits, with unacceptable values shaded. The unshaded area is the acceptable zone. Within that acceptable zone we may either select the centre for maximum ruggedness of formulation or process, or may look for a maximum (or minimum or target value) of the key response.

1. Graphical optimization of two opposing responses – 2 factors

The optimization of a gel formulation of ketoprofen has been described by Takayama and Nagai (4). They studied the effect of *d*-limonene and ethanol in their formulation on the penetration rate of the drug through the skin of the rat *in vivo*. They also measured the lag time and the skin irritation. A central composite design with 4 centre points was used and predictive second-order equations were obtained by multi-linear regression.

Contour plots are shown in figure 6.2a for irritation score and penetration rate, which appear to be correlated. It would not be possible to achieve a high penetration rate without unacceptable skin irritation.

Let the desired penetration rate be between 0.5 and 1 $mg.h^{-1}$, the lag time be less than 0.75 h, and the highest acceptable irritation score be 10. Figure 6.2b shows these limits superimposed, revealing a small acceptable zone around the point 1% *d*-limonene and 45% ethanol. In the original paper, the authors used a multi-objective optimization method.

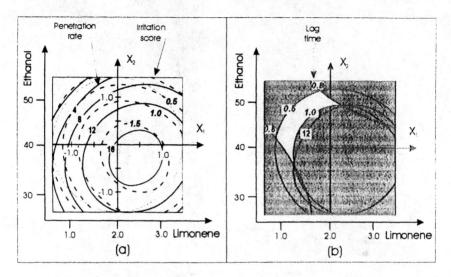

Figure 6.2 (a) Contour plots for penetration rate and irritation score. (b) Definition of optimum zone.

2. Optimization of 2 responses for 3 factors

The same approach may be used where there are more factors, but it is now necessary to select the one which seems to have least influence and to plot response surfaces of "slices" at different levels of this factor.

Perturbation analysis

This is a complementary method that can be useful for examining the form of the functions of the responses. The response is calculated from a given point (usually the centre), varying each factor in turn while keeping the others constant. The effects of a number of factors on a response may be seen on the same graph. This kind of plot does not indicate interactions. Perturbation analysis assists in the selection of contourplots to examine.

Senderak *et al.* (5) studied the effects of polysorbate 80 concentration, propylene glycol, and invert sugar on the turbidity and cloud point of a formulation of a slightly soluble drug, using a central composite design. Both the design and the determination of the model are described in some detail in chapter 5.

Figure 6.3 shows perturbation plots for the two responses. Figure 6.3a suggests that to lower the turbidity, the polysorbate concentration (X_1) should be increased, whereas an intermediate concentration of propylene glycol (X_2) should be used. Sucrose (X_3) has very little effect on turbidity. On the other hand, in figure 6.3b, sucrose appears to be the only factor affecting the cloud point to any important extent.

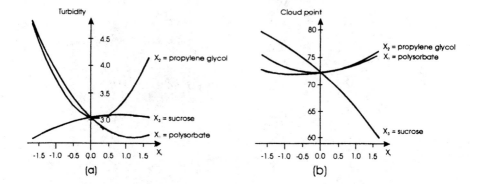

Figure 6.3 Perturbation plots: (a) turbidity; (b) cloud point.

Graphical analysis for formulation of an oral solution

The objective was to reduce the turbidity as far as possible, and to obtain a solution with a cloud point less than 70°C, at a level of invert sucrose that is as high as possible (in spite of its deleterious effect on the cloud point). Figure 6.4 shows slices taken in the propylene glycol, sucrose plane (X_2, X_3) at different levels of polysorbate X_1 (coded levels $x_1 = -\frac{1}{2}, 0, \frac{1}{2}, 1$). Examination of the response surfaces indicates an optimum compromise formulation at 60 mL sucrose medium, 4.3% polysorbate 80 and 23% polyethylene glycol.

Optimization of a microsphere formulation

Zeng, Martin, and Marriot also used a central composite design to optimize an albumin microsphere formulation for delivery of tetrandine to the lungs (6). The variables were the pH and the concentrations of albumin and of tetrandine. The level of drug trapped in the microspheres and the diameter of the particles were the selected responses and a quadratic model was determined for each response.

A diameter of 11 - 12 µm was sought as being optimal for pulminary delivery, with better than 10% drug incorporated in the microspheres. Overlaid contour plots of diameter and yield were drawn as functions of pH and albumin at different concentrations. Only at terandrine concentrations of about 40 mg mL^{-1} was there a region satisfying the requirements (figure 6.5).

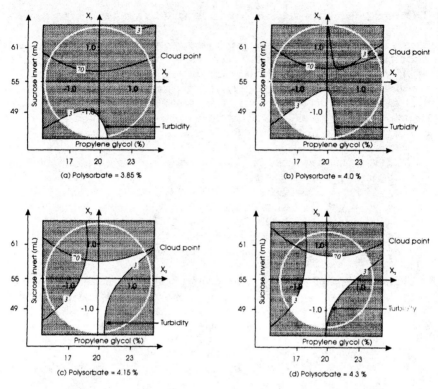

Figure 6.4 Acceptable zones in formulation of an oral solution.

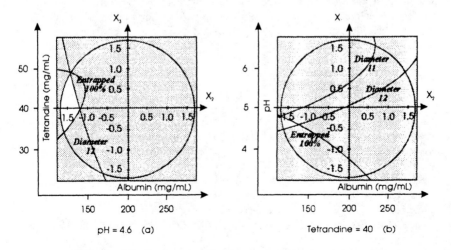

Figure 6.5 Optimum zones for tetrandine microspheres.

3. Optimization of polyglutaraldehyde nanoparticles synthesis (5 factors)

There could be difficulties in using the above approach for multiple responses and where there are 4 or more variables. For example, McLeod *et al.* optimized the conditions of synthesis of polyglutaraldehyde nanoparticles (7) using a central composite design for 5 factors: monomer, surfactant concentration, pH, oxygen level, and stirring rate. The most important responses were particle size and yield. Optimization involved finding a region of high yield and low particle size and the two responses tended to be opposed, that is, regions of a high yield also gave large particles.

This is an example of the kind of case where, when used alone, the graphical method can become totally impracticable, in spite of its apparent simplicity and ease of use. Each of the 5 variable factors may be set at 5 levels. Levels of -1, -0.5, 0, 0.5, 1 would be usual in the case of a composite design. If all combinations are treated there are 150 diagrams to be plotted! One possibility is to select 2 factors as variables for the diagrams, and take all possible combinations of fixed factors of the other 3; thus up to 125 diagrams must be displayed or printed.

It would therefore be difficult to find an optimum for this problem, using the simple graphical approach of taking slices though the factor space. The optimum path method could help in the choice of diagrams. In the following section we shall indicate how a purely numerical method (canonical analysis) may be used sometimes to help choose diagrams to study. The authors of the paper used the method of *desirability*, described in section VI.

B. Process Study for a Granulated Tablet Formulation

We now demonstrate the simultaneous graphical optimization of a number of responses in more detail, taking an example of 5 factors. Two of the factors are qualitative, at only 2 levels each, thus reducing the complexity of the problem considerably. The example is one of a process optimization study carried out by Benkerrour *et al.* (8).

1. Process and experimental domain

The formulation was a conventional tablet containing 33% soluble drug, lactose, microcrystalline cellulose, hydroxypropylmethylcellulose as binder, crospovidone as disintegrant, silicon dioxide and magnesium stearate. The process to be optimized was one of wet granulation, followed by tableting. Five process variable factors were studied. They are given, along with their variation limits, and a summary of the process, in table 6.2. Two of the five variable factors are *qualitative* (X_3, X_4) and were therefore *fixed* at the defined levels (2 levels). The other three factors are *quantitative* and *continuous*. The granulation factors were allowed to vary about their centre of interest (circular domain). The lubrification time could take any value between 1 and 6 minutes.

Table 6.2 Summary of Granulation/Tableting Process

Operation		Factor	Experimental domain and factor levels
Mixing of drug substance with diluents in a high speed mixer-granulator			constant
Addition of binder solution (wetting)	X_1	volume of binder solution (L)	centre = 2750 g [limits ± 250 g]
Mixing of wet mass (kneading)	X_2	granulation time	centre = 4.5 min [limits ± 1.5 min]
Drying of granulate in an oven to a given humidity	X_3	residual humidity	level -1 = 1% level +1 = 3%
Calibration	X_4	sieve calibre	level -1 = 1 mm level +1 = 1.25 mm
Mixing with crospovidone and with silicon dioxide			constant
Lubrification (mixing with magnesium stearate)	X_5	lubrification time	centre = 3.5 min [limits ± 2.5 min]
Compression on rotary tablet machine			constant

2. Mathematical model and experimental design

A special model was postulated for the responses:

$$y = \beta_0 + \beta_1 x_1 + \beta_2 x_2 + \beta_3 x_3 + \beta_4 x_4 + \beta_5 x_5 + \beta_{11} x_1^2 + \beta_{22} x_2^2 + \beta_{55} x_5^2 \\ + \beta_{12} x_1 x_2 + \beta_{45} x_4 x_5 + \varepsilon \qquad (6.2)$$

justified by the following reasoning.

The full second-order model consists of 19 terms, (β_{33} and β_{44} being impossible to evaluate, the associated variables X_3 and X_4 being fixed at 2 levels only). It was considered to require too many experiments to determine. The granulation conditions were expected to have the most effect on the compression characteristics of the granulate and quality of the tablets; therefore a full second-order model was selected for X_1 and X_2. It was considered likely that there is an optimum lubrification time; this justified the introduction of a square term, x_5^2. It was expected that interactions other than X_1 with X_2 would not be significant, so they were omitted, except for the $x_4 x_5$ term (calibration × lubrification time). This latter term was added in case the effect of lubrification were particle size dependent.

A design of 18 experiments was set up (table 6.3). Details of the construction of this non-standard design, which required the use of an exchange algorithm, will be given in chapter 8. Since we are considering only the optimization of the process here, all that is required is knowledge of the domain and an adequate model for each response. Other details of the experimental design used to determine the estimates of this model thus do not concern us at this point.

Consider first of all the continuous variables, X_1, X_2 and X_5. The domain is (unusually) *cylindrical*. It is circular in the granulation variables (X_1, X_2) because the experimental points form a pentagon and cubical in the lubrification time (X_5) which takes 3 levels only. The two remaining variables are discrete and treated as qualitative variables. They may take any of the four factorial combinations of the 2 settings available to each factor. The cylindrical domain is therefore repeated 4 times, once for each combination of the fixed values of X_3 and X_4.

3. Interpretation of the fitted model – important effects

The experiments were performed and the properties of the resulting granulate and tablets, as well as the compression characteristics were measured for each experiment. Results are listed in table 6.3. The model coefficients for each response were estimated by linear regression and are shown in table 6.4. Coefficients which seem to be important are printed in bold type. For reasons which will be explained in chapter 8, it was not possible to determine the statistical significance of the coefficients by analysis of variance.

The objective was to obtain tablets of low friability and satisfactory hardness. Good compression properties were also needed, that is a high transmission ratio and also a high cohesion index (see below). A granulate with at least 20% fine particles was also considered desirable, for the powder to flow correctly.

Only X_1, X_2, and X_5 are continuous variables. The mathematical model may be used to predict responses inside the experimental domain and possibly even to extrapolate outside it, but this involves a great deal of risk.

Table 6.4 shows that a high proportion of the coefficients are important. Exceptions to this are b_4, the effect of the calibration diameter, which does not appear significant even for the percentage of fine particles, and the interaction b_{45}, which is not significant either. This simplifies the analysis to some extent.

There is no active coefficient for X_4 (calibration diameter). It was arbitrarily fixed at +1. Except in the case of tablet hardness, the lubrification time X_5 had no effect either; neither for the main effect b_5, nor for the square term b_{55}, nor the interaction b_{45}. So in choosing the level of X_5 we need only consider the response y_3.

This greatly simplifies the choice of graphs for plotting. All that is needed is to fix the factor X_3 at its most favourable level and study each response in the X_1, X_2 plane.

Table 6.3 Design, Plan and Experimental Response Values for a Granulation/Tableting Process Study

No.	Design (coded variables)*					Plan (natural variables)*					Responses**				
	X_1	X_2	X_3	X_4	X_5	U_1 g	U_2 s	U_3 %	U_4 mm	U_5 min	y_1 %	y_2 %	y_3 kP	y_4 %	y_5
1	1.000	0.000	1	-1	-1	3000	270	3	1	1	11.3	0.33	10.34	83.7	690
2	1.000	0.000	1	-1	1	3000	270	3	1	6	12.8	0.40	11.57	86.8	846
3	1.000	0.000	-1	1	0	3000	270	1	1.25	3.5	8.7	0.14	9.63	87.2	611
4	0.309	0.951	-1	-1	-1	2827	360	1	1	1	9.4	0.20	8.82	86.9	573
5	0.309	0.951	-1	-1	1	2827	360	1	1	6	8.9	0.2	8.57	86.4	559
6	0.309	0.951	1	1	0	2827	360	3	1.25	3.5	12.8	0.3	15.46	83.8	717
7	-0.809	0.588	1	-1	0	2548	326	3	1	3.5	38.2	0.3	13.99	75.3	882
8	-0.809	0.588	1	1	-1	2548	326	1	1.25	1	37.1	0.3	12.74	83.9	768
9	-0.809	0.588	-1	1	1	2548	326	1	1.25	6	37.3	0.3	10.89	82.9	668
10	-0.809	-0.588	-1	-1	0	2548	214	1	1	3.5	38.6	0.4	14.40	74.7	843
11	-0.809	-0.588	1	1	-1	2548	214	3	1.25	1	42.0	0.5	15.21	70.8	929
12	-0.809	-0.588	1	1	1	2548	214	3	1.25	6	40.2	0.7	13.23	67.8	830
13	0.309	-0.951	1	-1	-1	2827	180	3	1	1	28.6	0.36	12.19	83.7	723
14	0.309	-0.951	1	-1	1	2827	180	3	1	6	26.4	0.38	11.54	84.9	699
15	0.309	-0.951	-1	1	0	2827	180	1	1.25	3.5	24.4	0.34	10.79	85.2	664
16	0.000	0.000	-1	-1	0	2758	270	1	1	3.5	21.3	0.42	10.64	85.7	651
17	0.000	0.000	1	1	-1	2758	270	3	1.25	1	19.2	0.48	12.30	84.5	737
18	0.000	0.000	1	1	1	2758	270	3	1.25	6	23.3	0.47	12.10	86.2	748

* See table 6.2 for a description of the variables
** See table 6.4 for identification of the responses

Table 6.4 Response Variables and Estimates of the Model Coefficients for the Optimization of a Tableting Process

Coeff.	y_1 % fines	y_2 friability (%)	y_3 hardness (kP)	y_4 transmission ratio	y_5 cohesion index
b_0	20.96	0.413	12.350	85.0	719.3
b_1	**-16.45**	**-0.083**	**-1.750**	**6.6**	**-93.5**
b_2	**-6.30**	**-0.085**	-0.065	**1.8**	-27.6
b_3	**1.14**	0.044	**1.354**	**-1.8**	**60.1**
b_4	-0.11	0.008	0.298	0.2	-7.4
b_5	0.11	0.023	-0.308	0.2	-5.8
b_{11}	**6.04**	-0.079	0.782	**-6.0**	**92.3**
b_{22}	2.17	**-0.118**	0.734	**-2.7**	-23.2
b_{55}	-0.05	0.039	**-1.763**	1.5	-37.2
b_{12}	**-5.71**	0.109	0.800	**-6.6**	23.2
b_{45}	0.31	0.008	-0.363	-0.6	-25.5

4. Response surface modeling of individual effects

Particle size - percentage of fine particles (y_1)

The particle size distribution of the granulate was measured and the percentage of fine particles defined as the percentage of the mass passing through a 125 μm sieve.

The granulation conditions had most influence on this response (as was the case for all other response variables). In particular an increased volume of the granulating liquid X_1 led to a sharp decrease in the percentage of fine particles. In the case of a "dry" powder (less than 2.750 kg granulating liquid added), increasing the granulation time had little effect on y_1, whereas for a "wet" powder (more than 2.750 kg granulating liquid) the percentage of fine particles decreased with increasing granulation time. In terms of the model coefficients, there is a large negative interaction between X_1 and X_2. No other variables appear to be active.

A fairly fine granulate was required, with between 20 and 35% of the powder having a diameter less than 125 μm. The shaded area in the graph (figure 6.6a) is therefore excluded, that on the right hand side representing overgranulation, and that on the left undergranulation. The same response surface is plotted three dimensionally in figure 6.7a.

Tablet friability (y_2)

This was measured as the percentage loss in weight of a number of tablets agitated for a certain fixed time (25 rpm for 15 min). It was to be minimized and must not in any case exceed 0.5% loss in weight.

The experimental results varied from 0.14 to 0.7%. The most friable tablets resulted from a low volume of granulating liquid and a short mixing (granulation) time, as shown in figures 6.6b and 6.7b. The friability may be reduced on increasing the granulation time, at the same time increasing the amount of binding liquid. This would allow a value of between 0.25 and 0.3% to be obtained while staying in the experimental domain.

Tablet hardness (y_3)
The hardnesses of tablets obtained at constant compression force were measured. Harder tablets were obtained for lower amounts of binder, but results are satisfactory over all the domain (figures 6.6c and 6.7c). A lubrification time of about 3 minutes appears to be the optimum (figure 6.6d). 3% humidity gave significantly harder tablets than the powder with humidity below 2% .

Compression characteristics
These are defined in terms of the transmission ratio and cohesion index, as suggested by Guyot (9). The *transmission ratio* (y_4) is defined as:

$$y_4 = Y_1 / Y_2 \times 100$$

where Y_1 is the resultant of the forces applied by the upper punch in the tablet machine, being transmitted by the powder bed from the force Y_2 applied by the lower punch. It was measured at $Y_1 = 15$ kN.

Sticking or gripping of the tablets was seen when the transmission ratio was below 80%. It was again the granulation conditions that had the major effect on this response (figures 6.6e and 6.7d), but a higher humidity of the granulate also had a slight influence, giving a slightly lower transmission ratio (figure 6.7d).

The *cohesion index* (y_5) is defined as the ratio of the hardness of the tablet to the value of the force applied on the lower punch. The larger the index, the harder are the tablets obtained for a given applied force (9). Values were obtained at constant applied force (15 kN).

Higher cohesion indices were obtained for "dry" granulates. Mixing time had no significant effect. For granulates of the higher humidity, the cohesion index was significantly increased, comparatively less force being needed to achieve a given tablet hardness (figure 6.6f).

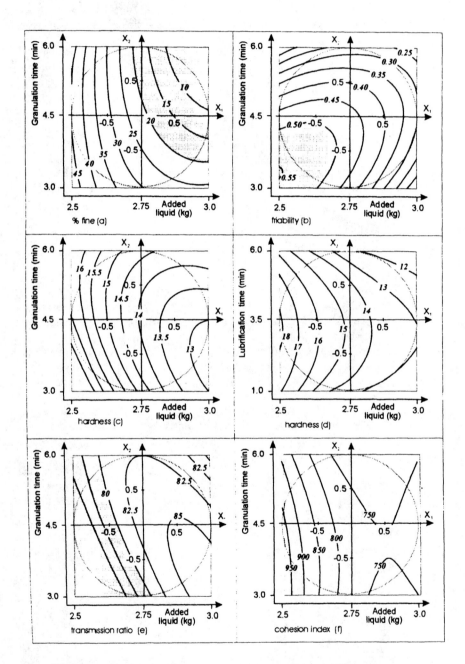

Figure 6.6 Response surfaces: (a) % fine particles;   (b) friability;
(c)-(d) hardness;   (e) transmission ratio;   (f) cohesion index.

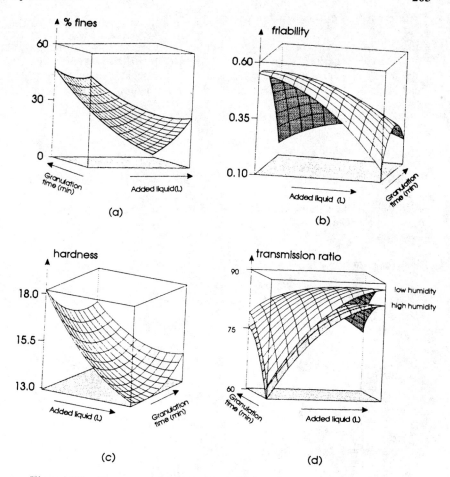

Figure 6.7 Response surfaces: (a) % fine particles; (b) friability; (c) hardness; (d) transmission ratio.

5. Superposition of the response curves

We select 3% residual humidity as the optimum level for X_3 on the basis of tablet hardness and cohesion. Calibration had no effect, so 1.25 mm is chosen. Lubrification time also had little effect and 3 minutes seemed to be optimum for tablet hardness. This value was therefore selected for plotting. We take therefore:

fines	25% - 35%
transmission	$\geq 80\%$
friability	$\leq 0.3\%$

These limits are plotted in figure 6.8. For clarity we have not included tablet hardness or cohesion index. Comparison with table 6.3 and with figures 6.6 and 6.7 will show that the hardness and the cohesion index were both fairly constant and, at satisfactory levels, in the part of the domain that was acceptable according to the above criteria.

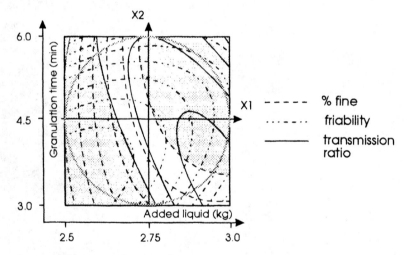

Figure 6.8 Graphical optimization of a tableting process.

The process study, using RSM experimental design, allows identification of the main tendencies and also the choice of optimum manufacturing conditions at the pilot scale.

III. OPTIMIZATION OF MULTIPLE RESPONSES USING DESIRABILITY

A. The Problem of Multiple Responses

We have noted the impossibility of optimizing several responses unless these are highly correlated, and correlated as we would like them to be. One way of overcoming this difficulty is to define an acceptable range for each response. The satisfactory zone is that part of the domain for which the value of each and every one of the calculated responses is acceptable. This was the approach of graphical optimization.

This method has the disadvantage that, in reality, for those variables where we have defined an acceptable range, all values within the range are acceptable but some are more acceptable than others. In formulating a tablet we could perhaps live with a friability of 1%, if it were the only means of obtaining the desired value of a more essential parameter, but a friability of 0.1% would be much preferred.

We recall that the optimization criteria for the granulation example (8) of section II.B, were:

particle size:	at least 20% and if possible between 25% and 30% fine particles in the granulate
friability:	no more than 0.5% and if possible less than 0.1%
tablet hardness:	at least 11 kP and if possible more than 16 kP
transmission ratio:	minimum 80%, with a target of over 90%
cohesion index:	600 minimum, with a target value of 1000 or above

Derringer and Suich (10) described a way of overcoming this difficulty. In the optimization, each response i is associated with its own *partial desirability function* (d_i). This varies from 0 to 1, according to the closeness of the response to its target value. For the friability, we would like as low a value as possible. The target is therefore 0% friability and the desirability at this point is equal to 1. If no formulation with a friability of more than 2% can be considered acceptable, than the desirability is equal to zero for all values of 2% and over. Between 0% and 2% the desirability decreases linearly or in a convex or concave form.

Just as each response variable can be calculated over the experimental domain using the model and the calculated coefficients, so can the corresponding desirability be calculated for that variable at all points in the domain.

The r individual desirability functions are then combined together, usually as the geometric mean, to obtain the overall *desirability function* (D) for the system whose maximum value can then be looked for within the domain.

B. Partial Desirability Functions

We will examine the different kinds of partial desirability functions available using the same example of a granulation process study (8) that has already been used to demonstrate graphical analysis and optimization.

It is assumed that the experimenter, having carried out his experiments and analysed the results, will now have a satisfactory mathematical model for each response that is to enter into the global optimization. He can therefore calculate a partial desirability for each response in all parts of the domain, and obtain from these an overall desirability D. The problem becomes one of numerical optimization of the function D within the domain.

1. Linear partial desirability functions

Maximization

For the desirability of a function requiring maximization, it is assumed that there is a *target value* for the response, above which we are totally satisfied. There also exists a lower *threshold* below which the result is not acceptable.

In this case the *cohesion index* required maximization. Values below 600 were considered unacceptable. The target value was 1000 and it was not considered that increasing the index above this value would improve the result in any way. The resulting partial desirability function d_5 for the response y_5 is shown in figure 6.9a.

Minimization

We assume there is a *target value* for the response to be minimized, and that values lower than this are entirely satisfactory. There also exists an upper limit above which the result is not acceptable.

The *friability* of the tablets was to be minimized. Values above 0.5% (the threshold) were to be rejected, but all friabilities below 0.1% (the target value), on the other hand, were considered completely acceptable. The partial desirability function d_2 for the response y_2 is shown in figure 6.9b.

Bilateral desirability function

The values of certain variables in an optimization need to be held close to a particular value, that is the target. Any deviation from this value leads to a decrease in quality. There are thus two thresholds, not always symmetrical about the target, above and below which the satisfaction is rated as zero (figure 6.9c).

There are also cases where the response is totally satisfactory provided it lies between the two threshold values: this is called a *target interval*. The desirability function is written:

$$d_i = 0 \quad \text{if} \quad y_i < y_i^{(min)}$$
$$d_i = 1 \quad \text{if} \quad y_i^{(min)} \leq y_i \leq y_i^{(max)}$$
$$d_i = 0 \quad \text{if} \quad y_i^{(max)} < y_i$$

and illustrated in figure 6.9d. Equivalent functions may be defined for maximization and minimization. Their use is equivalent to that of the graphical optimization method.

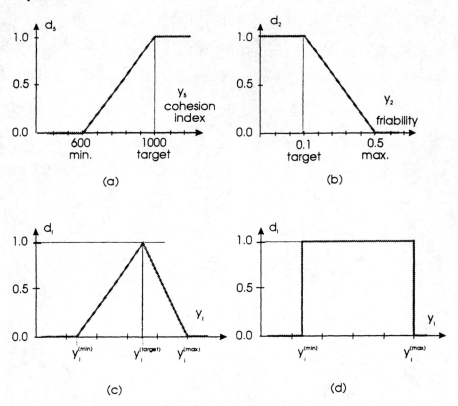

Figure 6.9a-d Simple desirability functions.

2. Non-linear partial desirability functions

The different possibilities given above and in figure 6.9a-d may be refined or developed further. For example, the experimenter may penalize response values near the thresholds and correspondingly favour those near the targets.

Maximization and minimization
We take the example of the maximization of the transmission ratio in the granulation process study. The minimum is 80% and the target 90%, but it was desired to favour high transmission ratios still further by using the following function:

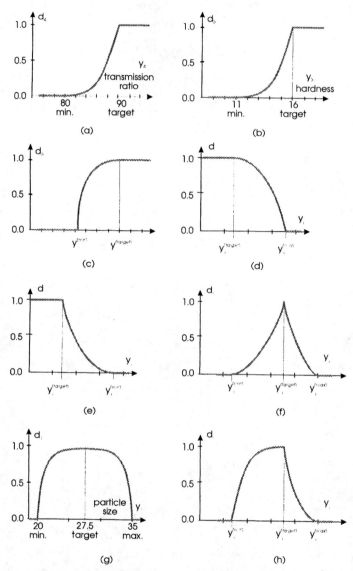

Figure 6.10a-h   Non-linear desirability functions.

$$d_i = 0 \qquad\qquad\qquad \text{if } y_i \leq y_i^{(min)}$$

$$d_i = \left(\frac{y_i - y_i^{(min)}}{y_i^{(max)} - y_i^{(min)}} \right)^s \qquad \text{if } y_i^{(min)} \leq y_i \leq y_i^{(max)}$$

$$d_i = 1 \qquad\qquad\qquad \text{if } y_i^{(max)} \leq y_i \qquad\qquad (6.3)$$

A value $s = 3$ gave the curve shown in figure 6.10a. A similar relationship was selected for the tablet hardness (figure 6.10b), where the minimum value 11 kP was penalized still further, as a higher exponent $s = 5$ was selected. In the case of the linear partial desirability function, the exponent s is unity.

In the same way it is possible to design a function where values exceeding the threshold, but still rather less than the target, are little penalized, by choosing $0 < s < 1$. This kind of function may be selected for a response which has less importance in the optimization. An example is shown in curve 6.10c.

The functions for minimization are similar (figure 6.10d-e).

Bilateral functions
On combining the functions for maximization and minimization we define a bilateral desirability:

$$d_i = 0 \qquad\qquad\qquad \text{if } y_i \leq y_i^{(min)} \text{ or if } y_i^{(max)} \leq y_i$$

$$d_i = \left(\frac{y_i - y_i^{(min)}}{y_i^{(target)} - y_i^{(min)}} \right)^{s_1} \qquad \text{if } y_i^{(min)} \leq y_i \leq y_i^{(target)}$$

$$d_i = \left(\frac{y_i^{(max)} - y_i}{y_i^{(max)} - y_i^{(target)}} \right)^{s_2} \qquad \text{if } y_i^{(target)} \leq y_i \leq y_i^{(max)} \qquad (6.4)$$

As before, the curves may be made biconcave (figure 6.10f), biconvex [as for particle size (figure 6.10g)], or mixed (by choice of s_1 and s_2). One side of the interval may be made linear by setting either s_1 or s_2 in equation 6.4 equal to 1.

3. Global desirability function

In combining partial desirability functions, it is possible to take a weighted mean. However it is far more frequent to use a geometric mean of the desirabilities, of the form:

$$D = (d_1 \times d_2 \times d_3 \times \ldots d_r)^{1/r} = \left[\prod_{i=1}^{r} d_i \right]^{1/r} \qquad (6.5)$$

The largest value that the desirability can take is 1 and the desirability function for all parts of the domain where an individual response is outside the acceptable range is therefore zero.

An alternative more general form of the function is:

$$D = \prod_{i=1}^{r} d_i^{\,p_i} \tag{6.6}$$

where p_i is the weighting of the ith response, normalised so that $\sum_{i=1}^{r} p_i = 1$.

This allows optimization to take into account the relative importance of each response, while selecting the most appropriate form of the partial desirability function. Equation 6.5 is the special case of equation 6.6 for equal weightings and is the form most often used at present. It has the disadvantage that weighting can only be done by adjusting s.

The corresponding desirability is calculated over the domain for each response, using the model equations. The overall function is then determined and is graphically mapped over the domain. The form of the function is investigated, just as for an individual response, using response surface analysis. Examination of the various contourplots of the desirability surface will give a reliable general picture of the acceptable region. Computer programs use standard search methods to determine the point of maximum desirability, which in the case of systems with more than 3 variables makes a good starting point for taking slices.

The shape of the function tends not to be as smooth as the response surfaces.

The desirability function is a guide in optimizing a process or a formulation using multiple response data from a statistically planned experiment. It should be used with caution. Always try a number of different individual desirability functions for the different responses, and when an apparently optimum point or region has been found, go back to the plots of the original responses to check that it is really what is required.

C. Examples of Multi-Criteria Optimization

1. Optimization of a process study

The aims of the optimization are summarised in table 6.5. All weighting factors were equal to one another ($p_i = 0.2$) – equation 6.6 is used.

Table 6.5 Optimization Criteria and Desirability Functions for the Optimization of a Wet Granulation Process (8)

Response	Partial desirability	$y_i^{(min)}$	$y_i^{(target)}$	$y_i^{(max)}$	Exponent s lower	upper
% fine particles	d_1	20%	27.5%	35%	0.1	0.1
Friability	d_2	-	0.1%	0.5%	-	1
Hardness	d_3	11 kP	16 kP	-	5	-
Transmission ratio	d_4	80%	90%	-	3	-
Cohesion index	d_5	600	1000	-	1	-

Numerical optimization leads to the following experimental conditions, expressed in coded variables:

X_1	X_2	X_3	X_4	X_5
-0.290	1	1	1	-0.660

The optimum is found to be insensitive to changes in X_5 around about -0.5, corresponding to a lubrification time of about 2.25 min. Taking this point as a "working" optimum, we calculate an overall desirability of 0.26. The calculated individual responses are:

fine particles:	$\hat{y}_1 = 24.5\%$	$d_1 = 0.95$
friability:	$\hat{y}_2 = 0.24\%$	$d_2 = 0.64$
hardness:	$\hat{y}_3 = 14.64$ kP	$d_3 = 0.20$
transmission ratio:	$\hat{y}_4 = 80.83\%$	$d_4 = 0.025$
cohesion index:	$\hat{y}_5 = 801.46$	$d_5 = 0.38$

Figure 6.11 shows the response surface of the global desirability in the $\{X_1, X_2\}$ plane, with factors X_3 and X_4 both fixed at +1, and X_5 at -0.5. Comparison of figure 6.11 with figures 6.6 and 6.8 is instructive; the two optimization methods, graphical and by desirability, lead to very similar conclusions, with an extensive zone of acceptable granulation conditions defined.

Note that X_3 and X_4 are *qualitative factors* and their values are limited to ±1. An optimization method that could distinguish between discrete and continuous variables and which would thus only search the space of permitted values of the qualitative, or discrete quantitative variables, would be preferred. It can also be seen that the maximum desirability, on the edge of the square defined by $\{X_1, X_2\}$ at ±1, is actually just outside the cylindrical experimental domain. The optimum should, strictly speaking, be displaced very slightly to lie on the edge of the design space.

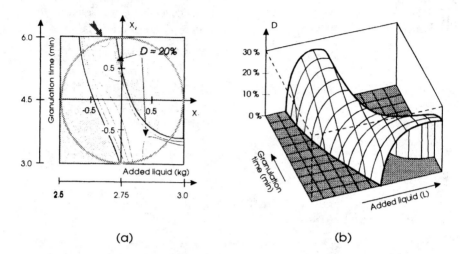

(a)   (b)

Figure 6.11 Response surface of global desirability as function of granulation time and volume, with fixed residual humidity 3%, sieve 1.25 mm, and lubrification time 2.25 min.

2. Formulation of an oral solution

Graphical optimization of this solution, using the data of Senderak *et al.* (5) was also described earlier. Numerical optimization may be carried out, using the desirability functions of table 6.6. The exponent $s = 1$ in all cases (c.f. figure 6.9a). Note that one of the factors (sucrose) is also being treated as a response for optimization (that is: $y = x_3$), and is maximized for reasons of taste.

Table 6.6 Optimization Criteria and Desirability Functions for the Optimization of an Oral Solution (5)

Response	Partial desirability	Goal	$y_i^{(min)}$	$y_i^{(max)}$	Result	Partial desirab.
turbidity ppm	d_1	minimize	1.8	4	2.070 ppm	0.88
cloud point °C	d_2	maximize	60	80	70.4 °C	0.52
invert sucrose mL	d_3	maximize	44	65	60.1 mL	0.77

The optimum is found at 4.43% polysorbate, 21.4% propylene glycol, and 60.1 mL sucrose. Results are very similar to those for the graphical optimization. It would also be possible to take into account the concentrations of the remaining constituents, propylene glycol and polysorbate 80, perhaps minimizing them.

3. Synthesis of nanoparticles

Introducing the problem
McLeod *et al.* optimized the conditions of synthesis of polyglutaraldehyde nanoparticles (7) in terms of 5 factors, monomer and surfactant concentrations, pH, oxygen level, and stirring rate. They used a central composite design, the factorial part of the design being a half fraction factorial of resolution V. Four responses were measured, the most important (the ones that we consider here) being the particle size and yield.

Models
The reader is referred to the original paper for the experimental data. The authors used a full second order design and the original untransformed variables. We have applied Box-Cox analysis (described in the following chapter), which indicated that the data should be transformed for multi-linear regression, the transformation being logarithmic in the case of yield and reciprocal for the size. The resulting regression equations for the yield y_1 and the particle size y_2 are:

$$\log_{10}\hat{y}_1 = 2.662 + 0.510\ x_1 + 0.004\ x_2 + 0.078\ x_3 + 0.001\ x_4 - 0.051\ x_5$$
$$- 0.096\ x_1^2 + 0.001\ x_2^2 - 0.164\ x_3^2 - 0.002\ x_4^2 - 0.000\ x_5^2$$
$$-0.027\ x_1x_2 - 0.075\ x_1x_3 - 0.032\ x_1\ x_4 + 0.026 x_1 x_5 - 0.033\ x_2 x_3$$
$$-0.062\ x_2 x_4 + 0.007\ x_2 x_5 - 0.046\ x_3 x_4 + 0.014\ x_3 x_5 - 0.016\ x_4 x_5 \qquad (6.7a)$$

$$\hat{y}_2^{-1} = 0.633 - 0.364\ x_1 + 0.077\ x_2 + 0.099\ x_3 + 0.006\ x_4 - 0.012\ x_5$$
$$+ 0.067\ x_1^2 + 0.030\ x_2^2 + 0.129\ x_3^2 + 0.005\ x_4^2 + 0.041\ x_5^2$$
$$+ 0.000\ x_1x_2 + 0.124\ x_1x_3 - 0.004\ x_1x_4 + 0.026 x_1 x_5 - 0.104\ x_2 x_3$$
$$- 0.014\ x_2 x_4 + 0.018\ x_2 x_5 - 0.093\ x_3 x_4 + 0.109\ x_3 x_5 + 0.081\ x_4 x_5 \qquad (6.7b)$$

A large number of the terms are statistically significant and numerically important, main effects, square terms and interactions. Significant effects ($< 5\%$) are shown in bold type.

Treatment of the models and lack-of-fit
Analysis of variance showed considerable lack of fit of the regression equations 6.7a-b even with the Box-Cox transformations. This is not surprising, as there is a wide range for both responses and the repeated experiments at the centre showed the process to be highly repeatable. At the end of chapter 5 we listed the various actions that may be taken on rejection of the model:

- The domain could be contracted, rejecting areas where the yield was poor or the particles produced were too large. However, it is not very easy to see

at once how this should be done.
- A higher order model may be used. However, no extra terms should be added to the model without further experimentation.
- The model may be used as it stands, in spite of the significant lack of fit.

The last possibility allows us to define a point of interest, which may not be a true optimum if the system shows deviations from the model. The approximation may be adequate for practical purposes, and failing that it is likely to be a good centre point for a new experimental design in a reduced domain.

Optimization

It has already been mentioned that graphical analysis of this problem, which had 5 factors and a number of responses, would prove difficult. Attempts to optimize by canonical analysis also failed because the important responses, yield and size of nanoparticles, proved to be "opposed" (that is, changes in any of the factors to maximize the yield would also give an increase in particle size). In such a case, numerical optimization by desirability appears to be the best, and probably the only, method of finding a compromise. Therefore this was the method used by the authors.

We set up partial desirability functions, as in table 6.7, using the same limits as the authors (7) for the size and slightly wider limits for the yield. Allowance also had to be made for the fact that the response data were transformed. The range for the size was relatively narrow, so the transformation would have little effect on the partial desirability function. On the other hand, the wide range for the yield means that smaller yields would tend to be penalised less than if there was no transformation. For this reason we increased the value of the exponent to $s = 5$ (c.f. figure 6.10b), in order to favour higher yields and compensate for this effect.

Table 6.7 Optimization Criteria and Desirability Functions for the Synthesis of Nanoparticles [from reference (7)]

Response	Partial desirability	Goal	$y_i^{(min)}$	$y_i^{(max)}$	Predicted response	Partial desirability
Size (µm)	d_1	minimize	0.5	1.0	0.71	0.31
Yield (mg)	d_2	maximize	100	1000	750	0.51

The solution found is shown in figure 6.12, with the size and yield as functions of the monomer concentration and the pH. The global desirability function is superimposed on this. This is very close to the optimum found by McLeod *et al.*, who took other responses into account at the same time.

It was necessary to carry out a number of runs for finding the optimum. A secondary maximum also existed within the domain, with less good predicted responses and a number of the searches found this one instead.

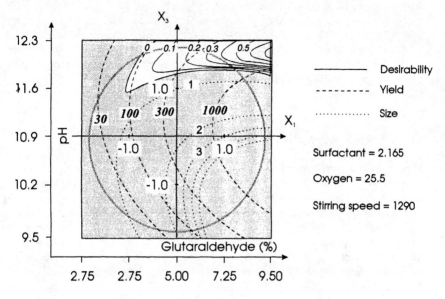

Figure 6.12 Optimization of nanoparticles.

4. General considerations when using *desirability*

Numerical optimization by the desirability method leads to a single point, but the experimenter is usually equally interested in the zone where the global desirability is greater than zero. Thus all criteria are acceptable in the neighbourhood of this optimum point. It is necessary to complete the determination of the optimum by drawing the contour plot of the desirability surface.

Always compare the above desirability surface with the surfaces for the original factors. Carry out a number of optimizations with the same parameters, models and limits but different starting points. The program may often find a secondary maximum, especially where there are a large number of factors.

In case of difficulty, usually when there are a large number of factors and responses, start with wide optimization limits for the different responses, and then when an approximate optimum has been found, make the limits narrower. If exponents are introduced in the partial desirability functions, it may be useful to test several values of these.

If the responses are transformed, remember that the desirability functions are on the transformed responses. It may be necessary to compensate by adding an exponent to the partial desirability function.

IV. OPTIMIZATION BY EXTRAPOLATION

The possible experimental domain is not always known at the beginning of a study. On the one hand, there is a risk of choosing too extensive an experimental region, and then the responses may vary over so wide a range that it becomes impossible to fit the simple quadratic model to them. On the other hand, the response surface may show discontinuities, such as in a tableting experiment, where there may be whole regions where tablets cannot be obtained because of incompatibility with the machine, or lack of cohesion. To avoid this problem a much smaller domain may be selected, with the result that the optimum may then be discovered to be outside it.

There are two main model-based methods for extrapolating outside the domain, steepest ascent (first-order model) and optimum path (second-order). In addition, we will briefly describe the well-known *sequential simplex* methods, which are model-independent.

A. Steepest Ascent Method

1. Circumstances for use

In a sense, this is a direct optimization method. Only when the domain has been established with a reasonable degree of certainty can the final experimental design be set up to enable response surface modelling and optimization. Very often the initial experiments to situate the region of interest will be part of a screening study, hence used to determine a first-order model.

When the optimum is outside the domain, we need to arrive at it rapidly. Changing one factor at a time will not work well, especially if there are interactions. One possibility is to use the sequential simplex, but it is here also that the steepest ascent method comes into its own. The procedure is simple. Assuming that the response (such as the yield) is to be maximized, we determine this response as a function of the *coded* variables x_1, x_2, ... and find the position and direction of maximum rate of increase (steepest ascent) in terms of these coded variables.

A straight line may be drawn along that direction, from the centre of the domain, and experiments performed at a suitable spacing along this line, measuring the response. If a maximum or near target value, or at least a much improved response were found at a certain point, this point could be used in turn as the centre for a new experimental design (figure 6.13).

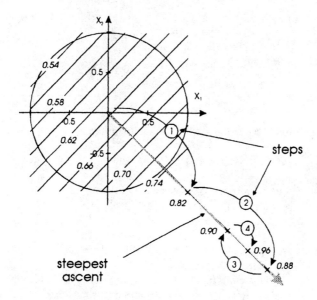

Figure 6.13   Steepest ascent method.

2. The first-order design and determining the direction of steepest ascent

Box, Hunter, and Hunter (1) recommend a design which allows the first-order model to be fitted, but which also includes test points (for example at the centre) and repeated experiments to allow the error to be estimated.

The example described below allows the model to be fitted, but without any check on how good the fit is. (At least one experiment at the centre point is to be preferred, so that there is some assessment of the linear model's validity.) Huang *et al.* (11) described optimization of the conditions for iontophoretic transdermal delivery of pilocarpine. They studied the effects of 4 factors, initially using a 2^{4-1} factorial design of defining relation $I = 1234$. The factors and the domain are given in table 6.8.

Table 6.8 Domain for a 2^{4-1} (Resolution IV) Factorial Design in Iontophoretic Transdermal Transport of Pilicarpine (8)

Factor	Associated variable	Lower level (coded -1)	Upper level (coded +1)
pH	X_1	3.0	5.0
ionic strength (mal/L)	X_2	0.004	0.020
pulse frequency (Hz)	X_3	1000	3000
current duty cycle (%)*	X_4	40	60

* % duty is the percentage of time the current is *on*

Let the first-order model within the domain be:

$$\hat{y} = b_0 + b_1 x_1 + b_2 x_2 + b_3 x_3 + b_4 x_4$$

For this model, the iso-response surfaces are parallel hyperplanes in 4 dimensional factor space. In general, for k variables, the iso-response surfaces are $k - 1$ dimensional hyperplanes. The path of steepest ascent is a straight line orthogonal to these planes, usually starting from the centre of the domain. The step-size for each variable is proportional to the estimate of the corresponding coefficient in the model.

The fitted relationship for the response (amount in µg transported during the first hour), determined from the results of the 8 experiments, is:

$$\hat{y}_i = 379.5 + 210.3\, x_1 - 141.8\, x_2 - 1.5\, x_3 + 71.9\, x_4$$

X_3 may be neglected; the frequency is thus held constant at 2000 Hz. It is X_1 which has the greatest effect. As we have seen, step-sizes of the coded variables for the steepest ascent path are proportional to the coefficients b_i of the fitted equation. The path of steepest ascent from the centre is:

$$x_1 = -1.48x_2 = -140x_3 = +2.92x_4$$

and the extrapolated calculated response (permeation) along that line, in terms of x_1, is:

$$\hat{y}_{i(steep)} = 379.5 + (210.3 - 141.8/(-1.48) + 71.9/2.92)\, x_1 = 379.8 + 330.7x_1$$

Huang *et al.* carried out experiments along this line at a spacing equivalent to 0.5 in X_1, that is 0.5 pH units. The steps sizes in the coded (X_i) and natural (U_i) variables were:

$$\Delta x_1 = +0.5 \qquad\qquad \Delta U_1 = +0.5 \text{ pH units}$$
$$\Delta x_2 = -0.5/1.48 = -0.34 \qquad \Delta U_2 = -0.0027 \text{ mol/L}$$
$$\Delta x_3 = 0$$
$$\Delta x_4 = 0.5/2.92 = +0.17 \qquad \Delta U_4 = +1.7\%$$

The experiments, in terms of coded and natural variables are shown in table 6.9. The results obtained are listed in the final column. (In general, it is not necessary for the experiments to be evenly spaced. On seeing that points 1 and 2 gave an improvement, the step size might have been increased and for certain problems, though not this one, the optimum could be arrived at more rapidly (1).)

Experiment 3 in table 6.9 showed a considerably increased permeability. On continuing along the calculated steepest ascent line, the response decreased. This experiment therefore served as centre of interest for the next step of the optimization. This would normally be a second-order design for RSM. In actual fact, the authors first carried out another 2^{3-1} fractional factorial design of 4 experiments, to test whether a further steepest ascent operation gave an improved response. It did not, so experiment 3 could be considered as being in the optimum region. They then constructed a second-order design centred on this point, augmenting the fractional factorial design with 6 axial experiments and additional centre points. The result was a small central composite design (see chapter 5, section V.A.4), which could thus be used for response surface modelling and determining optimum conditions. The variation of iontophoretic permeation within the new domain was small and the regression was not significant. However, conditions for a further slight improvement in the response were determined. The true optimum was still perhaps outside the domain. The optimum path method which follows could be used for further optimization, though, because the slope is so slight, it is unlikely that there would be significant further improvement.

Table 6.9 Steepest Ascent Optimization of Pilocarpine Iontophoresis [adapted from reference (8) by permission]

No.	Coded variables			Natural variables			Amount permeated in 1 h µg
	X_1	X_2	X_4	pH	ionic strength	duty	
1	0.5	-0.34	0.17	4.5	0.0095	52	660
2	1.0	-0.68	0.34	5.0	0.0070	54	829
3	1.5	-1.01	0.51	5.5	0.0045	56	950
4	2.0	-1.35	0.68	6.0	0.0020	58	726
5	2.5	-1.69	0.86	6.5	0.0000	60	452

B. Optimum Path

1. Introduction and conditions for use

If the design was for a second-order model and examination of the contour plots or canonical analysis (see below) showed that the optimum probably lay well outside the experimental domain, then the direction for exploration would no longer be a straight line, as for the steepest ascent method. In fact, the "direction of steepest ascent" changes continually and lies on a curve called the *optimum path*. The calculations for determining it are complex, but with a suitable computer program the principle and graphical interpretation become easy.

We demonstrate this method using the results of Vojnovic *et al.* for the optimization of a spheronization by granulation (12). These data have already been presented in chapter 5: the data for the median granulate size in terms of impeller speed, amount of liquid and spheronization time being listed in tables 5.18 and 5.19. The following mathematical model was estimated:

$$\hat{y} = 1594 - 97.4x_1 + 117.6x_2 + 16.7x_3 - 20.0x_1^2 - 379.2x_2^2 - 70.9x_3^2 - 222.8x_1x_2 + 119.6x_1x_3 + 82.8x_2x_3$$

We look first at the direction for maximizing the particle size. Figure 6.14a shows a "cut" at $x_3 = -0.2$, where there is only one possible direction for increasing the response. On the other hand, when the curves are branches of hyperbolae there are always two possible directions, as shown in figure 6.14b. This is a "slice" taken at $x_2 = -0.29$ (the only reason for drawing this contourplot being to show this particular form of isoresponse curve). The two directions for maximizing are not symmetrical with respect to the centre. Although the predicted response is the same at points A and B, A is much closer to the origin. B, being outside the domain (shown by the circle), is an extrapolation and its prediction is highly uncertain. The optimum path would tend to go in the direction of A.

2. Principle and maximization of the response

The principle of the method is as follows. Spheres of increasing radius are drawn about the origin of the domain. The position of maximum response is determined on each sphere. Thus the upper curve, figure 6.15a, gives the maximum value of the response on the sphere (as a function of its radius ρ). The lower curves, figure 6.15b, give the coordinates of that point (again as a function of the radius). For example, for a radius of 1 unit (the radius of the experimental domain in this case) the upper graph shows that at that distance from the centre the largest diameter is 1743 μm, at a point whose coordinates are $x_1 = -0.92$, $x_2 = +0.33$ and $x_3 = -0.21$.

It is clear that in following this path from the centre of the domain, x_1 (impeller speed) must be strongly decreased, x_2 (amount of liquid) slightly increased and x_3 (spheronization time) slightly decreased. This optimum path is traced on figure 6.14a in the X_1, X_2 plane. The path in X_3 must be imagined as being below the surface of the paper.

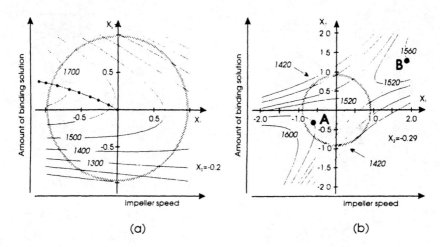

(a)   (b)

Figure 6.14 Median particle size as function of (a) impeller speed and binding solution and (b) impeller speed and spheronization time [from the data of reference (9)].

3. Minimization of the response

Although the above maximization illustrates the optimum path method, a more probable pharmaceutical objective would be to *decrease* the particle size, either to a value of 1000 μm, or under certain circumstances to a much smaller particle size. Figure 6.15c-d show the optimum path for the minimization, 6.15c giving the minimum value, and 6.15d the corresponding coordinates. It is evident that the size diminishes very steeply as the variable X_2 (binding solution) decreases. The other two variables must be held around zero. Therefore, if the centre of the domain corresponds to the present manufacturing conditions, a decrease in the percentage of binder liquid, while maintaining the other conditions constant, will lead to a decreased particle size.

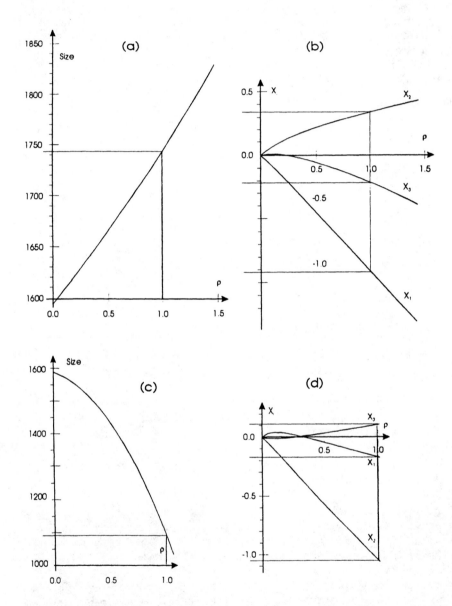

Figure 6.15 Optimum path for maximizing or minimizing the particle size, from the data of reference (12): (a) Maximum particle mean diameter as function of the distance ρ from the centre; (b) Dependence of coordinates of maximum point, as coded variables X_i, on ρ; (c) Minimum particle mean diameter as function of the distance ρ from the centre; (d) Dependence of coordinates of minimum point, as coded variables X_i, on ρ.

C. Sequential Simplex Optimization

1. Introduction

The simplex method has been used widely over the past 30 years, its success as much owing to its simplicity as to its efficiency. Unlike the other methods described in this chapter and most of the others in this book, it assumes no mathematical model for the phenomenon or phenomena being studied. The often long and costly phase of determination of a model equation may therefore be avoided, and the method is thus economical in principle. It is *sequential* because the experiments are analysed one by one, as each is carried out. Because the method is not model-based we will not describe it in detail, but well indicate how it can "fit in" and complement statistical experimental design.

The *simplex* which lends its name to this optimization method is a convex geometric figure of $k+1$ non-planar vertices in k dimensional space. For 2 dimensions it is a triangle, for 3 dimensions it is a tetrahedron.

The basic method originally described by Spendley, Hext and Himsworth (13) is quite powerful and also very easy. A modification employing a variable step-size was proposed in 1965 by Nelder and Mead (14) frequently performs even better and is more generally employed. It is this method that we will describe briefly in the following section.

2. Optimization by the Extended Simplex Method

Assume that we wish to optimize a response depending on 3 to 5 factors, without assuming any model for the dependence, other than the domain being continuous. We choose an initial domain and place a regular simplex in it, as described in chapter 5. The experiments for the initial simplex are then carried out and the response measured. In the basic simplex method an experiment is done outside the simplex in a direction directly opposite to the "worst" point of the simplex. The worst point is discarded and a new simplex is obtained, the process being repeated. The simplex therefore moves away from the "poor" regions towards the optimum, with a constant step size. In the extended simplex method on the other hand, if the optimum is outside the initial experimental domain, we may leave it rapidly while expanding the simplex for a region with an improved response. As the simplex approaches the optimum, it is contracted rapidly.

Principle of the method
We take the example of 3 factors. Of the experiments of a given simplex let the "worst" be (W), "next worst" (N) and "best" (B) points of the initial simplex. Here, depending on the value of the point R relative to the "worst" (W), "next worst" (N), and "best" (B) points of the initial simplex, it may be *expanded* to arrive quickly at the region of the optimum, and then be *contracted* around the optimum. The various possibilities are shown in figure 6.16. "$R > W$" means that point R is better than point W, etc.

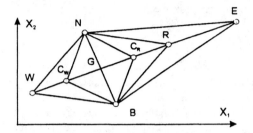

Figure 6.16 Summary of the extended simplex method of Nelder and Mead.

R replaces W if: $N \le R \le B$	Reflection
or: $R > B$ and $E \le B$	
E replaces W if: $R > B$ and $E \ge B$	Expansion
C_R replaces W if: $W < R \le N$	Contraction (exterior)
C_W replaces W if: $W > R$	Contraction (interior)

Optimization of a tablet formulation

Bindschaedler and Gurny (15) have demonstrated the use of the extended simplex method in optimizing the level of disintegrant and the compression force in a directly compressed tablet (figure 6.17). The responses to be optimized were the hardness (maximized) and the disintegration time (minimized), which were normalised and then combined to give a weighted average Φ, which was minimized. The reader is referred to the original paper for the full details. An upper limit of 24 kN compression force was imposed, and so certain experiments were rejected and assigned a low response without being carried out.

The majority of operations are reflections R (for example points 4, 5, 6) but there are also expansions E (point 7), contractions outside the simplex C_R (point 14) and inside the simplex C_W (point 11, result of a boundary violation), as indicated in table 6.10.

3. Other sequential simplex methods and approaches

The basic (fixed step) method

This works in the same way as the extended method, except that the only operation is the reflection R. Thus the simplex remains regular (in terms of the coded variables) and the same size, throughout the optimization. Special rules apply when R is worse than W, and for stopping the simplex. See references (13) and (16) for the general rules and reference (17) for a pharmaceutical example, where it is used by Mayné, again to optimize a tablet formulation.

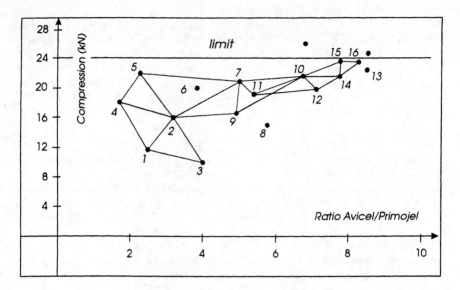

6.17 Optimization of a tablet formulation by the extended simplex method [adapted from reference (15), by permission].

Table 6.10 Optimization of a tablet formulation by the extended simplex method: response data for figure 6.17 [adapted from reference (15), by permission]

No	Hardness kg	Disint. time s	Φ	Operation	No	Hardness kg	Disint. time s	Φ	Operation
1	5.7	7.5	4.98	*Initial*	10	10.3	12.8	2.66	R
2	8.7	9.2	3.30	*Initial*	-	-	-	*	E
3	5.4	6.0	5.04	*Initial*	-	-	-	*	R
4	7.3	18.3	4.90	R	11	9.6	9.7	2.80	C_W
5	9.2	15.5	3.52	R	12	10.0	11.1	2.68	R
6	9.7	12.3	2.98	R	13	10.0	12.2	2.76	R
7	9.6	10.1	2.84	E	14	10.5	10.6	2.34	C_R
8	8.2	6.5	3.40	R	15	10.3	10.8	2.50	R
9	9.1	10.0	3.14	C_R	16	10.6	12.0	2.40	R

* Points not tested as they are outside the compression force limit. Φ set to 1000.

An initial simplex from a factorial design
The starting simplex is usually regular, but does not have to be so. It is quite possible to select points from a factorial or fractional factorial design, with reduction of the number of factors (taking only those that are active) and using either the basic or extended sequential simplex methods to move from the region of the factorial design to a more favourable point. However under such circumstances the steepest ascent method could equally well be used, and this should normally be preferred.

Optimization using a large initial simplex
The simplex method was first developed to look for an optimum outside, and frequently far from the initial domain. The variable step simplex method has also been used successfully to look for a maximum *inside* an experimental domain covered by the original simplex (16). Almost all steps are contractions, C_R or C_W. Shek, Ghani and Jones (18) have described the simplex optimization on 4 factors of a capsule formulation - the percentage drug substance, disintegrant, lubricant, and the total capsule weight - using this technique. Use of response surface methodology appears to be a better approach, provided that a polynomial equation can be fitted to the response surface.

4. Continuing the study

If the experimenter is confident in having achieved his objective, he will consider the study as *complete*, and do no further work on it. Otherwise after stopping the first series, the *simplex method* may be used to continue the optimization, perhaps by decreasing the simplex size or by changing to the extended simplex method.

If the experimenter requires a more certain or more precise identification of the optimum, or to know how the response or responses vary about the final point, he should *not* try to obtain a mathematical model at the end of the optimization by multi-linear regression over the points (not even limiting the points to those about the optimum). The variance inflation factors are often very high, a measure of the lack of reliability of the regression. Rather, he should carry out a *response surface study* around the supposed optimum.

The various criteria on deciding on such a design have been discussed in detail in chapter 5. However a design can also be constructed by taking the most interesting of the simplex experiments, usually those close to the optimum, and adding further experimental points, chosen so as to optimize the quality of the estimators. This is normally done according to the *D-optimal criterion*, using the exchange algorithm method of chapter 8. We also recall that a *Doehlert design* can be set up by adding experimental points to a *regular* simplex, normally one resulting from the basic simplex optimization.

V. VALIDATION AND OPTIMIZATION

Validation today is a key issue, and we may ask whether the work we do in formulating and developing a process has anything to do with industrial validation. The validation of a formulation or process, or for that matter, an apparatus or analytical method, is the demonstration (documented) that it behaves as it is intended to. It is therefore logical to conclude that if the documented development work (including that carried out using statistical design) supports the process and formulation, it must be considered a part of validation.

Validation can be considered to start with the conception of the formulation. Thus the presence of each component may be justified as a result of the experimental designs, carried out at the preformulation and formulation stages of the project, such as the screening and compatibility testing described in chapter 2. The quantitative composition is supported by the response surface and optimization studies carried out according to the methods of this and the previous chapter and, more specifically, for mixtures in chapters 9 and 10.

There may be other experiments carried out at the end of development to confirm the validation of the process or formulation. We have already examined designs for this kind of problem in chapter 2, where the use of a Plackett-Burman design for testing the ruggedness of an analytical method was demonstrated. Different aspects of these will be demonstrated in this final section. Further design methods for improving quality and ruggedness will be presented in the following chapter.

A. Validation of Clinical Trial Formulations

Registration authorities have become more demanding with respect to the formulations used for clinical trials than they were in the past. These now require validation.

In general, it is the final formulation that is used in Phase III clinical trials. If the materials are manufactured on the pilot scale, then results of the process study, of the kind described in chapter 5 and in sections IV and VI of this chapter, could be used to validate this programme, assuming the same equipment is used for the process study and for clinical trials manufacturing.

On the other hand, clinical trials for new drugs in phases I and II often involve formulations different from the final registered and marketed formulation. These are validated in a number of ways, but there are essential differences between their validation and that of the marketed formulation:

- Manufacturing conditions are unlikely to be constant. Different machines and scales may be involved.
- Raw material (drug substance) is not always consistent at this stage.
- A number of related formulations – different dosages, minor changes in formulation – are manufactured.

To validate these formulations there are the following rules, and possibilities:

- The batch records must include all results of very detailed testing. For example, the whole of the dissolution profile of solid dosage forms should be given for each unit tested, not just a single point. Studies of the characteristic final and intermediate properties of these batches should be even more detailed than for a validated production batch.
- Consistent manufacturing conditions should be employed so that the process can be shown to be reproducible.
- If changes are required they should be planned. If possible they should be part of an experimental design so that they give information for improving the process or demonstrate its robustness. EVOP (see below) could be a useful strategy here, for validating clinical trial formulations.

B. Validation of In Vitro Tests on Solid Dosage Forms

Many solid dosage forms are designed for controlled dissolution or liberation of the drug substance, controlled in the sense that its release is slow compared with the rate of systemic absorption. Dissolution testing using pharmacopoeial and other methods is an essential element of the pharmaceutical development process. There are still arguments about its relevance to the *in vivo* situation and there are still those who maintain that it cannot be anything other than a quality control tool to ensure batch to batch consistency.

However, if a quality parameter is to be controlled then the parameter measured should have some influence, direct or indirect on the performance. In the case of dissolution testing it is quite rare that the *in vitro* profile follows exactly or the function calculated as the absorbance profile from deconvolution of plasma levels in man after ingestion of the dosage form. There is, however, often some sort of direct relationship, depending on the kind of dosage form and on the dissolution test chosen.

The requirement to justify the choice of method may be satisfied by factorial experiments and especially by the so-called topographical studies: measuring the drug release profile as a function of pH (19). However, other factors, agitation, and buffer concentration are almost as important and the response surface methods of chapters 5 and 6 and also the non-classical exchange algorithm methods of chapter 8 can be used to map their effects. These methods have been employed to select conditions for dissolution testing controlled release products, by comparing the results of *in vitro* studies using factorial design (20) or response surface methodology (21) with calculated *in vivo* absorption profiles.

When a method has been selected on the basis of such studies, it can be validated for robustness, as described in chapter 2. But specifications, or limits are also required, and these limits must also be validated. If there is no close relationship between the *in vitro* dissolution profile and the *in vivo* calculated absorption profile, then it is necessary to test batches with dissolution profiles at (or possibly outside) those proposed limits in humans. Measurement of the blood plasma profiles given by the upper (more rapid dissolution) and lower (slower dissolution) batches and comparison with the intermediate target profile will show whether or not batches with dissolution profiles within those limits are equivalent or not, or alternatively show a correlation, which may be used to establish limits.

How are these upper and lower dissolution profiles chosen? One simple criterion is that of target percentage ± 10%. Another is to look at the results of the process study experimental design and to select a range according to the different profiles that were obtained. These may be some difficulty in obtaining batches with sufficiently different release rates (side batches), if the formulation and process are robust. The results of the process study experimental design may indicate methods for manufacturing them.

C. Experimental Strategy for Validation of Industrial Processes

1. The pilot scale process study and validation

We have already indicated that the stages of formulation and development of a pharmaceutical product must be considered as an integral part of its validation. This is in spite of the fact that validation of the industrial process requires the same equipment and materials as the industrial manufacturing process and the earlier development work is usually done on a smaller scale. Firstly, an adequate process study at the laboratory and pilot scales, will identify the *critical factors*, those most likely to influence the process and the quality of the finished product. Critical factors are:

- identified tentatively at the screening stage
- and/or confirmed by factor-influence studies
- and their limits determined in the RSM process study, normally at the pilot level

In spite of differences in the scale, and perhaps also in the design of the equipment used, a careful examination of the data will show which key factors in particular need to be controlled. It is easier to modify the method and scale of manufacturing where one is dealing with a properly and systematically studied process. Identification of critical factors will have shown which factors in scale-up are most likely to affect the finished product. Similar considerations apply to a change in manufacturing site. Designs used in scale-up are reviewed in chapter 7.

2. Validation of the industrial process

If the formulation is robust to variations in the process on the pilot or laboratory scale, this information supports the industrial scale validation. Consider, for example, a simple tablet formulation, with wet granulation, where the process study indicated that the process was fairly robust, but affected by the granulating liquid volume, granulation time, and lubrification time. There are two possible approaches.

The first one is to control these factors carefully, proposing allowable and reasonable limits. The object of validation is to show that the process works, when carried out under the prescribed conditions. It should give a product of the desired characteristics and should be reproducible. Most regulatory authorities ask for at least three validation batches. On this basis a suitable experimental design would be:

Validation batch no.	Granulating liquid volume	Granulation time	Blending (lubrification) time
1-3	0	0	0

in terms of the coded operating levels, where the normal level is coded as zero.

Nothing could be simpler! However, we might well want to know what happens if the granulating process is allowed to overrun by a minute, or if too much granulation liquid is added. The regulatory authorities also ask for this as part of the validation. If the process study has shown that a number of factors affect the intermediate and final products, it *may* be necessary to test the effects of these on a production scale.

3. Ruggedness testing on the validation batches

The second approach is to study the effects of the factors on an industrial scale. We may draw up a standard 2^{3-1} reduced factorial design for them (below). The 0 level in each case stands for the "normal" process condition and +1 for the presumed worst case. Using this design the effects of the excess volume and times can be assessed. This would give sufficient information for "worst case" considerations. If all experiments gave acceptable results, this would demonstrate a certain ruggedness in the process. The normal process is repeated:

Validation batch no.	Granulating liquid volume	Granulation time	Blending (lubrification) time
1	0	0	0
2	+1	0	+1
3	0	+1	+1
4	+1	+1	0
5	0	0	0

Five validation batches for a simple process, optimized and validated at the pilot scale could be considered by some to be an excessive number. An alternative approach would be to prepare two batches with the standard process. The protocol would then state that if the batch 1 were satisfactory the next batch would challenge the formulation and process using the worst case conditions *selected on the basis of the results of the process study*. If all batches are satisfactory, the process may be considered to be validated.

Validation batch no.	Granulating liquid volume	Granulation time	Blending (lubrification) time
1	0	0	0
2	0	0	0
3	+1	+1	-1

The validity of this approach rests on the existence of complete and reliable data obtained at the process stage. For a simple process it is perfectly adequate. If the worst case scenario gives an unacceptable product then further work is required.

D. Optimization of Industrial Processes by Evolutionary Operation (EVOP)

We have assumed that the process study is performed on a laboratory or on a pilot scale, and this is almost invariably the case. This is because it would normally be prohibitively expensive to realise such studies on an industrial scale. Also, the process for a pharmaceutical product is defined in the registration dossier, and it is also validated. There is not a great deal of scope for changing either the conditions or the formulation.

However robust the process and however successful the scale-up has been, there may still be some room for improvement. In addition, with time the process may "drift". Fine tuning of the process may be advantageous and also justifiable.

Evolutionary operation (EVOP) was proposed by Box and Draper to answer this problem (22). Any number of variables may be treated, but in general it is limited to the two or three *critical factors* known already, from the pilot scale process study, to have an influence on the properties or yield of the product. By very slightly altering the values of these variables in a systematic manner – they should remain within the limits already defined as acceptable – the dependence of the product on the operating conditions can be assessed, and in some cases the process can be improved. Two level factorial designs, or the simplex (1), are most commonly used.

We take the example of 3 variables, for a manufacturing or factory scale process whose performance and reproducibility is well known. The factors' effects may be investigated by a 2^{3-1} design (4 experiments) plus centre point. A cycle of the 5 experiments is performed and the effect of each variable determined as described in chapter 2. Since its standard deviation σ is already known, each measured effect may be compared with the calculated standard deviation for each effect, $\sigma/2$.

If differences are observed, but they are not statistically significant the design can be repeated a number of times, and the effects calculated using the data of all the cycles. The effects are then estimated with rather more precision, the standard deviation being $\sigma/2\sqrt{n}$, n being the number of cycles.

Should the result warrant it, a statistically significant improvement having been found after a sufficient number of cycles, the manufacturing conditions may be changed accordingly and a new set of EVOP cycles begun about the new improved set of operating conditions.

References

1. G. E. P. Box, W. G. Hunter, and J. S. Hunter, Statistics for Experimenters, J. Wiley & Sons, N. Y., 1978.
2. D. S. Montgomery, Design and Analysis of Experiments, Second Ed., J. Wiley & Sons, N. Y., 1984
3. M. F. L. Law and P. B. Deasy, Use of canonical and other analyses for the optimization of an extrusion-spheronization process for indomethacin, *Int. J. Pharm.*, **146**, 1-9 (1997).
4. K. Takayama and T. Nagai, Simultaneous optimization for several characteristics concerning percutaneous absorption and skin damage of ketoprofen hydrogels containing d-limonene, *Int. J. Pharm.*, **74**, 115-126 (1991).
5. E. Senderak, H. Bonsignore, and D. Mungan, Response surface methodology as an approach to optimization of an oral solution, *Drug. Dev. Ind. Pharm.*, **19**, 405-424 (1993).
6. X. M. Zeng, G. P. Martin, and C. Marriot, Tetrandine delivery to the lung: the optimisation of albumin microsphere preparation by central composite design, *Int. J. Pharm.*, **109**, 135-145 (1994).
7 A. D. McLeod, F. C. Lam, P. K. Gupta, and C. T. Hung, Optimized synthesis of polyglutaraldehyde nanoparticles using central composite design, *J. Pharm. Sci.*, **77**, 704-710 (1988).
8. L. Benkerrour, J. Francès, F. Quinet, and R. Phan-Tan-Luu, Process study for a granulated tablet formulation, unpublished data.
9. J-C. Guyot, L. Tête, S. Tic Tak, and A. Delacourte, Practical interest of the cohesian index for the technological formulation of tablets, *Proc. 6th Int. Conf. Pharm. Tech.* (APGI), **3**, 246-254 (1992).
10. G. Derringer and R. Suich, Simultaneous optimization of several response variables, *J. Qual. Tech.*, **12**(4), 214-219 (1980).
11. Y. Y. Huang, S. M. Wu, C. Y. Wang, and T. S. Jiang, A strategy to optimize the operation conditions in iontophoretic transdermal delivery of pilocarpine, *Drug Dev. Ind. Pharm.*, **21**, 1631-1648 (1995).
12. D. Vojnovic, P. Rupena, M. Moneghini, F. Rubessa, S. Coslovich, R. Phan-Tan-Luu, and M. Sergent, Experimental research methodology applied to wet pelletization in a high-shear mixer, *S. T. P. Pharma Sciences*, **3**, 130-135 (1993).
13. W. Spendley, G. R. Hext, and F. R. Himsworth, Sequential application of simplex designs in optimization and evolutionary operation, *Technometrics*, **4**,

441 (1962).

14. J. A. Nelder and R. Mead, A simplex method for function minimization, *Comput. J.*, **1**, 308 (1965).

15. C. Bindschaedler and R. Gurny, Optimization of different pharmaceutical formulations by the simplex method with a TI-59 calculator, *Pharm. Acta. Helv.*, **57**(9), 251-255 (1982).

16. S. L. Morgan and S. R. Deming, Simplex optimization of analytical chemical methods, *Anal. Chem.*, **46**, 1170-1181 (1974).

17. F. Mayné, Optimization techniques and pharmaceutical formulation: example of the sequential simplex in tableting, *Proc. 1st Int. Conf. Pharm. Tech.* (APGI), **5**, 65-84 (1977).

18. E. Shek, M. Ghani, and R. E. Jones, Simplex search in optimization of capsule formulation, *J. Pharm. Sci.*, **69**, 1135-1142 (1980).

19 J. P. Skelly, M. K. Yau, J. S. Elkins, L. A. Yamamoto, V. P. Shah, and W. H. Barr, *In vitro* topographical characterization as a predictor of *in vivo* controlled release quinidine gluconate bioavailability, *Drug Dev. Ind. Pharm.* **12**, 1177-1201 (1986).

20. C. Graffner, M. Sarkela, K. Gjellan, and G. Nork, Use of statistical experimental design in the further development of a discriminating in vitro release test for ethyl cellulose ER-coated spheres of remoxipride, *Eur. J. Pharm. Sci.* **4**, 73-83 (1996).

21. K. Ishii, Y. Saitou, R. Yamada, S. Itai, and M. Nemoto, Novel approach for determination of correlation between *in vivo* and *in vitro* dissolution using the optimization technique, *Chem. Pharm. Bull. Tokyo,* **44**, 1550-1555 (1996).

22. G. E. P. Box and N. R. Draper, Evolutionary Operation: a Statistical Method for Process Improvement, John Wiley & Sons, N. Y., 1969.

Further reading

• R. H. Myers and D. C. Montgomery, Response Surface Methodology, Wiley Interscience, N. Y., 1995

7

VARIABILITY AND QUALITY

Analysing and Minimizing Variation

I. THE EFFECT OF VARIABILITY ON DESIGN

Associated with each measured value of an experimental response is an *experimental error*, which is the difference between this measured value and the unknown "true value". Variation in the results of replicated experiments carried out under "identical conditions" may be ascribed to fluctuations in the experiment conditions. These fluctuations are either in controlled factors which not perfectly fixed or, in other factors, left partially or totally uncontrolled. Added to these are the errors from imprecision of the measurement method. The overall result is a dispersion of values about a mean, this variation being the experimental *repeatability*.

  Up to now we have assumed that this dispersion, characterised by the standard deviation σ, is more or less constant over the domain; in any case, sufficiently uniform to be able to neglect changes in σ. We will now question this assumption, at the same time exploring and extending the concept of experimental variation.

Part of the basis of the scientific method is that any experiment may be reproduced by another worker and/or another laboratory, provided the original protocol is rigidly adhered to. We have seen already that this is an ideal that the experimenter seeks to approach. However, a single worker – in a fixed spot, using the same apparatus or machine, the same batch of material, the same reagents – is faced with variability. Once we add the effects of geography, changed materials and equipment, and different operators, this variation is likely to be still greater. This global variation in the value of the response is known as the *reproducibility*.

Variation has many causes. There are certain factors which cannot be controlled, or which one chooses not to control, and these can have non-negligible effects on the response or on its measurement. Variation in these factors (temperature, humidity, wear of the equipment, the operator's learning curve, etc.) is not always random. It is because all of these increased variability that we have stressed the necessity of randomizing experiments whenever possible.

In pure and applied research and also in industrial development, an experiment is usually carried out in order to verify a theory, to quantify a phenomenon or to demonstrate that certain factors influence certain responses. The experimenter is often keen to understand and explain, as well as to describe the phenomenon (especially in the case of factor-influence studies described in chapter 3). We therefore try to carry out the experiment and also analyse the data so that the postulated model is determined with the greatest possible precision and the estimates affected as little as possible by the dispersion in the responses. In following two sections of this chapter we will describe techniques most commonly used for achieving this, where there is heterogeneity of the experimental variance.

The objectives of industrial production experiments are often rather different. The responses are usually characteristics of the *final* product. According to quality control principles, the product is considered acceptable provided these characteristics are within the established specifications or norms. These specifications consist of an upper and a lower limit, or alternatively they comprise a nominal value with a tolerated range on either side. This notion of *tolerance limit* where a product just in the interior is accepted and one just outside it is rejected is too abrupt. Taguchi (1) was one of the first to introduce the idea of a **continuous loss function** where any difference between the measured response and its target value leads in some way to a decrease in the product's value – to a loss in quality. The greater the difference between a product's measured characteristics and the target properties, the higher is the cost to the company producing the product – in increased cost of after-sales service, returned products, replacements, compensation, and possible desertion by customers.

Taguchi also showed that variability in a product is itself a measure of quality. A product of totally consistent quality may in fact be preferred to one whose average characteristics are better, but more dispersed. So, in an experiment the aim is no longer to eliminate the consequences of the variability from the analysis of the effects and of the mathematical model we are studying by means of weighted regression or transformation of responses. It is rather the variability of the performance of the product that must be studied and minimized.

Another consideration peculiar to industrial production is that not only must the variability of the performance characteristic be minimal at the time of manufacture, but it must stay low throughout the product's lifetime. The conditions under which the product is kept and used are likely to be very variable. The conditions of use are uncontrolled factors which must have as little effect as possible on the product's quality, that is, on the variability of its characteristics with time.

II. ANALYSIS OF RESIDUALS

A. The Sources of Residual "Error"

Predictive models, determined by multilinear regression, are tested by ANOVA. We also saw, in chapter 5, that they might also be tested by analysis of the residuals, the differences between the measured response and that calculated by the model, $y_i - \hat{y}_i$. This analysis is usually graphical, by normal or half-normal plots of the residual and by plotting the residuals against the value of each factor, against time, and against the response. Analysis of the residuals is only useful if there is an adequate number of degrees of freedom (at least 5). It should be used for example in analysing RSM designs such as central composite or Doehlert designs.

The residuals may be distributed in various different ways. First of all they may be scattered more or less symmetrically about zero. This *dispersion* can be described by a standard deviation of random experimental error. If this is (approximately) constant over the experimental region the system is *homoscedastic*, as has been assumed up to now. However the analysis of residuals may show that the standard deviation varies within the domain, and the system is *heteroscedastic*. On the other hand it may reveal systematic errors where the residuals are not distributed symmetrically about zero, but show trends which indicate model inadequacy.

In the following section we describe some of these methods and how they may show the different effects of dispersion and systematic error. Then in the remaining two sections of the chapter we will discuss methods for treating heteroscedastic systems. In the first place, we will show how their non-constant standard deviation may be taken into account in estimating models for the kind of treatment we have already described. Then we will describe the detailed study of dispersion within a domain, often employed to reduce variation of a product or process.

B. Graphical Analysis of Residuals

The use of all of these methods is recommended. Each may possibly reveal systematic errors due to inadequacy of the model or may show that the system is heteroscedastic. In the latter case a transformation of the response may sometimes be appropriate, as described in section III, or the influence of the factors on the variability of the system may be investigated, as in section IV.

1. Normal and half-normal plots

The residuals, if they are a measurement of the error, would be expected to be normally distributed. Therefore they may be expressed as a cumulative frequency plot, just as was done for the estimated coefficients in chapter 3. If the plot is a straight line, this supports the adequacy of the model. If there are important deviations this may indicate an inappropriate model, the need for a transformation, or errors in the data. The deviating points should be looked at individually.

Just as for the normal plot of the coefficients, the residuals must be *studentized* – that is, divided by the standard deviation of prediction at that point.

2. Dependence of the residual (error) on the factors

It is useful to plot the residuals, or the studentized residuals, against the values of the coded variables X_i in turn. These should be evenly distributed, with no obvious dependence on the factor. Figure 7.1 gives the example of the response of cloud point, in the case of the formulation of an oral solution (3) already discussed in chapters 3, 5, and 6. The studentized residuals are plotted in turn against the polysorbate 80 concentration (X_1), the propylene glycol concentration (X_2), and the invert sucrose medium (X_3).

The design (central composite) was for a second-order model, so this model was used for the calculation. However, analysis of variance showed that the second-order coefficients were not statistically significant, and there was (possibly) some lack of fit (significance 0.067). Graphs (a) and (c) show no trends; the points are scattered evenly. However, graph (b) of cloud point against propylene glycol concentration, shows a plot clearly cubic in type. The fit of the model may be improved greatly if an x_2^3 term is added to the first- or second-order model. In view of the lack of significance of the quadratic model, and the high variability of the replicated experiments, it is unlikely that other cubic terms would be needed (see chapter 5, section VII).

The new model could be fitted without doing further experiments, but we would advise against this. Further experiments might be selected along the X_2 axis, for example at points $x_2 = -1.68, -0.56, +0.56, +1.68$, in order to confirm the model.

There may be cases where there are no systematic differences between predicted and experimental values, but where the scatter of the residuals depends on the value of the factor. This could be because the standard deviation is dependent on the response (see section 4). Otherwise, if this scatter of residuals proves not to be response-dependent, but a function only of the level of the factor, then the methods of section IV could prove useful for reducing the variability of the formulation or process being studied.

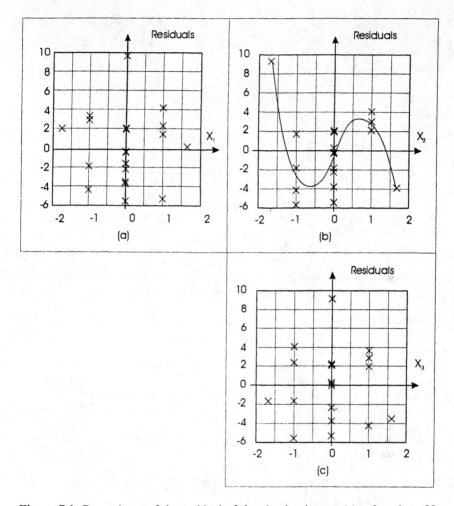

Figure 7.1 Dependence of the residual of the cloud-point on: (a) polysorbate 80, (b) propylene glycol, and (c) sucrose invert medium concentrations

3. Dependence of the residuals on combinations of 2 or more factors

Systematic dependance of the residuals may be observed for randomization restrictions, where for example the batch resulting from the first stage in the process is split into sub-batches for the remaining stages. The experiments are not independently carried out and the errors of experiments corresponding to each sub-batch in a given batch are correlated.

We take the example of a granulation experiment (see chapter 6, section IV.B, and chapter 8, section III.A). Let the granulation variables (time, amount of liquid, agitation, etc.) be represented as X_{1i}. Each batch of granulate is split into a number of sub-batches for the following stages (drying, sieving, lubrification, tableting, etc.) the variables for which are represented by X_{2j}. The random error in the granulation stage is represented by ε_1 with an expectation of zero and standard deviation σ_1, and the random error of the remaining stages is ε_2, also with expectation zero and with standard deviation σ_2. If a first-order model for the responses is postulated, it is:

$$y = \beta_0 + \Sigma\beta_{1i}x_{1i} + \Sigma\beta_{2j}x_{2j} + \varepsilon_1 + \varepsilon_2$$

The coefficients are estimated by multi-linear regression. Analysis of residuals only gives useful information if there are sufficient degrees of freedom. If the random error of the granulation step (ε_1) is to influence the residuals, there must be more granulation batches prepared than there are granulation terms in the model. In such a case, if the residuals are grouped according to the granulation batch from which the sub-batch is taken and they appear to be correlated (and if ANOVA did not show lack of fit), we might conclude that the random variation of the first stage is comparable or greater than the random error of the second stage. This is why, although the best values of the coefficients are those determined by multi-linear regression, it is not possible to use analysis of variance as described in chapter 4 to estimate the significance of the model.

The problem is a common one in industrial experiments (4). The design is known as a *split-plot* (the name has an agricultural origin). Failure to recognise this situation is likely to lead to coefficient estimates which appear to have a higher statistical significance than is really the case. In the above example, if there are no degrees of freedom for the granulation stage, the residuals will be estimates of the random errors only of the second stage in the process and the true error will be underestimated. If, on the other hand, there are sufficient degrees of freedom for the first step, it will be possible, by an appropriate analysis of variance, to estimate the standard deviations of the errors of both stages and to test for the significance of the model and the individual coefficients.

It may happen that the *dispersion* of residuals without systematic effects is observed for certain combinations of factor levels where the experiments in the design have been carried out independently. If the scatter is not dependant on the response value, then the results may usefully be analysed in order to reduce variability, as explained in section IV.

4. Dependence of the residuals on the time

When the residuals are plotted in the order in which the experiments were carried out, they should be scattered evenly about zero. However, there may be a slope demonstrating a trend in an uncontrolled factor that is affecting the results of the experiment. Provided the plan has been randomized, this will affect the estimations of error and significance, but will not greatly affect predictions. [See chapter 3, section IV (*time-trend*)].

Secondly, there may be a trend to increased, or decreased scatter, indicating a change of variability with time. Decreasing residuals in a long series of runs may point to the operator's increasing expertise. Increasing residuals are more worrisome!

5. Dependence of the error on the response value

The experimenter should test for this possibility, especially where there is wide variation of a response over the experimental domain. If the response range is relatively narrow, changing by less than a factor of 3 or 4, it is unlikely that the dependence of the variance on the response will have much effect on the significance of the factor study or the predicted response surfaces. If the response varies by a more than an order of magnitude, it is often found that transformation improves the analysis.

This situation, where the variance of a response function depends on the value of the response, should be distinguished from the phenomenon where the variance changes over the experimental region as a function of the *underlying process or formulation variables* (see II.3 above). For this, see section IV of this chapter.

III. TREATMENT OF DATA WITH NON-CONSTANT VARIANCE

A. Weighted Multilinear Regression

In determining a mathematical model, whether by linear combinations or by multi-linear regression, we have assumed the standard deviation of random experimental error to be (approximately) constant (*homoscedastic*) over the experimental region. Mathematical models were fitted to the data and their statistical significance or that of their coefficients was calculated on the basis of this constant experimental variance. Now the standard deviation *is* often approximately constant. All experiments may then be assumed equally reliable and so their usefulness depends solely on their positions within the domain.

If, however, the system is *heteroscedastic* and the standard deviation does indeed vary within the domain so that certain experimental conditions are known to give less precise results, this must be taken into account when calculating the model coefficients. One means of doing this is by **weighting**. Each experiment i is assigned a weight w_i, inversely proportional to the variance of the response at that point. Equation 4.5, for least squares estimation of the model coefficients, may thus be rewritten as:

$$\mathbf{B} = (\mathbf{X'WX})^{-1}\mathbf{X'WY}$$

where $\mathbf{W}$ is the weights matrix:

$$W = \begin{bmatrix} w_1 & 0 & 0 & . & 0 \\ 0 & w_2 & 0 & . & 0 \\ . & . & . & . & . \\ 0 & 0 & 0 & . & w_N \end{bmatrix}$$

The weights must be known or assumed. One method, expensive in time and resources, is to repeat each experiment several times and to estimate the variance at each point. If the variance cannot be predicted in any way, this is the only possible method.

Analysis of the residuals $(y_i - \hat{y}_i)$ as a function of the *independent variables* X_i may reveal a correlation between the residues and one or more of the variables. This could indicate an increase in the variance of the experimental results as the value of the factor increases (which could be the case, for example, if not all levels of the factor can be controlled with the same precision). One might then suggest a weighting w_i which decreases as the value of the natural variable U_i increases. An example might be $w_i = U_i^{-1}$, provided all U_i are greater than zero.

Other possibilities are described in the literature, which deal with cases where the variance depends on the response, or where the errors are correlated, or where the errors depend on the order in which the experiments are carried out. For the most part these are outside the scope of this book. However the use of *transformations* of response values to correct for a non-constant variance, as well as non-normality, is widespread, ever since it was proposed by Box and Cox (2), and this is the subject of the following two sections.

B. Variance Depending on the Experimental Response

1. Examples of variance depending on the response values

There are certain situations where for physical or physico-chemical reasons, or because of the methodology, we might expect the variance to depend on the response in specific ways.

(a) The relative standard deviation may be constant. For example in a solubility experiment where the measured solubility varied from 0.05 to 10 mg/mL, the error in the solubility, due perhaps to analytical error and variation in sample preparation, was approximately constant at 8%. The standard deviation is thus proportional to the response.

(b) In the measurement of ultraviolet absorbance A, at least part of the error results from a constant electrical noise σ_T in the transmittance measurement T. Since $A = \log_{10} T$ the standard deviation of the absorbance σ_A is given by:

$$\sigma_A = \left(\frac{dA}{dT}\right)\sigma_T = \frac{\sigma_T}{2.303\,T} = \frac{10^{-A}}{230.3}\sigma_T$$

(c) The Poisson distribution applies where there is a constant, relatively small possibility of an event occurring. A possible example is that of the number of broken tablets in a friability test. For this distribution, the variance is equal to the mean, so the standard deviation varies with the square root of the response.

(d) A simplification which is sometimes useful is to assume that the standard deviation varies according to a power α of the mean value of the response at a given point:

$$\sigma = y^{\alpha}$$

The response is usually measured directly on the most physically accessible quantity. Flowability, for example, is normally measured by recording the time for a certain quantity of powder to flow. It could equally well be expressed as a flow rate. The variance dependance will be different according to how the data are expressed.

It is quite rare for there to be enough data to show deviations from normality. Least squares regression is usually adequate and cases requiring "robust" regression methods are infrequent.

2. How to recognise the need for a transformation

Transformation of the response data is often advantageous where there is variation in the response of an order of magnitude or more. A transformation will not always be necessary, but should be tested for.

(a) Plotting either absolute or studentized values of residuals against the corresponding calculated values may show a relationship between them, usually increasing values of the residuals with increasing values of the calculated response. However, such trends are not always visible, even when a transformation is required because of error in the prediction at low-response values. Analysis by the method of Box and Cox, given at the end of this section, is preferred.

(b) There are theoretical or physico-chemical arguments for such a transformation.

3. How to choose the appropriate transformation

A transformation of the data giving an improved fit may sometimes be obtained by trial and error, but one or both of the following approaches is recommended. One may make a *theoretical choice* of transformation and test if it gives improved results – a regression that is more significant in analysis of variance, reduction in the number of outliers, residuals normally distributed. Alternatively the *Box-Cox transformation* method may be used.

4. Principle of the Box-Cox transformation

Box and Cox (2, 5) showed that, provided the experimental errors were independent, transformation of the initial responses might correct for non-normality and non-constant experimental variance. The function

$$y^{(\lambda)} = \frac{y^\lambda - 1}{\lambda \dot{y}^{\lambda-1}}$$

is calculated for each experimental point y for values of λ from -2.5 to 2.5. $\dot{y}$ is the geometric mean of y. $\lambda = 0$ is a special case where:

$$y^{(0)} = \dot{y} \log_e y$$

The $y^{(\lambda)}$ are analysed according the model equation and residual sums of squares. S_λ are then calculated for each value of λ and plotted against λ. $\lambda \dot{y}^{\lambda-1}$ is a normalising factor in the above equation, which allows for the change of scale on transformation, so that the sums of squares may be compared. The best value of λ is that for which the sum of squares is a minimum and y may be transformed accordingly to y^λ. Note that certain values of λ give rise to particular transformations (shown in table 7.1).

The value of λ for the minimum sum of squares is not necessarily exactly a multiple of ½. But there will normally be a choice of transformations that are nearly as good, at least not statistically worse. An approximate 95% confidence interval may also be calculated (1, 2), and any transform within the confidence interval may be selected.

The Box-Cox transformation method is valid only for all $y > 0$. If some $y \leq 0$, a small constant term may be added to all the data to give positive values. In our experience this is rarely useful when there are negative values, but may be contemplated when a few responses are equal to zero. A value equal to the smallest detectable response could be added to each response before transforming.

Table 7.1 Transformations Corresponding to Particular Values of λ

λ	Transformation $(y \rightarrow y^\lambda)$	Example or type
1	none	data with constant standard deviation
0.5	square root	Poisson distribution
0	logarithm	constant relative standard deviation
-0.5	reciprocal square root	
-1	inverse	rates

Possible advantages of a successful transformation are:

• The sensitivity of the experiment is improved, with analysis of variance

showing the model to be more significant.
- Apparent interaction terms in the original transform are no longer significant if they were the result of an inappropriate scale.
- The residuals are often normalised.

C. Example of a Box-Cox Transformation

1. Description of the problem, experimental domain, model and design

To illustrate the method we examine the data of Wehrlé, Palmieri, and Stamm (6), who report the optimized production of theophylline pellets by a simple one-step process in a high speed Stephan granulator. The formulation consisted of 20% theophylline drug substance, 30% lactose, and 50% microcrystalline cellulose. It was granulated with a 15% hydroalcoholic solution of hydroxypropyl-methylcellulose.

The process was considered to depend on the amount of granulating liquid added, and on the kneading or granulation time. The responses were the yield of pellets between 200 and 630 µm, the mean diameter, and the flowability, their dependences on these factors were investigated. The speed of the granulator was also varied, but in this example we take only those results obtained at a speed of 1000 rpm and the only response considered is that of yield.

The limits for added liquid were between 190 and 210 mL and the kneading time was between 0 and 20 minutes. The domain was therefore not spherical but cubic. Having carried out the 5 experiments of a 2^2 factorial design with a centre point, the authors chose to carry out 4 further experiments to obtain the 9 experiments of a full factorial design at 3 levels, 3^2 (chapter 5, section VI.A). The design and the experimental results are given in table 7.2.

Table 7.2 Full Factorial Design 3^2 for Granulation [data taken from reference (6), by courtesy of Marcel Dekker, Inc.]

No.	X_1	X_2	liquid (mL)	time (min)	yield (%)
1	-1	-1	190	0	48
2	0	-1	200	0	74
3	+1	-1	210	0	82
4	-1	0	190	10	84
5	0	0	200	10	86
6	+1	0	210	10	56
7	-1	+1	190	20	84
8	0	+1	200	20	50
9	+1	+1	210	20	21

2. Analysis of transformations

Figure 7.2 shows the results of the transformation according to the method of Box and Cox. The minimum is found at $\lambda = 0$, corresponding to a logarithmic transformation. (The sharpness of this minimum is however highly unusual.)

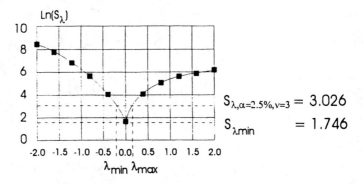

$$S_{\lambda,\alpha=2.5\%,v=3} = 3.026$$

$$S_{\lambda min} = 1.746$$

Figure 7.2 Box-Cox transformation of the data of table 7.2.

3. Analysis of results

The analysis of the variance of the regression is shown on table 7.3. Table 7.4 gives the estimates of the coefficients of the model:

$$\log_{10} y = \beta_0 + \beta_1 x_1 + \beta_2 x_2 + \beta_{11} x_1^2 + \beta_{22} x_2^2 + \beta_{12} x_1 x_2 + \varepsilon$$

The yield, transformed back to the original variables, is given as a contour plot in figure 7.3 (solid lines). The process may be optimized by reference to the diagram, to give maximum yield.

Table 7.3 ANOVA of the Regression on the Transformed Response

	Degrees of freedom	Sum of squares	Mean square	F	Sign.
Total	8	0.31826			
Regression	5	0.31797	0.06359	639.66	***
Residual	3	0.00030	0.00010		

Table 7.4 Coefficients of the Model for $\log_{10}$(yield)

Coefficient		Sign.		Coefficient		Sign.
$b_0 =$	1.924	***	$b_{11} =$		-0.081	***
$b_1 =$	-0.091	***	$b_{22} =$		-0.134	***
$b_2 =$	-0.086	***	$b_{12} =$		-0.209	***

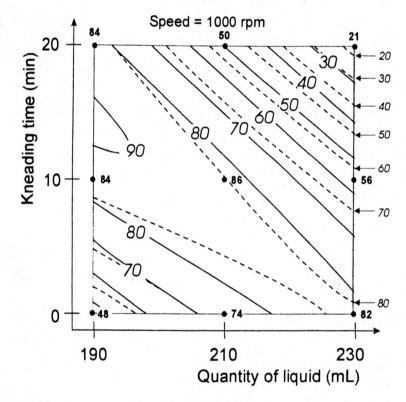

Figure 7.3 Response surfaces for yield, calculated by regression on transformed data (solid lines) and untransformed data (dotted lines) of reference (6).

4. Comparison with untransformed results

Figure 7.3 shows that the model using the logarithmic transformation of the yield is better than that using the regression on the untransformed response. The numbers in bold type are the experimental values obtained, the dotted lines represent the response surface calculated for the model $y = f(x_i)$ and the solid contour lines show the response surface calculated for the model $\log_{10} y = g(x_i)$.

D. Non-Power Transformations

Apart from the power transformations, usually selected after a Box-Cox analysis, certain other transformations may sometimes be useful.

1. The *logit* transform

This transformation may be applied to response data which fall within a finite range. Common examples are percentages between 0 and 100, size data such as a yield, or the percentage dissolved at a given time. Consider the dissolution testing of a number of formulations. The formulations may be divided, approximately, into 3 groups with slow, intermediate, and fast dissolution. Those slow to dissolve might have percentages dissolved clustered between 0 and 20%. For the second group of intermediate dissolution rate, the percentage dissolved would vary from about 20% to 80%. If there were just those 2 groups, a power transformation like the one we have discussed might be adequate. A reciprocal transformation, for example, would transform them to rates. But if there were a third group with rapid dissolution clustered near 100%, a different kind of transformation could be useful, one which spreads out the values near the boundaries:

$$\text{logit } (y) = \log_e \left(\frac{y - y_0}{y_\infty - y} \right)$$

where y is the response and y_0 is the lower value for the response (for example 0%) and y_∞ is the maximum value (for example 100%). This is known as the logit transformation.

2. Binomial data and the arcsine transformation

Here the response is one of a fraction of results that "pass" or "fail". If the fraction that passes is p then the following transformation to a new variable P may be made (as suggested by Fisher) before analysis with the chosen model:

$$\sin P = \sqrt{p}$$

$$P = \arcsin(p^{\frac{1}{2}})$$

Obviously, P must be between 0 and 1. Unlike the logit transformation, P can also take those limiting values 0 and 1. Possible examples are (as for the previous example) the fraction of drug dissolved or liberated at a given time in a dissolution test or the fraction with a particle size below a given limit.

IV. VARIABILITY AND QUALITY

A. Introduction

1. Quality control and quality assurance

The term *quality* applied to a product includes all the properties and characteristics by which it satisfies the needs of those who use it. This concept is not new to pharmaceutical scientists. They are used to specifications, whether these are in pharmacopoeias or otherwise. Specifications are fixed and these are designed to ensure the quality. It is verified after manufacture that the product enters within these specifications, and so, historically, the concept of *quality* is tied up with the concept of *control*. The elimination of products that do not conform to commercial specifications will satisfy external clients that the quality criteria are being met, but this is at a cost. *In no way does control directly improve the quality of the product that is actually being manufactured.*

It is necessary to introduce structures, procedures and methods which ensure quality – that is *quality assurance* (7). Quality control is only one aspect of this, although in general it is a necessary verification.

The cost of quality may be broken down into:

- the cost of conformity, which is the cost of prevention plus the cost of control, and
- the cost of non-conformity, which is the cost of internal plus external failure.

Internal failures must be thrown out or recycled. These should all be detected in products where a batch can be sampled and tested. External failures are much more serious and costly, although it is not always easy to quantify the cost. In the case of a "normal" product, the customer will return it. If he is really unhappy he will not buy it again. If a batch of medicine is faulty, the F.D.A. (to name but one national authority) may recall it. If the agency is really unhappy about the company's structures, procedures, and methods, it may well take stronger action.

There is therefore an advantage in making sure that no defective product leaves the factory and for this there are two main strategies – increased control, so that all defective products are eliminated, and improved quality, so that there are no defective products.

If we reduce the number of actual failures, we will automatically reduce the number of both internal and external failures. This modern approach to quality, spearheaded by Japanese workers (1) but now formally applied over a much wider area (8), consists of prevention of failure rather than its detection. The idea is a simple one – to avoid refusing a product it is enough at the production stage to control the process in such a way that the manufactured product is as close as possible to the desired product.

2. The loss function

When we manufacture a product, be it chemical, mechanical, electrical, or pharmaceutical, we look first of all for properties of the actual product that are within the specifications that were previously fixed. They were either fixed by some external authority (for example a pharmacopoeia) or internally, at levels considered necessary by the company, and then registered. In any case, they are not negotiable, the only possible exception being the distinction between official specifications and internal company specifications, which may be rather narrower.

Each measured property is known as a *performance characteristic*. If we can give this property a number, it can be expressed or "graded" as y. The ideal value is the *target value*, represented by τ. So let y be the performance characteristic's measured value, η its expectation $E(y)$. The ideal result is:

$$E(y) = \eta = \tau$$

where the expected (mean) value of the property that is being measured is also the target value. This alone is not enough, as individual items can still fall outside specifications. The performance characteristic y is certain to vary, with a variance σ^2 representing the variability:

- of the measurement method,
- due to the manufacturing process,
- due to the manufacturing environment.

The traditional approach to quality is dominated by quality control. A permissible range for the performance characteristic y about the target value is defined. If y is within specification the product is accepted, but if y is outside specification, the product is refused (figure 7.4a). The weakness of this approach is the implication that even when the performance characteristic of a product is close to the specification limit, this product is still of equivalent quality to one where y is very close to the target value. In spite of this it appears evident that the closer y is to τ the "better" is the product.

The Japanese approach takes this into account. Taguchi (1) states that any product whose performance characteristics are different from the target values, suffers a loss in quality. He quantifies this by the following loss function $L(y)$:

$$L(y) = K(y-\tau)^2$$

where K is a constant, see figure 7.4b. The objective of Taguchi and his successors has been to find ways of minimising this loss in quality.

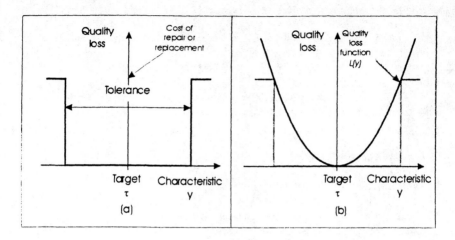

Figure 7.4 Quality loss functions: (a) classical and (b) as suggested by Taguchi.

3. Variability

In research, development, and above all in production, increased variability of a product implies reduced quality. Due to this fact a product's variability is itself a response. Therefore, the problem of choosing an optimum formulation, or conditions for manufacturing it, is not only one of obtaining the best values of the responses (whether maximum, minimum or target values), but also that of finding conditions where those characteristics vary as little as possible.

For this purpose, Taguchi classified factors influencing the responses. We will consider the two main classes:

- *control* factors, those factors which may normally be controlled and whose levels may be fixed,
- *noise* factors, which are difficult, impossible, or very expensive to control. These may be manufacturing factors (that depend on machines, differences in speed, non-homogeneous temperature, etc.), to which may be added external variations in their conditions of use (over which the manufacturer has no control). We have already seen the effects of such factors in discussing blocking and block factors.

Taguchi's approach consists of identifying the effects of all factors on both the response and on its variability by means of experimental design. Then, since the noise factors cannot be controlled (except possibly during a particular experiment), he would try to minimize their effects. Thus, diminishing the variability of the response is equivalent to removing the effects of uncontrolled factors on the response. This is a considerable advance on the traditional approach, which is to try to *control the variation* in these noise factors.

    To do this it is absolutely necessary to determine the interactions between noise and control factors. Thus, the levels of certain control factors will be fixed at levels that minimize the effect of noise factors with which they interact, whereas the other control factors will be used to adjust the mean response to its nominal value.

    There are three main types of experimental design approaches to this. All tend to require a large number of experiments.

B. Experimental Strategies

1. Random variation in the noise factors

The first method is to construct a classical experimental design in the controlled factors, and to repeat it a large number of times, hoping that there is enough random variation in the uncontrolled noise factors to cover the domain. Mean and variance may be determined for each experimental point and the variance may then be considered as a response and accordingly minimized.

    The disadvantages of this approach are twofold. It requires a very large number of experiments. Also, one is never sure that the noise (non-controlled) factors have varied enough over the period of the carrying out of the experimental design, so that they are representative of the variation to be expected over months or years, under manufacturing conditions. This being said, it is the only possible method when it is technically impossible to control the noise factors.

2. Taguchi's matrix products

Rather than rely on chance variation of the noise factors, Taguchi (1) proposes that at each point of an experimental design set up to study the control factors the noise factors are also allowed to vary following an experimental design. Suppose, for example, there are two control factors F_1 and F_2 and two noise factors, F_3 and F_4. For each of the noise factors choose two real but extreme levels. One may then construct a factorial design 2^2 with the control factors and repeat each of the 4 experiments using each possible combination of levels of the noise factors, F_3, F_4. We thus obtain the design of table 7.5.

    The design comprising the control factors, F_1, F_2, on the left hand side of table 7.5 is known as the *inner array* and the design formed by the noise factors, F_3, F_4, is the *outer array*. If there are N_1 experiments in the inner array and N_2 in the outer array, then there are $N = N_1 \times N_2$ experiments in all.

    This may quickly result in a prohibitively large number of runs. Because of this, Taguchi proposes using only screening designs. Although their R-efficiency approaches 100%, they do not allow interactions to be calculated. Taguchi generally neglects interactions, but this may be a source of considerable error.

Table 7.5 Matrix Product of Two Factorial Designs

Inner array		Outer	array			
F_1	F_2	-1	+1	-1	+1	F_3
		-1	-1	+1	+1	F_4
-1	-1	y_1	y'_1	y''_1	y'''_1	
+1	-1	y_2	y'_2	y''_2	y'''_2	
-1	+1	y_3	y'_3	y''_3	y'''_3	
+1	+1	y_4	y'_4	y''_4	y'''_4	

When there is a considerable number of noise factors to be studied, certain authors suggest studying them by groups, as in group screening, allowing them to vary in pairs or triplets rather than individually.

3. Use of classical designs

Before eliminating the possible effect of noise factors on the variability, it may be interesting to study these effects. One may therefore include these factors in the design, just as the controlled factors are included (7, 8, 9, 10). This involves two kinds of design.

The first are designs for factor influence studies (chapter 3). They may be ones set up for special models, where interactions between the noise factors are not included, but interactions between noise factors and controlled factors are studied in detail. However even when the design has not been set up for the express purpose of studying variability, Box and Meyer (11) and Montgomery (12) have shown that it is possible to detect factors affecting the variability of the response. A recent example of the use of this technique in studying a fluid-bed granluation process has recently been described (13). This used a 2^{5-1} resolution V design.

Other authors use RSM (chapter 5) with graphical methods to reveal favourable manufacturing conditions. This is illustrated below.

C. Choice of response

What variables do we use to describe quality? Taguchi (1) takes into account:

(a) the difference between the measured response and its target value,
(b) the variability of the measured response.

The response y has a target value τ which is neither zero nor infinity but equal to a nominal value τ_0 – "*nominal is best*" or "target is best". We need to obtain a performance characteristic as close to τ_0 as possible, at the same time reducing the variation. For n experiments at a given setting of the control factors we obtain responses y_1, y_2,... y_i,..., y_n. In this case, Taguchi recommends maximizing the function:

$$Z = S/N = +10 \log_{10} \left(\frac{\bar{y}}{s} \right)^2 \tag{7.1}$$

where $\quad \bar{y} = \frac{1}{n} \sum_{i=1}^{n} y_i \qquad s^2 = \left(\frac{1}{n-1} \sum_{i=1}^{n} (y_i - \bar{y})^2 \right)$

by minimizing its variability s^2.

There are other possibilities, according to whether the target value for y is as small as possible, or as large as possible. If the desired value is zero, the objective, $\tau = 0$, is *"smaller is better"*. The performance characteristic, y, is positive and the loss function increases with y:

$$L(y) = Ky^2$$

It is suggested that the function:

$$Z = S/N = -10 \log_{10} \left(\frac{1}{n} \sum_{i=1}^{n} y_i^2 \right) \tag{7.2}$$

should be maximized, the sum of squares of the y_i being minimized. An example might be the presence of an impurity in a synthesized starting material or finished product.

We look next at the similar case, where the target value is as large as possible ("infinity") – *"larger is better"*. Maximizing y is equivalent to minimizing y^{-1}. Taguchi proposes maximizing:

$$Z = S/N = -10 \log_{10} \left[\frac{1}{n} \sum_{i=1}^{n} y_i^{-2} \right] \tag{7.3}$$

Other authors prefer using the initial response (with possible transformation, as we saw in chapter 2), the study of the dispersion being carried out on $\log(s)$.

D. Example: Granulation in a High-Speed Mixer

1. Problem and experimental domain

Wehrlé, Palmieri, and Stamm have demonstrated this method, which they used for optimizing the production of theophylline pellets in a one-step process in a high speed granulator (6). The process was considered to depend on the amount of water added, between 190 and 210 mL, and the kneading time, between 0 and 20 minutes.

These are the *control factors* and the domain is cubic shaped.

We want to know the variability of the responses as well as the shapes of their response surfaces. Variability is very difficult to measure experimentally. In order to estimate it by replicate measurements a very large number of experiments would be required. Here it was thought that variability might be due to small differences in the speed of the impeller blade of the mixer granulator. The blade speed was thus allowed to vary by a small amount about its normal value of 1100 rpm. This is therefore a *noise factor*. The two levels for the speed, 1000 and 1200 rpm, are believed to cover the range of random variation of the speed.

2. Experimental design, plan, and results

The authors decided to use Taguchi's strategy, presented in section IV.B.2. The design used was a product matrix between an *inner array*, which is a complete 3^2 factorial design, and an *outer array*, which is a 2^1 design. The dependence of the responses, which were the yield of pellets between 200 and 630 μm, and also the flowability on the control factors was studied by the inner array. The 18 experiments and their results are given in tables 7.2 and 7.6. Note that for the design in table 7.2, the level of the coded noise factor X_3 is -1, throughout.

Table 7.6 Additional Results at 1200 rpm Mixer Speed [data taken from reference (6) by courtesy of Marcel Dekker, Inc.]

No.	X_1	X_2	X_3	liquid (mL)	time (min)	speed (rpm)	yield (%)
1	-1	-1	+1	190	0	1200	66
2	0	-1	+1	210	0	1200	83
3	+1	-1	+1	230	0	1200	85
4	-1	0	+1	190	10	1200	76
5	0	0	+1	210	10	1200	88
6	+1	0	+1	230	10	1200	36
7	-1	+1	+1	190	20	1200	79
8	0	+1	+1	210	20	1200	75
9	+1	+1	+1	230	20	1200	53

3. Global analysis of the overall results

Although the design was constructed according to Taguchi's method, it is in fact a full factorial $2^1 3^2$ design, which may be analysed in the usual way. The authors of the original article showed the necessity of introducing quadratic terms in X_1 and X_2. The postulated model therefore contains the 6 terms of the quadratic model, the single noise factor term, in X_3, and 5 cross-terms between the noise factor and the remaining terms in X_1 and X_2:

$$y = \beta_0 + \beta_1 x_1 + \beta_2 x_2 + \beta_3 x_3 + \beta_{11} x_1^2 + \beta_{22} x_2^2 + \beta_{12} x_1 x_2 + \beta_{13} x_1 x_3$$
$$+ \beta_{23} x_2 x_3 + \beta_{113} x_1^2 x_3 + \beta_{223} x_2^2 x_3 + \beta_{123} x_1 x_2 x_3 + \varepsilon \qquad (7.4)$$

A similar calculation to that carried out in section III.C shows that a Box-Cox (logarithmic) transformation gives an improved fit, but the improvement is not statistically significant. However it was decided to use this transformation because of the much improved result demonstrated for the half plan at 1000 rpm, above, and the fact that the significance of the model with transformation of the data is 2.5%, and without transformation it is 6.7%.

Table 7.7 gives the analysis of variance of the regression in the two cases and estimates of the coefficients (logarithmic tranformation) are given in table 7.8.

Table 7.7 ANOVA of the Regression

	Degrees of freedom	Sum of squares	Mean square	F	Sign.
(a) Untransformed response					
Total	17	6546			
Regression	11	5673	515.7	3.5	6.7%
Residual	6	873	145.4		
(b) Logarithmic transformation					
Total	17	0.4595			
Regression	11	0.4174	0.0374	5.4	2.5%
Residual	6	0.0421	0.0070		

Table 7.8 Coefficients of the Model for $\log_{10}$(yield)

Coefficient		Sign.	Coefficient		Sign.
$b_0 =$	78.9	***	$b_{12} =$	-17.7	**
$b_1 =$	-8.7	*	$b_{13} =$	0.8	81.3%
$b_2 =$	-6.3	11.7	$b_{23} =$	1.8	62.1%
$b_3 =$	-1.4	82.1	$b_{123} =$	6.5	17.6%
$b_{11} =$	-11.8	9.6	$b_{113} =$	-4.3	50.4%
$b_{22} =$	-4.3	50.4	$b_{223} =$	11.2	11.2%

The rotation speed has apparently little effect on the response and can only give rise to slight variations. There may, however, be some effect on the interaction time b_{12} and the curvature b_{22}. Although the effects are small, we continue the treatment to demonstrate how these methods may be used.

4. Graphical analysis

Interesting information can be shown in different ways. Firstly, the above equation allows response surfaces to be traced, for the yield, on the planes $x_3 = -1$ (speed = 1000 rpm) and $x_3 = +1$ (speed = 1200 rpm). These curves are super-imposed in figure 7.5. The shaded zones correspond to yields greater than 70% (light shading) and 80% (heavier shading) for the two extremes. Note that the two response surfaces are identical to those that would have been obtained if the two 3^2 designs had been treated separately by a second-order model.

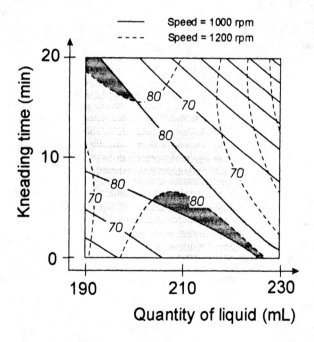

Figure 7.5 Response surfaces for different mixing speeds, showing region with over 80% yield (plotted using the coefficients of table 7.8).

This analysis on its own does not allow us to minimize the effects of variations in the speed. The two curves are shown in perspective in figure 7.6. It may be seen that the curvature on the X_2 axis (kneading time) is modified. The intersection of the two surfaces includes all the conditions where the calculated response is the same for the two rotation speeds, and it is on this intersection of the two curves where the variability due to fluctuations in the speed is minimized. Its projection onto figure 7.5 will allow any target value to be selected, minimizing the variability at the same time. It is evident that the *maximum* yield would be chosen here; nonetheless the method is valid for a response where the optimum is a target, not a maximum or minimum.

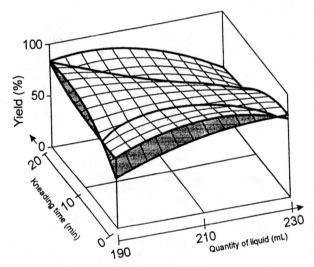

Figure 7.6 Response surfaces for different mixing speeds, showing intersection of the two surfaces (plotted using the coefficients of table 7.8).

Another graphical method is to calculate the difference between the results at each speed, Δy_i, for each experiment of the inner array. Δy is treated as a new response, and analysed by multilinear regression, according to the second-order model:

$$\Delta y = \beta_0 + \beta_1 x_1 + \beta_2 x_2 + \beta_{11} x_1^2 + \beta_{22} x_2^2 + \beta_{12} x_1 x_2 + \varepsilon \qquad (7.5)$$

The values of the coefficients, which are not highly significant, are given in table 7.9, and the response surface of $\Delta \hat{y}$ is plotted in figure 7.7. As before, we may conclude that the variation in the yield depends only on the curvature on the X_2 axis and the interaction between the two variables.

Table 7.9 Coefficients for the Model: $\Delta \hat{y} = f(x_1, x_2)$

Coefficient		Sign.		Coefficient		Sign.
b_0	-2.9	71.5 %		b_{11}	-8.7	29.9 %
b_1	1.7	70.2 %		b_{22}	22.2	*
b_2	3.6	42.8 %		b_{12}	13.0	7.5 %

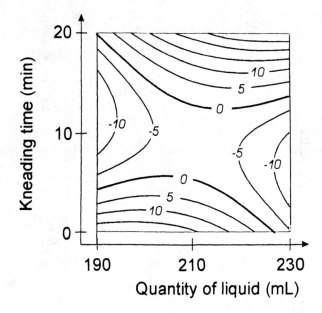

Figure 7.7   Response surface for $\Delta \hat{y}$ (difference in yield at different speeds).

5. Analysis using Taguchi's method

The criterion for the yield is "larger is better" (equation 7.3). The performance characteristic Z_Y:

$$Z_Y = -10 \, \log_{10} \left[\frac{1}{n} \sum_{i=1}^{n} y_i^{-2} \right]$$

is therefore maximized. The second-order model was postulated, and so the coefficients of the model, where S/N replaces Δy in equation 7.5, are those given in table 7.10 and the contour lines in figure 7.8. The area of maximum performance follows quite closely the area of maximum yield.

For the flowability, where the response is the flow time, the criterion is "smaller is better" (equation 7.2). The performance criterion Z_F is maximized:

$$Z_F = -10 \ \log_{10} \left[\frac{1}{n} \sum_{i=1}^{n} y_i^2 \right]$$

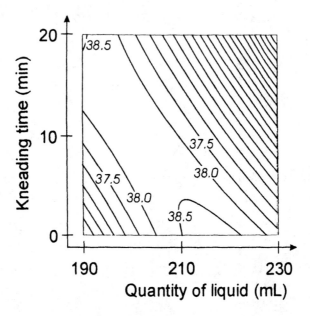

Figure 7.8 Response surface for Taguchi performance characteristic Z_Y, calculated from the coefficients of table 7.10.

Table 7.10 Coefficients for the Model: S/N(yield) = $f(x_1, x_2)$

	Coefficient	Sign.		Coefficient	Sign.
b_0	37.95	***	b_{11}	-2.19	7.0 %
b_1	-1.86	*	b_{22}	-0.90	34.4 %
b_2	-1.44	5.1 %	b_{12}	-3.26	**

6. Conclusions

As we have just shown, the various approaches do not lead to exactly the same conclusions. This being said, the predicted optimum conditions are quite close, whether it is the characteristic response of the product (the yield), the absence of variability in the yield, or the *S/N* ratio, that is treated. We superimpose the different results in figure 7.9:

- The shaded regions show maximum yields for the two mixer speeds.
- The lightly shaded lines represent the points where the yield depends least on the variations in the speed.
- The full lines are the optimum contour lines of the *S/N* ratio for the yield.
- The two circles represent regions where the different objectives may best be achieved.

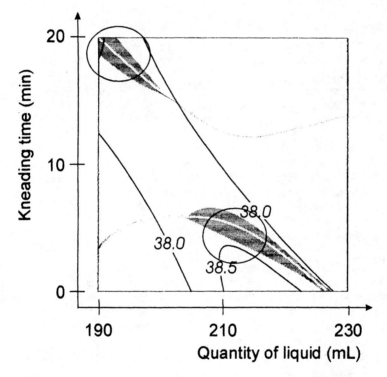

Figure 7.9 Robust granulation conditions predicted by the different methods.

E. Scaling-up and Other Applications

1. Identifying possible noise and control factors

Many pharmaceutical processes are well controlled and repeatable. However, properties may change for no apparent reason or sometimes a necessary modification of the manufacturing method may provoke an apparent instability. Dissolution profiles of controlled release formulations may cause problems, or the variation may be in the stability of the dosage form. Control charts may show the variation to be part of a constant trend or it may be random fluctuation.

The first stage is to examine the existing data, screening studies, factor influence studies, optimization, and scale-up (see below). These may give a clue as to what factor is causing the variation. They should also indicate what factors may be varied in order to change the properties and perhaps improve the robustness of the product or process.

After that, factors difficult to control are noted. Possible ones (according to the actual problem) might be:

- different batches of drug substance,
- the batch of excipient,
- the ambient temperature,
- the ambient relative humidity,
- the machine used,
- the operator,
- the exact granulation time or speed,
- the rate at which liquid is added.

Some of these, like the granulation time, may also be control factors. When the noise and control factors have been chosen, a design can be set up in the control factors. This should be based on the results already obtained in a process study, if available. It is then multiplied by a design in the noise variables, generally factorial or fractional factorial.

2. Scaling-up of pharmaceutical processes

Scale-up and technology transfer
A general problem and source of variation in pharmaceutical development and production is that of scaling-up and of process transfer. Formulations and processes are normally first developed on a small scale because of the need to test a large number of formulations and the carrying out of a large number of experiments for optimization of the processing factors, as described in chapter 6. The problems are that a large mass of material may well behave differently from where there are only a few kilograms and the production scale equipment is likely to have changed characteristics from that used at the pilot scale.

Optimum levels of the process variables for a pharmaceutical form manufactured using one piece of equipment will not be the same as for another. The correspondence between the process variables needs to be established. This is obviously easier if the equipment used on the laboratory, or pilot scale, is similar to the production scale equipment and the development scientist will try to be equipped with material similar to that which is used in the factory.

A related problem is one of process transfer, where a process in production at one factory is transferred to another, using different equipment. The process which has been optimized for one set of equipment has to be modified or changed completely. This is especially frequent for multinational companies, extending or transferring production across international borders. The modifications may sometimes be very considerable, such as a pharmaceutical form originally developed for a one-step granulation process where the granules are dried and lubricated within the same mixer used for the initial granulation is transferred to a multi-step process, with granulation, sieving, drying, and lubrification in different equipment. Or the transfer may be in the opposite sense.

Scale-up and quality
Some of the problems to be tackled in scale-up are very similar to those of variability. There are in fact two possible approaches, both using rather similar designs. Most in tune with the philosophy of design for quality is the approach of considering the scale of manufacturing or type of apparatus as a *noise variable* and optimizing the process or formulation so that the resulting dosage form should be not only optimum, but as robust as possible to the apparatus used and the scale. The methods described in sections C and D may be used.

Another approach is to treat the manufacturing scale as a normal qualitative variable, and optimize the process and/or formulation at each level. Alternatively it might also be treated as a *quantitative discrete variable*, as the volume of the apparatus, or possibly better, its logarithm. Literature examples of the use of experimental design in scale-up take this approach (14, 15, 16).

Designs and models
The designs are similar to those involving control and noise variables. A design of the appropriate order for the process variables (c.f., control variables), is multiplied by a full factorial design for the scale or apparatus variables. It is thus a product design and may be considered as consisting of an inner and an outer array if the "quality/variability" method of analysis is to be used.

The models for quality/variability consist of an appropriate model for RSM in the process factors, a term or terms for the effect of scale, and all possible interactions between scale and process. Take a simple example of a granulation at the 2, 5 and 20 kg scale. The process factors are the amount of liquid X_1 and the granulation time X_2. We will treat the scale as a quantitative variable, but will transform it to the logarithm, instead of using it directly. The levels are thus 0.30, 0.70, and 1.3. The levels of the associated coded variable Z_3 are therefore -1, -0.2, and +1, the middle level being slightly displaced from the centre.

A hexagonal design with one centre point would be suitable for studying the granulation. This may be carried out at each level of Z_3, giving 21 experiments in all. The model is obtained by multiplying the model for the granulation by the second-order model for the scale:

$$\hat{y} = [\beta_0 + \beta_1 x_1 + \beta_2 x_2 + \beta_{11} x_1^2 + \beta_{22} x_2^2 + \beta_{12} x_1 x_2] \times [\alpha_0 + \alpha_3 z_3 + \alpha_{33} z_3^2] + \varepsilon$$

There are 18 terms in the model, so the design is nearly saturated. The 3 fourth-order terms could probably be left out.

Literature examples of experimental design in scale-up
Surprisingly little has been published on this subject and all works appear to have taken the approach of optimizing individually at each level. The papers are mainly concerned with the problem of granulation.

Wehrlé *et al.* (14) have compared granulation with 5 different mixers. The formulation was a "placebo" mixture containing lactose and corn starch, with povidone as binder, granulated with a hydroalcoholic mixture. They investigated small and large planetary mixers (Ours 12 litre and Collette 60 litre size) and high shear mixers (Moritz Turbosphere, 10 and 50 litres, and the Lödige 50 litre mixer), in terms of the effect of granulation time and quantity of water added. For each piece of equipment they did experiments according to a 3^2 factorial design with 2 extra experiments at the centre. They were thus able to compare the properties of the resulting granules (flow properties, particle size distribution) and the resulting tablets (friability, disintegration time) both directly by plotting superimposed contour surfaces for pairs of mixers and also by using factorial discriminant analysis to reduce the number of responses.

The experiment gave two kinds of information, useful in scaling up. Firstly *optimum levels* could be found for each apparatus. For example, more liquid (compared with the powder mass) was needed for the smaller Turbosphere than the large. It *may* be possible to extrapolate this finding to other formulations. Secondly, *general characteristics* of the different mixers could be established, in terms of the principal components. (This method, deriving orthogonal combinations of the responses, and the whole subject of multi-response data analysis is outside the scope of this book.)

Ogawa *et al.* (15) compared results with small size mixer granulators (2 and 5 kg scale). The factors studied were the volume percentage of ethanol in the binder solution and the volume of binder solution. Mixing time, granulation time, and cross-screw rotation speed were constant and the blade rotation speed was adjusted so the speed at the end of the blade was the same in both apparatus. Granulation in each apparatus was studied by a central composite design, with 2 centre points, a total of 20 experiments. The total design was therefore a product design, product of the central composite with a 2^1 factorial, and the model consisted of the product of the two models.

The model was therefore a full second-order model in the granulation variables, but also included a first-order term in the third factor, the type of mixer. In addition there were interaction terms between the type of mixer and all the granulation terms. They also carried out experiments at the 20 kg scale.

Lewis *et al.* (16) compared granulation in Fielder mixers at the 25 litre and 65 litre scale for two sustained release tablet formulations. They also compared the effects of oven drying in trays and in a fluid-bed apparatus. Here the design was slightly different to those used above, as it was basically a central composite design where the factorial design was carried out at a small scale and the axial experiments at the larger scale, with centre points for both series of experiments. Thus fewer experiments were needed than for the full central composite design, replicated for each mixer-granulator. The design had the disadvantage that it was not possible to determine interactions between the mixer variable and either the square terms or the interaction terms in the granulation variables. However, the experiment allowed:

- determination of (average) differences in the responses at each scale,
- optimum conditions of granulation and lubrification to be identified for each scale of granulation and drying method,
- trends to be identified so that initial conditions could be selected for production scale manufacturing,
- critical variables to be identified, those which had a different effect at the two scales.

If the curvature of the response were significant and dependent on the type of apparatus used, it would have been necessary to use a full second-order design at each qualitative level of the mixer variable.

3. Closing remarks

Taguchi's philosophy of quality is a valuable addition to the range of methods available to us. However, these methods, whether those proposed by Taguchi himself or the more efficient designs described in this book, are little used in pharmaceutical development at the present. The designs are often difficult to set up; it is not always easy to identify noise factors, nor can the noise factor always be varied as we would wish. They require considerable experimentation. Nevertheless, failure is costly, and the possibility of designing quality into formulations so that they are robust enough to be manufactured on different pieces of equipment, under different conditions, and using drug substance and excipients from different sources, makes the approach potentially very attractive.

In this section the control variables have been process variables, but we may also wish to adjust the proportions of drug substance and excipients so that the formulation is insensitive to noise factors. The control factors are thus studied in a mixture design of the kind that will be described in the final two chapters.

References

1. G. Taguchi, System of Experimental Design: Engineering Methods to Optimize Quality and Minimize Cost, UNIPUB/Fraus International, White Plains, N.Y., 1987.
2. G. E. P. Box and D. R. Cox, An analysis of transformations, *J. Roy. Stat. Soc. Ser. B,* **26**, 211 (1964).
3. E. Senderak, H. Bonsignore, and D. Mungan, Response surface methodology as an approach to the optimization of an oral solution, *Drug Dev. Ind. Pharm.*, **19**, 405-424 (1993).
4. D. C. Montgomery, Design and Analysis of Experiments, 2nd edition, J. Wiley, N. Y., 1984.
5. G. E. P. Box, W. G. Hunter, and J. S. Hunter, Statistics for Experimenters, J. Wiley, N.Y., 1978.

6. P. Wehrlé, G. F. Palmieri, and A. Stamm, The Taguchi's performance statistic to optimize theophylline beads production in a high speed granulator, *Drug Dev. Ind. Pharm.*, **20**, 2823-2843 (1994).
7. R. N. Kacker, Off-line quality control, parameter design and the Taguchi method, *J. Qual. Technol.*, **17**, 176-209 (1985).
8. V. Nair, Taguchi's parameter design: a panel discussion, *Technometrics*, **34**, 127-161 (1992).
9. R. V. Leon, A. C. Shoemaker, and R. N. Kacker, Performance measures independent of adjustment, *Technometrics*, **29**, 253-285 (1987).
10. M. S. Phadke, Quality Engineering Using Robust Design, Prentice Hall, 1989.
11. G. E. P. Box and R. D. Meyer, Dispersion effects from fractional designs, *Technometrics*, **28**, 19-27 (1986).
12. D. C. Montgomery, Using fractional factorial designs for robust process development, *Quality Engineering*, **3**, 193-205 (1990).
13. A. Menon, N. Dhodi, W. Mandella, and S. Chakrabarti, Identifying fluid-bed parameters affecting product variability, *Int. J. Pharm.*, **140**, 207-218 (1996).
14. P. Wehrlé, Ph. Nobelis, A. Cuiné, and A. Stamm, Scaling-up of wet granulation: a statistical methodology, *Drug Dev. Ind. Pharm.*, **19**, 1983-1997 (1993).
15. S. Ogawa, T. Kamijima, Y. Miyamoto, M. Miyajima, H. Sato, K. Takayama and T. Nagai, A new attempt to solve the scale-up problem for granulation using response surface methodology, *J. Pharm. Sci.*, **83**, 439-443 (1994).
16. G. A. Lewis, V. Andrieu, M. Chariot, V. Masson, and J. Montel, Experimental design for scale-up in process validation studies, *12th Pharm. Tech. Conf.*, 1993.

Further reading

*   D. M. Grove and T. P. Davis, Engineering Quality and Experimental Design, Longman Scientific and Technical, Harlow, 1992.
*   S. R. Schmidt and R. L. Launsby, Understanding Industrial Designed Experiments, 3rd edition, Air Academic Press, Colorado Springs, 1993.
*   G. S. Peace, Taguchi Methods, A Hands-on Approach to Quality Engineering, Addison-Wesley, Reading, 1992.

8

EXCHANGE ALGORITHMS

Methods for Non-Standard Designs

I. INTRODUCTION TO EXCHANGE ALGORITHMS

It is only during the last 10 years that the techniques described in this chapter have begun to be widely used by experimenters. Even now they are often considered difficult to use and of marginal utility. We intend to show that this prejudice is unjustified and we will demonstrate, with the help of several examples, that, because of their flexibility and general applicability, they have an important place in the formulator's and developer's tool kit along with the more traditional and better known methods. The long and complex calculations needed are taken care of by specialised computer programs, of which a large number are now available.

A. Recognising the Exchange Algorithm Situation

The so-called "classical designs" exist independently of the problem being treated. They consist of a set of points in factor space, their arrangement being optimal for the determination of a given type of model. So in order to use these designs we need to carry out a *projection* between the numerical values corresponding to the coded variables (these are the formal values of the design) and the real levels of the true factors, corresponding to the experimental conditions. Thus this projection allows a temperature of 40°C to be associated with a value -1 of a factorial design, for example, or a mixer rotation speed of 1200 rpm to correspond to a level 0.866 in a Doehlert design, etc.

The main problem with these designs is that they do not take into account specific experimental constraints which may render them inefficient or impossible to use. We will now outline various situations where this might occur. Exchange algorithms are used in all of these.

1. Constraints on the experimental domain

The domains so far studied have been symmetrical – cubes, spheres and regular simplexes. Classical designs are used, provided that there are no limitations on the experimental conditions that prevent one or more of the experiments being carried out. It is by no means unusual that certain combinations of values of the experimental conditions are excluded for reasons of cost, stability, feasibility, or simply interest to the researcher. In a tableting experiment there might be whole regions where tablets cannot be obtained because of incompatibility with the machine, lack of cohesion, etc. Thus the experimental domain is transformed from a regular polyhedron to a more or less irregular convex one, in which the experiments may no longer be placed according to a standard design. Figure 8.1 shows two examples of this, where the initially regular domains [a cube in case (a), a simplex for a mixture in case (b)] have been reduced by the constraints to irregular shapes. It is of course possible to construct standard designs within these

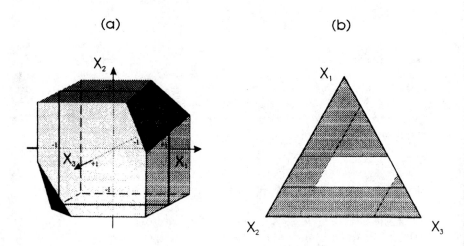

Figure 8.1 Experimental domains with constraints.

domains, but then considerable portions of the design space would remain outside the domain of the new design. The problem is well illustrated by Snee (1).

The use of exchange algorithms to construct D-optimal designs for *mixtures with constraints* is described in some detail in chapter 10.

2. Discontinuous experimental domain

Quantitative factors may sometimes be allowed only to take certain levels, not always evenly spaced. Examples are sieve sizes, volumes of mixer-granulators, speed controls, etc. These factors may be treated as qualitative, but this would cause potentially interesting information to be lost to the analysis. If these levels do not correspond to those of the standard designs, the methods of chapter 6 (RSM) would no longer be applicable.

3. Limitations on combinations of levels

Only certain combinations of levels of some of the variables may be possible. Two examples of this are given later in this chapter, where there were 5 factors in a process study, but only 6 combinations of the variables X_1, X_2 were allowed. This precluded all standard RSM designs.

4. Limited number of experiments

For reasons of cost, time, or amount of material available, the number of possible experiments might be limited, and the standard design might require too many. We show examples of this for a process study and for a screening study.

5. Incorporation of previous experiments

When preliminary experiments have already been done (under reliable conditions), and when the cost of an experiment is very important, it is often a more economical and better strategy to build up a design by adding experiments to those already carried out, rather than to start completely afresh with a standard design. All the same, it must be noted that the results of the complementary experiments are assumed to be homogenous with the preliminary ones. Otherwise, a block effect variable must be introduced into the model.

6. Use of an "incomplete" mathematical model

So far we have generally assumed complete mathematical models: first-order, synergistic, or second-order polynomials. In chapter 5 we supposed either that the model contained only first-degree terms, or it included *all* the second-order terms (square and rectangular), and the designs then proposed corresponded to those models. The experimenter may be led to postulate the presence or absence of certain terms in the model. For example, during a preliminary study he may have tested a factor at several levels, keeping other conditions constant, and seen a curvature in the response. He would then postulate a second-degree term in that variable. On the other hand, by analogy with related studies, he might have good reason to expect that another factor has a purely linear effect. He would then include only this first-order term in the model. The number of terms in the model is therefore less than in the second-order model and the use of classical experimental designs thus becomes expensive, with respect to the limited number of coefficients in the model to be determined.

We saw in chapter 7 that models for minimizing variability include interaction terms between noise variables and control variables, but omit interactions between two noise variables. They also are incomplete models. The experimenter may occasionally wish to use a "mechanistic", non-polynomial model, one resulting from, or inspired by a thermodynamic, or kinetic equation.

7. Complement to a design

In spite of careful reflection and discussion before carrying out the experiments, it may well happen that the phenomenon appears more complex than was initially believed. Interaction or curvature effects that were not considered significant *a priori*, may appear during or after the experimentation. A more complex mathematical model must then be postulated and it is then almost invariably necessary to add the most pertinent (informative) experiment(s) to those carried out already. As we have already mentioned, one would not normally consider starting afresh.

An example is the addition of third-order terms to a second-order model (chapter 5, section VII.B, and chapter 7, section III.A.2).

8. Repairing a design

The most careful attention must be paid to the choice of experimental domain, to the levels of each factor, and to their ranges of variation. The whole experimental plan (expressed in terms of the real factor levels and not the coded variables) must be examined closely before starting the experiments, in case certain experimental combinations are not feasible. In fact it is often wise to carry out any "doubtful" experiment first.

In spite of this, one or more experiments may be found to be impossible, or the results unreliable. In general, an experimental design can only be analysed if the whole of it has been carried out. Lacking a single result it *could* be totally unusable. There are statistical methods for replacing *missing values* under certain circumstances, but there is no treatment which can really replace an experimental datum. A design is repaired by adding one or several experiments that provide the same quantity of information as the failed experiment would have done (1).

II. FUNDAMENTAL PROPERTIES

The objective of using an exchange algorithm is to construct the "best" design within the proposed experimental domain for determining the proposed mathematical model. We will first clarify and quantify these intuitive ideas.

A. Choice of Mathematical Model and Model Discrimination

The experimenter does not usually know which is the most adequate mathematical model – that which best represents the phenomenon being studied in the experimental domain. From time to time, his theoretical knowledge might lead him to postulate certain special models (kinetic, mechanical, or thermodynamic models, for example). He must also take into account any partial knowledge that he has on the system, to postulate a curvature or a particular interaction between two factors. And by analogy with other systems already tested, he may put forward models which "work" in similar cases. Otherwise he may continue to postulate one of the empirical models we have used before, the first-order, second-order, or synergistic model, according to the circumstances. The models used for mixtures and formulation problems will be introduced in the following chapters.

There is no universal experimental design allowing us to carry out the experiments and then to select the best mathematical model.

The only exception to this rule would be to cover the domain with a sufficiently dense network of experiments, but this leads very rapidly to an excessive number. If s is the number of levels to be tested on each axis, and k the number of factors, it requires a complete factorial s^k design.

It is an illusion to imagine that one might choose the mathematical model on the basis of experiments carried out without any mathematical structure. The question is not, in general, one of finding a mathematical model which represents the experimental results already obtained, but rather to choose a model to treat the experiment that one intends to carry out.

We must therefore select experiments that are the most *representative* of all the possible experiments within the domain and which provide the maximum of information about the coefficients of the model. Now, it is well known that not all experiments provide the same amount of information. For example, the centre points provide very little information for a first-degree model, points midway between two axes are rich in information on the first-order interactions of two factors, etc.

To be optimal, a design must be constructed **after** *choosing the mathematical model.*

And, as we have already seen a number of times, the experimenter does not know what will be the best model. He therefore has the following dilemma: to set up the design he must postulate a model, but if he makes the wrong choice of model, the design loses much of its interest.

Techniques exist allowing identification of the experiments which provide the most information *distinguishing* the various models. These would therefore allow us to postulate the best model. They do not allow that model to be determined; for this the design must be completed by carrying out complementary experiments. These *model discrimination* methods (2, 3) are as yet little used and we will not discuss them here. The computer programs they use are not yet of very high performance, nor are they available generally.

There remains the traditional solution to the problem, adopted here: to postulate several models, the most likely ones, and either:
- determine the best design for each and combine them or try to find a compromise between them, or
- construct the optimum design for the most probable model and complete it so that the other models may be determined satisfactorily.

B. Candidate Points

Once the mathematical model has been postulated, there are two possible approaches.

- Use an algorithm to place the experiments directly in that position of the factor space where they are expected to provide the most information. Programs for these quite recent methods – genetic algorithms (4), simulated annealing (5, 6, 7) – are again not widely available and we will not describe their use here.
- Cover the domain with an assembly of *possible* experiments, called candidate points or candidate experiments, from which the algorithm chooses the "best" ones. The simplest solution is to set up a network or lattice, more or less dense, according to the required precision and the means of calculation available. Fortunately, it is possible to set up this network intelligently, allowing us to impose the properties we seek on the final solution and at the same time limiting the number of candidate points and the amount of computation required (8).

C. Quality Criteria for an Experimental Design

The main quality criterion for choosing a design is that of *D-optimality*. This is combined with the *R-efficiency* and normalized to give the *D-efficiency*. However there are other properties that are useful for assessing the design quality, though they are not used directly in determining the design.

1. D-optimality (8, 9, 10, 11)

In chapter 4, section III.C.3, we introduced the calculation of the confidence interval for calculating the coefficient. We recollect that:

$$b_i - t_{v,\alpha/2} \times s_{bi} \leq \beta_i \leq b_i + t_{v,\alpha/2} \times s_{bi} \tag{8.1}$$

This widely used formula is only true for β_i if all the other estimations b_j are equal to β_j. Figure 8.2a shows the confidence interval calculated for 2 coefficients β_1 and β_2 according to the formula 8.1, for a given value of α. More generally, a *joint confidence region* must be considered. For a model with 2 coefficients this is an ellipse (figure 8.2b) and for more coefficients it is an ellipsoid (the value of α being assumed to be the same for all coefficients).

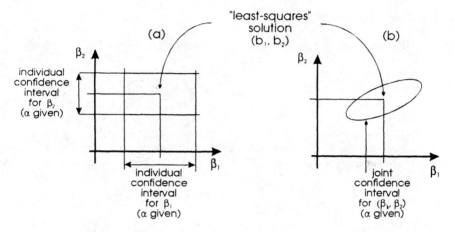

Figure 8.2 (a) Individual confidence limits. (b) Joint confidence limits or confidence ellipse (ellipsoid).

The smaller the volume of the confidence ellipsoid, the more precise will be the estimates of the coefficients. It may be shown that, for a significance level α and a given experimental variance, this volume is proportional to the determinant $|(X'X)^{-1}|$ of the dispersion matrix, defined in chapter 4, section II.C.4.

*An experimental design of N experiments is said to be **D** optimal within a given domain if the determinant of the dispersion matrix $|(X'X)^{-1}|$ is minimal, and thus the coefficients are estimated with maximum overall precision.*

The determinant of a matrix is equal to the reciprocal of the determinant of its inverse, and so:

$$|X'X| = \frac{1}{|(X'X)^{-1}|}$$

This property is a particularly useful one because it enables the quality of two designs to be compared, by comparing the determinants of the information matrices $X'X$, without inverting them. This leads to a considerable saving in computing time.

2. R-efficiency

We recollect that the R-efficiency is the ratio of the number of coefficients of the model to be determined, p, to the number of experiments in the design, N:

$$R_{eff} = \frac{p}{N}$$

This function is not particularly useful on its own. The objective in experimental design is not to do the minimum number of experiments, but to use the available resources in carrying out each experiment as effectively and as efficiently as possible, and thus to arrive at an optimum use of the allocated budget. There is no point in doing a minimum number of experiments, if the results obtained are unreliable, imprecise, and do not properly answer the question.

Let us consider a design of N experiments, to which we add one experiment, giving a design of $N + 1$ experiments. The determinant of the information matrix of the second design is greater than that of the first design.

$$|X'_{N+1}X_{N+1}| > |X'_N X_N|$$

Simply adding another experiment therefore leads to more precise overall estimates of the coefficients, because the volume of the confidence interval ellipsoid is diminished. So one method of improving a design is to add experiments to it. However these designs are not strictly comparable. Because the one design has one more experiments its cost is greater, and we may in effect ask ourselves whether it is worth carrying out that extra experiment. This depends on whether the information provided by that one experiment is great or small. When we wish to compare designs with different numbers of experiments (in the same domain) we "combine" the criteria of D-optimality and R-efficiency by the *D-efficiency*.

3. D-efficiency

The *moments matrix* M is defined by:

$$M = N^{-1} \times X'X$$

and its determinant is therefore:

$$|M| = |X'X|/N^p$$

A given design A has a greater D-efficiency than a design B if $|M_A| > |M_B|$.

This property may be considered as a description of the design's *"value for money"*.

To the above we may add several other criteria (10). They are not used directly for the construction of experimental designs in the way that the D-optimal and the D-efficiency criteria are used, but they are often useful for comparison once several D-optimal solutions, of the same or different numbers of experiments, have been found.

4. A-optimality

A-optimality refers to the *Average variance of parameters*.

The confidence ellipsoid is characterised by its volume $|(X'X)^{-1}|$, but also by its shape – whether it is flattened or not and its direction with respect to the axes b_i. The *trace* of the dispersion matrix $(X'X)^{-1}$ is the sum of its diagonal elements Σc^{ii}. For a given volume of ellipsoid (given by the determinant), the trace takes its minimum value when all the terms c^{ii} are equal, corresponding to a spherical confidence ellipsoid, as in figure 8.3a, where the estimations of the coefficients are independent (uncorrelated).

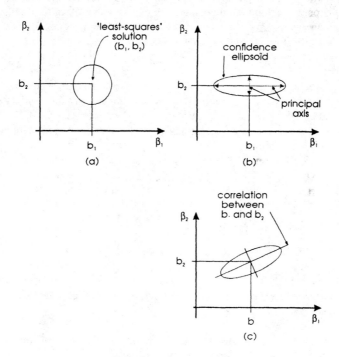

Figure 8.3 Confidence ellipsoids

Figure 8.3b shows a case where the estimations are uncorrelated, but of very different precisions. β_2 is estimated with a much greater precision than the coefficient β_1.

In figure 8.3c the estimations of the coefficients are correlated, as are the confidence intervals. This correlation will also become clear on examining the dispersion matrix, $(X'X)^{-1}$ by the presence of a non-zero, non-diagonal term, in this case $c^{12} > 0$. See chapter 4 for an example.

For equal $|(X'X)^{-1}|$ or $|X'X|$, we may select the design with minimum $tr(X'X)^{-1}$.

5. G-optimality

This refers to the *General variance*.

If the mathematical model has been determined, it is possible to calculate the theoretical value of the response and its confidence interval for every point in the domain. We saw in chapter 5 that d_A, the variance function at any point A, along with the experimental variance σ^2, describes the precision of prediction at that point. d_A depends only on the model and the experimental design and no measured response data are required. Within the domain there is at least one point where the precision of prediction of the response is the worst of all, so the variance function is maximum at this point: d_{max}. A design is called *G-optimal* when d_{max} is as small as possible.

In practice, it may be difficult to find the exact point where the variance function is maximal and it is more usual to investigate only the candidate points, and find which of these has the highest variance function. Another solution that is sometimes adopted is to calculate the variance function at a large number of points selected randomly within the domain.

The *G-efficiency* is given by:

$$G_{eff} = \frac{p}{N d_{max}}$$

D. Principle of the Exchange Algorithm Method

After the mathematical model has been chosen and the candidate experimental design (that is to say, all the experiments possible within the domain) has been generated we must be able to answer two questions:

- How many experiments are needed to obtain estimates of the coefficients as precisely as possible and at the least cost?
- Given we know the answer to the first question, which experiments do we choose?

It is in fact easier to reply to the second question than to the first! Therefore, we will first assume the number of experiments in the design to be *fixed*, equal to N.

Once we have described how the N experiments are chosen, we will be able to answer the first question.

1. Initial phase

The first stage is the choice of a sub-set of N experiments from the set of n_c candidate experiments. Knowing the mathematical model, the information matrix X'_0X_0 for this initial N-experiment design may be calculated and its quality can be quantified as $|X'_0X_0|$. This initial choice may be random, or using an algorithm to find an homogeneous distribution of points in factor space (9).

2. Iteration

The objective is to progressively replace experiments within the sub-set so as to increase the determinant of the information matrix.

Fedorov has shown how it is possible to choose a pair of points (A, B), A being within the design and B a candidate point, so that their exchange gives rise to a maximum increase in the determinant of the information matrix (9, 11). The details of the calculation are outside the scope of this book. Other methods for optimizing $|X'_1X_1|$ have been developed by Mitchell (12) and Wynn (13).

If $|X'_1X_1| > |X'_0X_0|$ the exchange of points is accepted and the iteration is begun again, at the end of which a new exchange is carried out.

3. Stopping condition

If at step i, $|X'_iX_i| \approx |X'_{i-1}X_{i-1}|$, that is the determinant has stopped increasing, the iteration is stopped.

4. Validation of the solution found

As is the case for many iterative methods where the starting point is chosen randomly, the final solution may depend on the initial choice. Although the algorithm is monotone, with the determinant increasing, it may still converge to a local maximum. It is then necessary to begin the treatment again, starting from several initial random starting choices.

If the solutions found are identical, or equivalent (that is, different experiments but each design having the same determinant) this increases the probability that the maximum that is found is the highest possible value. The design of the final solution is therefore the D-optimal design.

If the solutions are very different, this may indicate an instability, which must be analysed.

5. Protected points

Certain points in the candidate design are *protected*. This means that they are *selected* for each run and they are maintained in the design throughout the iteration, not being replaced. This is normally because:

- the experiments have already been carried out as preliminary exploratory experiments, (see section I.A.5);
- they have been done, or it is intended to do them, as a preliminary design (see section I.A.7 and also chapter 10),
- they are part of a (standard) design, also already carried out, but requiring "repair", either because of experiments which proved impossible, or because of the need to introduce extra terms in the model (see section I.A.8).

6. Exhaustive and non-exhaustive selection of points

Some computer programs give the possibility of 2 modes for choosing experiments from the candidate design (8).

Non-exhaustive selection
Once any point has been chosen, it remains within the candidate design and can therefore be selected a second or a third time. By this means the exchange algorithm gives the design with maximum $|X'X|$ without any further consideration. When the number of experiments is greater than the number of coefficients in the model, $N > p$, it is highly probable that some experiments will be replicated within the final D-optimal design. It is quite frequent that when, for example $N = 2p$, the final design consists of p duplicated experiments. Use of such a design and analysis by regression and ANOVA will give an unbiased estimation of the experimental variance. However the design is likely to be either saturated or nearly saturated, there being only as many *distinct* experiments as there are coefficients (or just a few more). It will not be possible to test for the goodness of fit of the model, as there can be no estimate of the variance by the regression.

    The method is also useful where a number replications of each experiment is required for reasons of precison, for example, *in vivo* experiments on formulations. The "best" experiments are thus replicated more times than those less well placed in the design space.

Exhaustive selection of points
Here any point chosen is eliminated from the candidate design. Thus *all* points in the final design are distinct from one another. If $N > p$, the determinant of $|X'X|$ will often be smaller than that obtained by non-exhaustive selection. Analysis of variance allows estimation of the variance about the regression equation, but the validity of the model cannot be tested because of the lack of replicated points. The model validity (lack of fit) can only be tested provided there are more distinct points than coefficients, and some points are repeated.

    This mode of running the exchange algorithm is especially recommended when the experimenter suspects that not all points belong to the same population, fearing a possible discontinuity or unstable zone. In such a case there is no point in giving excessive weight to experimental points which may, theoretically, be rich in information according to the information matrix $|X'X|$, but for which that information may be biased.

Combined selection

As elsewhere, it is generally better to find a compromise between the two solutions. A possible approach, allowing estimation of the experimental error, the significance of the regression and testing of the model's goodness of fit would be:

- non-exhaustive selection of enough points to allow a certain number of repetitions. Select for example N large enough (at least $p + 5$) to have 5 duplicated points in the design, to allow estimation of the pure error variance with 5 degrees of freedom,
- then protect the duplicated points, eliminating them from the candidate design, and continue the exchange algorithm in the exhaustive choice mode, to ensure that more distinct candidate points are chosen within the domain.

III. EXAMPLES OF EXCHANGE ALGORITHMS

A. Process Study for a Wet Granulation

We illustrate the setting up of a D-optimal design with the example of a process study (14) already discussed at length in chapter 6, where the experimental results of this example were analysed. We will now show how the design itself was obtained. The process was a wet granulation, which was followed by drying, sieving, lubrification, and compression (see table 6.2).

1. Constraints imposed by the problem

There are three reasons why this problem could not be solved using a standard design, and required the use of an exchange algorithm.

- The factors X_3 (residual humidity) and X_4 (sieve size) took only 2 levels and could be treated as qualitative factors. The mathematical model could not contain square terms in these factors.
- Even apart from this, the mathematical model (equation 6.2) was not a full second-order one, as only 2 interactions, X_1X_2 and X_4X_5, were postulated.

$$y = \beta_0 + \beta_1 x_1 + \beta_2 x_2 + \beta_3 x_3 + \beta_4 x_4 + \beta_5 x_5 + \beta_{11} x_1^2 + \beta_{22} x_2^2 + \beta_{55} x_5^2$$
$$+ \beta_{12} x_1 x_2 + \beta_{45} x_4 x_5 + \varepsilon$$

- The possible number of experiments was limited because a fixed quantity of active substance, a 30 kg batch, was available for the study. Since the product contained 33% active substance, this allowed 90 kg to be formulated. The mixer-granulator required about 15 kg of product for each granulation, so no more than 6 granulations could be carried out. On the other hand, lubrification could be done on a scale of only 5 kg. Each batch of 15 kg of granules might therefore be divided into 3 parts (sub-batches) for drying and sieving, followed by lubrification. Thus 18 experiments might be carried out, but with certain restrictions in the combinations of levels.

2. Candidate points

The construction of the candidate design is very important. An apparently simple solution is to postulate as many experiments as possible. The humidity X_3 and the sieve size X_4 are limited to 2 levels each, but the remaining 3 factors are continuous and may take any number of levels. If we allow 10 equidistant levels for each, there will be a total of 4000 combinations of factors. Between 11 and 18 experiments are to be selected from these 4000 possible ones. Under these circumstances, it is by no means certain that the algorithm would converge to the optimum solution, though it would probably be close. It is probable that certain programs will not accept so many data, and the calculation times may be long, particularly if we wish to determine several solutions for different numbers of experiments. The solution may not respect the different constraints. This approach should therefore not be used.

It is far better to propose only the candidate points which satisfy the conditions, and that is what we shall do here. Consider the factors X_1 and X_2 (granulation time and added liquid). The optimal design can only contain 6 different combinations (since only 6 granulations may be carried out). So the candidate design itself should contain only 6 combinations. Also, the model is second-order with respect to those 2 variables, so we must find an experimental design allowing us to determine a second-order model, with only 6 separate combinations of X_1 and X_2. Only two of the designs described in chapter 5 may be used. The first is a design of 6 experiments, for the cubic domain (BD206 in table 5.28, see appendix III). The second is an equiradial design, a regular pentagon with one centre point (table 5.8, column 3, and figure 5.5), and it is this one that was chosen.

The effect of the lubrification time X_5 includes a square term, so it must be set at 3 levels. The humidity X_3 and the sieve size X_4 are each allowed 2 levels. Thus each granulation batch is divided into 3 sub-batches of 4 kg, for the remainder of the processing, each with a different combination of levels of X_3, X_4, and X_5. There are 12 possible treatments for each of the 6 granulations. The candidate design is therefore the product of the pentagonal design and a $2^2 3^1$ factorial design and contains 72 experiments.

3. D-optimal design

The number of experiments is fixed at 18, the maximum possible. The D-optimal design thus obtained by running the exchange algorithm is described in table 6.3, and in figure 8.4. The experimental results for each run are given in chapter 6 (see table 6.3), where they were used for determination of the model and optimization of the process.

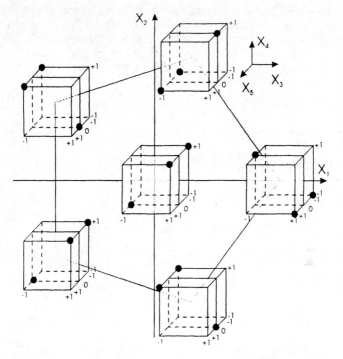

Figure 8.4 D-optimal design for wet granulation process study.

4. Determination of the optimum number of experiments

We have shown the principle of the exchange algorithm in paragraph II.D, for a fixed number of experiments. The next question to be answered is: how do we decide on the number of experiments? Subject to possible external constraints, this depends on the statistical properties of the design. We demonstrate this by determining D-optimal designs for different values of N, in the above process study, and examine trends in their properties.

There are 11 coefficients in the model so the design must contain at least 11 experiments. The maximum number of experiments is 18. D-optimal designs were therefore determined for 11 to 18 experiments. Twenty D-optimal designs were calculated for each value of N, each starting from a different random selection from the 72 experiments. The "best" design (that with the highest value of $|X'X|$) was selected in each case. The trends in the design criteria are shown in figures 8.5a-d, $|X'X|$ and $|M|$ being normalized as $|X'X|^{1/p}$ and $|M|^{1/p}$ respectively, where p is the number of coefficients in the model.

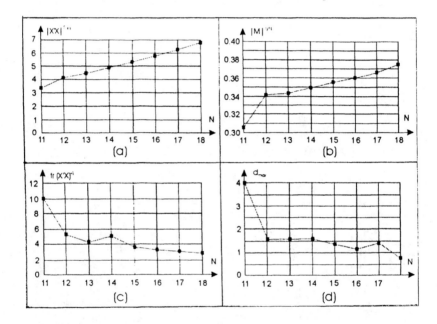

Figure 8.5   Design properties : (a) $|X'X|^{1/11}$; (b) $|M|^{1/11}$; (c) $tr (X'X)^{-1}$; and (d) d_{max}.

The determinant of the information matrix $|X'X|$

Figure 8.5a shows a steady increase in $|X'X|$ with increasing N. Addition of *any* point to a design increases the determinant of the information matrix. If the decision criterion is that the absolute precision of the coefficient must be maximized, the determinant $|X'X|$ must be maximal and 18 experiments must be carried out.

The determinant of the moments matrix $|M|$

$|M|$ represents the quantity of information per experiment. Figure 8.5b shows it also increasing with N, but it is clear that on going from 11 to 12 experiments the information brought in by each experiment is considerably increased. The value then continues to increase from $N = 12$ to 18. We conclude that at $N = 18$ each experiment provides the maximum of information for determining the coefficients.

The trace of the dispersion matrix tr $(X'X)^{-1}$

Figure 8.5c shows the trace of $(X'X)^{-1}$, diminishing by a factor of 2, when going from 11 to 12 experiments. The one additional experiment greatly improves the precision. Going from 12 to 18 experiments leads to very little further improvement. This signifies that the sum of the variance functions c^{ii} stays constant, whereas the volume of the confidence ellipsoid, defined by $|X'X|$, increases (figure 8.5a). It seems that the ellipsoid becomes more spherical and thus the precision of estimation of the coefficients becomes more homogenous and the estimations become more independent (orthogonal). The addition of experiments 12 to 18 improves the symmetry of the distribution of experiments chosen within the experimental domain.

The variance function d_{max}

d_{max} also diminishes very considerably between 11 and 12 experiments and remains practically constant up to $N = 17$ (figure 8.5d). With the addition of the 18th experiment we see a further decrease to a value of less than 1, which means that the variance of prediction of the model is equal to the experimental variance, since $\text{var}(\hat{y}_A) = d_A\sigma^2$.

5. Problems analysing the "split-batch" design

The part of the model describing the effect of the granulation parameters in this problem is determined by a saturated design. Thus, there are no degrees of freedom for determining its coefficients.

The 18 experiments are not independent. The errors in each experiment may be attributed partly to the granulation (and these are therefore considered the same in each of the 3 sub-batches of a given granulation batch) and partly to the succeeding operations and measurements. These latter errors are probably independent. Analysis of variance, therefore, gives an error that is representative only of the post-granulation steps in the process. It is for this reason, that although multi-linear regression is used correctly to estimate best values for the coefficients, analysis of variance may not be used here to estimate the significance of either the model or the coefficients for this problem, as already noted in chapter 6.

This is an example of the *split-plot design* (15), discussed briefly in chapter 7, with reference to correlated errors.

B. Another Wet Granulation Process Study

Another very similar process study was described by Chariot *et al.* (16). The same factors were studied and the constraints were very similar. It was possible to carry out up to 7 granulations and each granulation could be divided into up to 4 sub-batches. The model was also similar, except that it included one more interaction term and no square term in the granulation time. The solution chosen by the authors was to take a 2^13^1 factorial design for the granulation experiment, and multiply this by a 2^23^1 factorial, as in the previous example, to give a candidate design of 72 experiments.

D-optimal designs were determined between 12 and 28 experiments. There was a maximum in $|\mathbf{M}|$ for 22 experiments, so it was this design that was carried out. For 5 of the 6 combinations of granulating conditions (batches), there were 4 combinations (sub-batches) of the other conditions. The remaining batch was split into only 2 sub-batches. This is therefore another example of the split-plot design.

If the model had included a square term in X_2 it would have been possible to treat the problem using the pentagonal design, or (since there was material for 7 granulations, although only 6 were in fact done), the hexagonal design, of 7 experiments, or if it was desired to work in a cubic domain, the 6 experiment design of Box and Draper (BD306), tabulated in appendix III.

C. A Screening Design

We saw in chapter 2 that where factors took different numbers of levels it was not always possible to obtain a suitable design by collapsing of a symmetrical design as described in chapter 2, section V.A. This is especially the case where a limited number of experiments may be carried out. In chapter 2, section V.C we saw how a 12 experiment design (table 2.25) was proposed for treating a problem in excipient compatibility screening. There were 4 diluents, 4 lubricants, 3 levels of binder (including no binder), 2 disintegrants, glidant and no glidant (2 levels), capsule and no capsule (2 levels) to be studied. A 16-experiment design may be derived by collapsing A and B. If 16 experiments are too many, a D-optimal design must be sought.

1. Proposed model

This is a screening model, an additive (first-order) model for 6 factors, where the s_i levels of the ith qualitative variable are replaced by s_i - 1 independent (presence-absence) variables:

$$y = \beta_0 + \beta_{1A}x_{1A} + \beta_{1B}x_{1B} + \beta_{1C}x_{1C} + \beta_{2A}x_{2A} + \beta_{2B}x_{2B} + \beta_{2C}x_{2C}$$
$$+ \beta_{3A}x_{3A} + \beta_{3B}x_{3B} + \beta_{4A}x_{4A} + \beta_{5A}x_{5A} + \beta_{6A}x_{6A} + \varepsilon$$

There are thus 12 coefficients in the model. We therefore construct D-optimal designs with 12 or more experiments.

2. Candidate design

The full factorial $4^2 3^1 2^3$ design, of 384 experiments ($4 \times 4 \times 3 \times 2 \times 2 \times 2$), consisting of all possible combinations of the levels of the different variables, is a solution to the problem. This set of experiments is therefore used as candidates, from which we extract *exhaustively* a representative sub-set, of 12 to 20 experiments.

3. D-optimal designs

For each number of experiments $N = 12$ to 20 we determined the final design, to which the exchange algorithm converged. Figures 8.6a-d show the trends in the four properties of the design that we have already used. $|X'X|$ and $|M|$ are again normalised as $|X'X|^{1/p}$ and $|M|^{1/p}$, where $p = 12$. All of these show that the most efficient solution contains 16 experiments:

- The determinant $|M|$ (quantity of information per experiment) is maximal.
- The trace of $(X'X)^{-1}$ remains stable from $N = 16$.
- The determinant $|X'X|$ increases steadily.
- The value of d_{max} reaches a value of less than 1 (for each point of the full factorial design) at $N = 16$, but increasing the number of experiments beyond this does not lead to any further decrease in d_{max}.

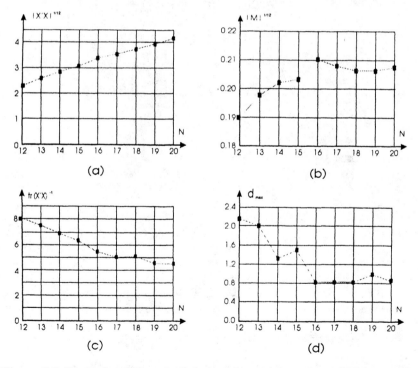

Figure 8.6 Properties of D-optimal designs for excipient compatibility screening problem: (a) $|X'X|^{1/12}$; (b) $|M|^{1/12}$; (c) $tr\ (X'X)^{-1}$; and (d) d_{max}.

In conclusion, a D-optimal screening design of 12 experiments is possible for treating this problem, but if the resources are available it is better to carry out 16.

Figures 8.7a-b show the distribution of the determinants $|X'X|$ obtained for the designs resulting from the 30 runs at $N = 12$ and $N = 16$. For $N = 12$, 30 % of the runs converged to give the same value of $|M|$. These were probably not exactly the same designs, but equivalent. The best solution for $N = 16$ was only obtained once! Nothing proves that those at $N = 12$ are the best solutions possible, but the probability of finding a better solution is small and they are sufficiently close to one another to be considered acceptable. The best result at $N = 16$ experiments is identical to the one derived from the $4^4//4^2$ symmetrical screening design, by collapsing A followed by B (see chapter 2, table 2.22).

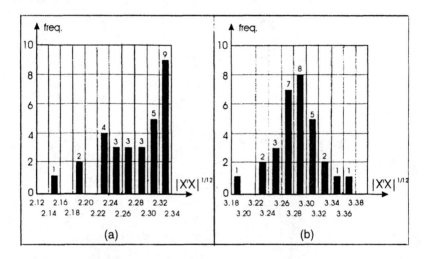

Figure 8.7   Distributions of $|X'X|^{1/12}$ for solutions to the screening problem: (a) $N = 12$ and (b) $N = 16$.

D. D-optimal Designs at 2 Levels for Factor Influence Studies

In chapter 3 we showed that a factorial design of resolution V is needed to determine the constant term, the main effects and the first order interactions in the model. We may thus refer to any design as being of resolution V, should it allow these effects to be estimated, even if it is not a fractional factorial design.

1. Designs for 4 factors

In chapter 3, we gave 3 different saturated or near saturated designs for determining a second order synergistic model for 4 variables:

- a Rechtschaffner design (11 experiments),
- a ¾ design of a 2^4 full factorial design (12 experiments), and
- a D-optimal design (12 experiments).

We will now show the construction of the D-optimal design, and compare the different designs.

Postulated model and candidate design
The model is the second order synergistic model of 11 coefficients:

$$y = \beta_0 + \beta_1 x_1 + \beta_2 x_2 + \beta_3 x_3 + \beta_4 x_4$$
$$+ \beta_{12} x_1 x_2 + \beta_{13} x_1 x_3 + \beta_{14} x_1 x_4 + \beta_{23} x_2 x_3 + \beta_{24} x_2 x_4 + \beta_{34} x_3 x_4 + \varepsilon$$

The candidate experiments are those of the complete 2^4 factorial design (16 experiments), which corresponds to the above model. We select the sub-set of the most representative experiments from these.

D-optimal designs
We have determined the D-optimal designs for 11 to 16 experiments, each experiment in the candidate design being selected no more than once. There were 10 trials for each value of N, each one starting with a random choice of points. Each converged to the same solution. Figures 8.8a-b show the trends in $|M|$ and d_{max}. The trace of $(X'X)^{-1}$ decreases steadily and $|X'X|$ increases with N.

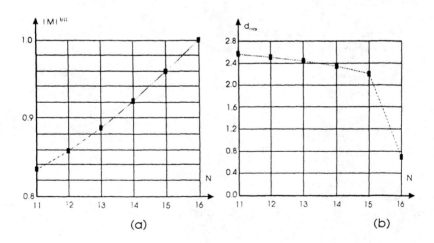

Figure 8.8 Properties of 2 level D-optimal designs for the 4 factor synergistic model of degree 2.

All criteria indicate that the full factorial design is the best solution. There is no design more efficient than this. Note that if experiments had been allowed to be selected more than once from the candidate matrix (non-exhaustive choice), and D-optimal designs had been constructed for $N > 16$, the plot of $|M|$ as a function of N would have shown a maximum at $N = 16$.

Comparison of the different designs at 11 and 12 experiments
Table 8.1 shows the characteristics of the different designs constructed. The D-optimal design for 11 factors is not given, as it is identical to the Rechtschaffner design (which is thus D-optimal).

Table 8.1 Designs for the 4 Factor Synergistic Model: Statistical Properties and Optimality Criteria

Design	N	$|X'X|^{1/11}$	$|M|^{1/11}$	$tr(X'X)^{-1}$	d_{max}	VIF
Rechtschaffner	11	9.172	0.834	1.486	2.56	1.52
D-optimal design	12	10.293	0.858	1.313	2.50	1.37 - 1.50
¾ of 2^4 design	12	10.293	0.858	1.313	2.50	1 - 1.50
2^4 design	16	16	1.0	0.688	0.688	1.0

The two designs for $N = 12$ have the exactly the same characteristics in $|X'X|$, $|M|$, $tr(X'X)^{-1}$, and d_{max} but they are not equivalent. It is not possible to obtain one from the other by permutating lines and columns. They also differ in their variance inflation factors, which are slightly more constant in the D-optimal than the ¾ design. However, they are very close and there is no reason to choose one rather than another on the basis of these properties.

2. Two level designs for 5 to 8 factors

The second order synergistic model (with main effects and all possible first-order interactions) is postulated in each case. As in the previous example, the candidate design is the full factorial, 2^5, 2^6, 2^7, 2^8.

Full and fractional factorial designs
A similar analysis to the above shows that the most efficient designs, according to the D-efficiency criterion, are the full and fractional factorial designs of resolution V and better:

- 2^5 and 2^{5-1} ($I = 12345$, resolution V: $N = 16$ runs)
- 2^6 and 2^{6-1} ($I = 123456$, resolution VI: $N = 32$ runs)
- 2^7 and 2^{7-1} ($I = 1234567$, resolution VII: $N = 64$ runs)
- 2^8 and 2^{8-2} ($I = 12347 \equiv 12568$, resolution V: $N = 64$ runs)

• 2^9 and $2^{9\text{-}2}$ (I = **134689** ≡ **235679**, resolution VI: N = 128 runs)

However, a number of these designs have low R-efficiencies, so alternative solutions may be sought if resources are limited and a minimal design is required.

Saturated and near designs

Of the saturated designs, we have already seen (chapter 3) that the Rechtschaffner design for 5 factors is identical to the resolution V $2^{5\text{-}1}$ fractional factorial. The 6 factor Rechtschaffner design, with 22 experiments to determine 22 coefficients, is also D-optimal, like the above 4 factor saturated design.

The Rechtschaffner designs from 7 factors onwards are no longer D-optimal, due to their lack of balance. In terms of precision of estimation of the coefficients and VIF values, the 7 factor Rechtschaffner design is only slightly inferior to the best design found using the exchange algorithm. Because it is easily constructed and all coefficients are estimated equivalently, it is probably to be preferred to the D-optimal design. In contrast to this, the D-optimal design for 8 factors has clearly better properties than the corresponding Rechtschaffner design.

For 7 and 8 factors, increasing the number of experiments from 29 or 37 to 64 gives a monotonic increase in the determinant of M, in both cases. There appears to be no ideal number of experiments, and no particular advantage in carrying more than the minimum number of experiments.

For 9 factors, the efficiency of the D-optimal design, quantified as $det(\mathbf{M})$, increases with the number of experiments from the saturated design of 46 runs, to reach a peak at 50 runs, and then steadily decreases. We would therefore advise use of the nearly saturated 50 experiment D-optimal design for treating this problem.

3. Number of runs for determining the D-optimal design

It should be noted, just as for the screening designs, that the carrying out of a large number of different runs with the exchange algorithm is highly recommended, if at all possible, in order to increase the probability of finding the true D-optimal design at least once.

References

1. R. N. Snee, Computer-aided design of experiments – some practical experiences, *J. Qual. Tech.*,**17**, 222-236 (1985).
2. A. C. Atkinson and D. R. Cox, Planning experiments for discriminating between models, *J. Roy. Statist. Soc., Ser. B*, **36**, 321-348 (1974).
3. A. C. Atkinson and V. V. Fedorov, Optimal design: experiments for discriminating between several models, *Biometrica*, **62**(2), 289-303 (1975).
4. A. Brodiscou, Contribution of genetic algorithms to the construction of optimal experimental designs, Thesis, University of Aix-Marseilles, 1994.
5. J. H. Holland, Adaptation in Natural and Artificial Systems, University of Michigan Press, Ann Arbor, 1975.

6. S. Kirkpatrick, C. D. Gelatt, and M. K. Vecchi, Optimization by simulated annealing: quantitative studies, *Science*, **220**, 671-680 (1983).

7. P. Franquart, Multicriteria optimization and methodology of experimental research, Thesis, University of Aix-Marseilles, 1992.

8. D. Mathieu, Contribution of experimental design methodology to the study of structure-activity relations, Thesis, University of Aix-Marseilles, 1981.

9. P. F. de Aguiar, B. Bourguignon, M. S. Khots, D. L. Massart, and R. Phan-Tan-Luu, D-optimal designs, *Chemom. Intell. Lab. Syst.*, 30, 199-210 (1995).

10. M. J. Box and N. R. Draper, Factorial designs, the $|X'X|$ criterion and some related matters, *Technometrics*, **13**(4), 731-742 (1971).

11. V. V. Fedorov, Theory of Optimal Experiments, Academic Press, N. Y., 1972.

12. T. J. Mitchell, An algorithm for the construction of "D-Optimal" Experimental Designs, *Technometrics*, **16**, 203-210 (1974).

13. H. P. Wynn, The sequential generation of D-optimum experimental designs, *Ann. Math. Stat.*, **41**(5), 1644-1664 (1970).

14. L. Benkerrour, J. Francès, F. Quinet, and R. Phan-Tan-Luu, Process study for a granulated tablet formulation, unpublished data.

15. D. C. Montgomery, Design and Analysis of Experiments, 2nd edition, J. Wiley, N. Y., 1984.

16. M. Chariot, G. A. Lewis, D. Mathieu, R. Phan-Tan-Luu, and H. N. E. Stevens, Experimental design for pharmaceutical process characterisation and optimization using an exchange algorithm, *Drug Dev. Ind. Pharm.*, **14**, 2535-2556 (1988).

Further reading

•   A. C. Atkinson, The usefulness of optimum experimental designs, *J. Roy. Stat. Soc. Ser. B*, **58**, 58-76 (1996).

9

MIXTURES

Introduction to Statistical Experimental Design for Formulation

I. EXPERIMENTAL DESIGNS FOR MIXTURES WITHOUT CONSTRAINTS

A. Introduction

1. Mixtures in pharmaceutical formulation

Pharmaceutical formulations, like other formulations, are mixtures, almost by definition. Once the excipients have been selected in the quite early stages of a product's development, probably subsequent to excipient compatibility studies of the kind described in chapter 2, the effects of different proportions of the excipients on characteristics of the formulation would normally be investigated. These characteristics include both the (final) properties of the dosage form and the ease of processing. They are defined in terms not of the *amounts* of the different constituents, but of their relative *proportions*.

There are some pharmaceutical mixtures where all components except one vary within quite narrow limits. The remaining component, normally a diluent or a solvent is considered as being a "filler", a "slack variable", or "*q.s.*" This situation has already been treated using factorial designs in chapter 3 and using the central composite design in chapter 5. The variables (except for the filler) can be varied independently and, therefore, the problem may be treated by the methods appropriate to independent variables.

We now consider the other extreme case, where there is no restriction on the proportion of each component, each being allowed to vary between 0 and 100%. This does not on the face of it appear very useful for pharmaceutical mixtures, but we will also see that mixtures where only lower limits are imposed on the proportions of each component have an identically shaped design space and can be treated using the same experimental designs.

We take a mixture of 3 components X_j and draw the design space first of all in the usual way, with orthogonal axes X_1, X_2, and X_3. There are two important restrictions. The fraction of a given component cannot be negative and cannot exceed 1. The total fraction of all components is 1 (or 100%).

$$0 \leq x_j \leq 1 \quad \text{and} \quad \Sigma x_j = 1$$

Because of these restrictions, the only possible experimental points lie on the equilateral triangular shaped plane whose corners are the three points representing the pure substances (1, 0, 0), (0, 1, 0), and (0, 0, 1). The figure (see figure 9.1a) is known as a simplex (no relation to the simplex optimization method). The design space may be expressed on paper as an equilateral triangle and the properties of the mixture may be plotted within it, giving a **ternary diagram** (figure 9.1b).

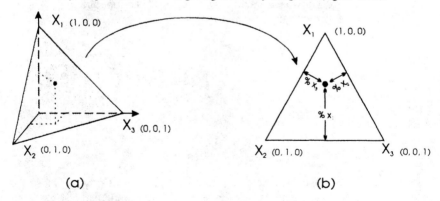

(a)   (b)

Figure 9.1 (a) Relationship between orthogonal axes and mixture factor space. (b) Ternary diagram.

It is improbable that any constituents will be allowed to vary from 0 to 100%, but we can well imagine the situation where several components form part of the same class, diluent, polymer for sustained release, disintegrant, etc., with total quantity fixed. Consider 3 polymers (A, B, and C) in a matrix tablet. The percentage of polymer is fixed at 25%. However, the composition of the polymer may vary freely, and one might use either polymer A, B, or C, or a mixture of all three. Within the polymer the proportion of each of the constituents may vary from 0 to 100%. For such a situation the experimental designs to be studied in this section are very useful. The following example of a placebo tablet, described by Huisman *et al.* (2), is a case in point: the diluent is fixed at a value of about 98% of the total weight and 3 individual diluents can be varied freely within it. At the other extreme, Frisbee and McGinity (3) described the preparation of a pseudolatex, where the total proportion of a mixture of up to three surfactants was fixed at 0.5% of the total emulsion, but the individual components were allowed to vary freely within this limit.

For a 4 component system, without any restrictions other than those given above, the experimental domain is a regular tetrahedron (figure 9.2a). For systems of 5 or more components, it is not possible to represent the design space or experimental region graphically. The properties can be plotted as response surfaces on ternary diagrams (figure 9.2b), which are slices through the domain, allowing 3 components to vary while holding the fraction(s) of the remaining component(s) constant.

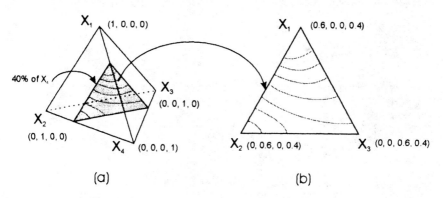

Figure 9.2 4 component factor space for mixtures.

2. Independent and non-independent variables

The restriction that the fractions of all components sum to 1 results in them being no longer independent. If one of them were to increase, the sum of all the others would decrease by the same amount. If we were to insist on continuing to work with independent variables the number of *independent* variables in the factor space would be reduced to one less than the number of components.

3. Mixtures and dosage units

For dosage units – tablets or gelatine capsules or other – it will be noted that if the total mass of the unit is allowed to vary, its composition may be described by the *amount* of the component in the unit rather than its proportion. However it is usually more convenient to keep the mixture description, and add the effect of the amount to the model. The problem is treated later in the chapter, when models and designs for the combined effects of mixture and process variables are introduced.

B. A Placebo Tablet with 3 Components

1. Formulation and measured responses

Huisman *et al.* (2) have described the use of mixture experimental design in formulating placebo tablets. The formulation they describe contained up to 3 diluents, α-lactose monohydrate, potato starch and anhydrous α-lactose. The tablets also contained magnesium stearate as lubricant but because this was fixed at a constant (low) level, the remainder of the formulation, consisting entirely of diluent, could be considered as being 100%. The measured responses were the crushing strength of the tablets, which we will consider first, and the disintegration time. The experimental domain (table 9.1) may be described using a ternary diagram, as in figure 9.1b.

Table 9.1   Experimental Domain for Placebo Tablet Formulation

Factor (component in mixture)	Associated variable	Lower limit %	Upper limit %
α-lactose monohydrate	X_1	0	100
potato starch	X_2	0	100
anhydrous α-lactose	X_3	0	100

2. Mathematical models

First-order canonical equation
We postulate a mathematical model which is first-order in the component fractions X_j, for the above domain:

$$y = \alpha_0 + \alpha_1 x_1 + \alpha_2 x_2 + \alpha_3 x_3 + \varepsilon$$

but as the variables are not independent, any single one of them can be eliminated from the equation by:

$$x_1 + x_2 + x_3 = 1$$

For example, if we decide to eliminate x_3, we obtain:

$$y = \alpha_0 + \alpha_1 x_1 + \alpha_2 x_2 + \alpha_3(1 - x_1 - x_2) + \varepsilon$$
$$= \alpha'_0 + \alpha'_1 x_1 + \alpha'_2 x_2 + \varepsilon$$

where $\alpha'_0 = \alpha_0 + \alpha_3$, $\alpha'_1 = \alpha_1 - \alpha_3$ and $\alpha'_2 = \alpha_2 - \alpha_3$.

But there is no obvious reason to leave out one component rather than another, particularly here, where there are no limits on the proportions of the components other than their sum being unity. It is more usual to eliminate the constant term α_0. Because the sum of all components is unity, $\Sigma x_j = 1$, we may replace α_0 by $\alpha_0(x_1 + x_2 + x_3)$:

$$y = (\alpha_0 + \alpha_1)x_1 + (\alpha_0 + \alpha_2)x_2 + (\alpha_0 + \alpha_3)x_3 + \varepsilon$$

or

$$y = \beta_1 x_1 + \beta_2 x_2 + \beta_3 x_3 + \varepsilon \tag{9.1}$$

where $\beta_j = \alpha_0 + \alpha_j$.

Second-order canonical equation

The second-order equations can be derived using a similar argument to the above:

$$y = \beta_1 x_1 + \beta_2 x_2 + \beta_3 x_3 + \beta_{12} x_1 x_2 + \beta_{13} x_1 x_3 + \beta_{23} x_2 x_3 + \varepsilon \tag{9.2}$$

There are no squared terms, $\beta_{jj} x_j^2$, in the canonical equation because $x_1(1 - x_2 - x_3)$ can replace x_1^2, etc., and the resulting expressions are then included in the first-order and rectangular terms.

3. Experimental design, plan and results

10 formulations were tested by Huisman *et al.* These points, which make up what is known as a *simplex centroid* design with 3 test points, are shown in figure 9.3 and table 9.2. In this and the following sections we simulate a sequential approach to the problem (similar to those of chapters 3, 4, and 5) studying polynomials of increasing complexity and testing their validity.

4. Calculation of coefficients and use of test points

Determination of the first-order model

For the first-order model, equation 9.1, the best design consists of points 1, 2, and 3. It is optimal with respect to the quality criteria studied in chapter 8 — determinant, trace, and R-efficiency. The coefficients may be estimated by the least squares method, but in fact, b_1, b_2, and b_3 in the fitted model are equal to the crushing strengths of the pure substances.

$$\hat{y} = b_1 x_1 + b_2 x_2 + b_3 x_3 = 55.8 x_1 + 36.4 x_2 + 152.8 x_3$$

Table 9.2 Placebo Tablets Design Plan and Experimental Responses (Tablet Crushing Strength and Disintegration Time) [adapted from reference (2), with permission]

	X_1	X_2	X_3	α-lactose monohydrate (%)	potato starch (%)	anhydrous α-lactose (%)	Crushing strength (N)	Disintegration time (s)
1	1	0	0	100.0	0.0	0.0	55.8	13
2	0	1	0	0.0	100.0	0.0	36.4	22
3	0	0	1	0.0	0.0	100.0	152.8	561
4	1/2	1/2	0	50.0	50.0	0.0	68.8	25
5	1/2	0	1/2	50.0	0.0	50.0	91.0	548
6	0	1/2	1/2	0.0	50.0	50.0	125.0	141
7	1/3	1/3	1/3	33.3	33.3	33.3	94.6	22
8	2/3	1/6	1/6	66.6	16.7	16.7	70.4	13
9	1/6	2/3	1/6	16.7	66.7	16.7	80.0	34
10	1/6	1/6	2/3	16.7	16.7	66.7	130.0	385

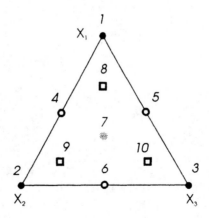

Figure 9.3 Simplex centroid design with test points.

The *test points* are the three binary mixtures 4, 5, and 6, chosen for the following reason: if the first-order model is invalid, the addition of these points will give a design which is optimal for the second-order model with respect to all the quality criteria.

This first-order equation may be used to calculate the crushing strengths of all mixtures. For example, the calculated value for point 4 ($\frac{1}{2}$, $\frac{1}{2}$, 0) is:

$$\hat{y}_4 = 55.8 \times \frac{1}{2} + 36.4 \times \frac{1}{2} + 152.8 \times 0 = \frac{1}{2}(55.8 + 36.4) = 46.1$$

which may be compared with the experimental value of 68.8. The predicted values for the three binary mixtures are given in table 9.3. The differences between calculated and experimental values are quite considerable. The tablets made with two of the binary mixtures are harder than predicted, indicating *synergism* or *synergistic blending*. Where the response is lower than predicted by the first-order model, blending is *antagonistic*.

Table 9.3 Placebo Tablet Formulation (2): Comparison Between Experimental and Predicted Values for Crushing Strength at Design and Test Points[*]

No.	X_1	X_2	X_3	Experimental value (N)	Predicted value (N)		
					1st order model	2nd order model	Reduced cubic model
1	1	0	0	55.8	55.8	55.8	55.8
2	0	1	0	36.4	36.4	36.4	36.6
3	0	0	1	152.8	152.8	152.8	152.8
4	1/2	1/2	0	68.8	**46.1**	68.8	68.8
5	1/2	0	1/2	91.0	**104.3**	91.0	91.0
6	0	1/2	1/2	125.0	**94.6**	125.0	125.0
7	1/3	1/3	1/3	94.6		99.4	94.6
8	2/3	1/6	1/6	70.4			**73.9**
9	1/6	2/3	1/6	80.0			**78.8**
10	1/6	1/6	2/3	130.0			**125.0**

[*] Predictions for test points are in bold type.

Determination of the second-order model
We can therefore go on to the more complex second-order model, equation 9.2. The additional coefficients may be calculated directly from the responses of the experiments 4, 5, and 6. Take point 4 as an example. The difference between the

calculated and the experimental value $y_4 - \hat{y}_4$ must be accounted for by the second-order term $b_{12}x_1x_2 = \tfrac{1}{4}b_{12}$ (as $x_1 = x_2 = \tfrac{1}{2}$). So $b_{12} = 4(y_4 - \hat{y}_4)$ and equation 9.2 becomes:

$$\hat{y} = b_1x_1 + b_2x_2 + b_3x_3 + b_{12}x_1x_2 + b_{13}x_1x_3 + b_{23}x_2x_3$$

$$= 55.8\,x_1 + 36.4\,x_2 + 152.8\,x_3 + 90.8\,x_1x_2 - 53.2\,x_1x_3 + 121.6x_2x_3$$

The design is saturated; there are no degrees of freedom for testing if the model can be used to predict responses within the domain, or whether a more complex model is needed. It is therefore usual to add test points.

Test points for the second-order model
A commonly used test point is the ternary mixture with equal proportions of all 3 components (point 7 in table 9.2). Let $\hat{y}_{123}^{(2)}$ be the calculated response for this point, using the second-order model.

$$\hat{y}_{123}^{(2)} = \frac{b_1 + b_2 + b_3}{3} + \frac{b_{12} + b_{13} + b_{23}}{9} = 99.4$$

This can be compared with the experimental datum, 94.6. It may be concluded that the difference is not very great and therefore the second-order model is accepted. If there are no data indicating the repeatability of the experiment the decision can only be made on the basis of the experimenter's desired precision for the model. Further test points may be added as described in the following section.

The fit may sometimes be improved by a transformation, for example a logarithmic one. Applying the Box-Cox method shows it to be unnecessary here.

5. The reduced cubic model and the simplex centroid design

The reduced cubic model
If the second-order model does not give a sufficiently good fit to the data (either by using test points, or by analysis of variance), a higher, normally third-order, model may be chosen. The canonical form of the complete third-order polynomial for a 3-component mixture such as the one we are considering is relatively complex:

$$\begin{aligned} y = &\beta_1x_1 + \beta_2x_2 + \beta_3x_3 + \beta_{12}x_1x_2 + \beta_{13}x_1x_3 + \beta_{23}x_2x_3 \\ &+ \gamma_{12}x_1x_2(x_1-x_2) + \gamma_{13}x_1x_3(x_1-x_3) + \gamma_{23}x_2x_3(x_2-x_3) + \beta_{123}x_1x_2x_3 + \varepsilon \end{aligned} \qquad (9.3)$$

and it is not possible to derive the most efficient design for it from the second-order design of the first 6 experiments of table 9.2. We shall discuss this model and the meaning of the various coefficients later in this chapter. There is, however, a *reduced cubic model*, where the terms in γ_{ij} are omitted:

$$y = \beta_1 x_1 + \beta_2 x_2 + \beta_3 x_3 + \beta_{12} x_1 x_2 + \beta_{13} x_1 x_3 + \beta_{23} x_2 x_3 + \beta_{123} x_1 x_2 x_3 + \varepsilon \qquad (9.4)$$

Design for the reduced cubic model

The simplex centroid design (figure 9.3) is the best design by all the criteria for determining the reduced cubic model. In the case of 3 components, a single point is added to the second-order design. It is that of the ternary mixture, with equal proportions of each component, each diluent being present at a level of ⅓ or 33.3% (test point 7, in table 9.2).

The reduced cubic model, equation 9.4, may be determined from the 7 data. The design is saturated. The estimations b_i, and b_{ij} of the first- and second-order coefficients β_i, β_{ij} are calculated as before. The method of estimating the third-order coefficient b_{123} can be demonstrated by the following argument.

The difference between the experimental value at point 7, y_{123}, and the value calculated by the quadratic model $\hat{y}_{123}^{(2)}$ at this point may be attributed to the term $b_{123} x_1 x_2 x_3$. Hence:

$$y_{123} - \hat{y}_{123}^{(2)} = b_{123} x_1 x_2 x_3 = \frac{b_{123}}{27}$$

$$b_{123} = 27 \, (y_{123} - \hat{y}_{123}^{(2)}) = 27 \times (94.6 - 99.4) = -129.2$$

The coefficients for the higher order terms in the mixture models are often numerically very large. This does not mean that they are significant, either statistically or physically. Unlike the factorial designs where the standard deviations of estimation of all coefficients, whether main effect or interaction are the same, in designs for mixtures the standard deviations of estimation of the second-order terms are higher than for the first-order terms, and those for the third-order terms higher still (see section I.C.4 and table 9.4).

Test points for a simplex-centroid design

Three further points, 8, 9, and 10, all of them ternary mixtures, can be added to the simplex centroid design between the existing points of the lattice to give a total of 10 points (figure 9.3, table 9.2). These enable testing of the reduced cubic model (or possible further testing of the quadratic model, equation 9.3). The reduced cubic model is determined from experiments 1 to 7 above, and used to calculate values for these 3 points, shown in the right hand column of table 9.3. There is good agreement between calculated and experimental values.

Although test point experiments can be carried out after doing the other experiments of the design, this carries the danger of experimental drift error and it is far better to include them, randomized, at the same time as the design experiments. The rules defining the choice of test points are given later (section I.C.5.)

Since in the latter case there are 3 degrees of freedom, the model coefficients may also be estimated by least squares multi-linear regression, instead of the direct method shown above. The values thus calculated are only slightly different from the previous estimates.

6. Multi-linear regression and analysis of variance for mixture models

We demonstrate below the alternative method of analysing the data. The first-order, second-order, and reduced cubic models (equations 9.1, 9.2, and 9.4) can be estimated from the 10 data points by multiple linear regression. The statistical significance and the goodness of fit of the models can then be determined.

Multi-linear regression
Best values of the coefficients are estimated for each model by the least squares method (see chapter 4) and listed in table 9.4. Estimates for the second-order and reduced cubic models are very similar to those obtained by the direct method above. Estimates for the first-order model are very different. Note that:

- a suitable computer program, designed for treating models without a constant term, is needed for these calculations (otherwise the models must be transformed to ones with independent factors, by elimination of one of the X_i);
- the relative standard error is higher for the second-order coefficients than for the first-order coefficients, and it is particularly high for the third-order coefficient b_{123}; and
- the variance inflation factors are greater than 1.

These are all consequences of the constraints on the model and the fact that the coefficients are not independent.

Table 9.4 Placebo Tablet Formulation (2): Models Determined by Multi-Linear Regression

Coeff.	First-order model			Second-order model			Reduced cubic model		
	Estimate (N)	Standard error	V.I.F.	Estimate (N)	Standard error	V.I.F.	Estimate (N)	Standard error	V.I.F.
b_1	55.5	9.9	1.15	54.9	3.2	1.96	54.7	3.1	1.97
b_2	57.1	9.9	1.15	36.8	3.2	1.96	36.6	3.1	1.97
b_3	158.8	9.9	1.15	154.2	3.2	1.96	154.0	3.1	1.97
b_{12}				81.8	14.7	1.98	88.9	15.7	2.38
b_{13}				-60.1	14.7	1.98	-53.0	15.7	2.38
b_{23}				117.3	14.7	1.98	124.3	15.7	2.38
b_{123}							-113.9	103.3	2.47

Analysis of variance
The analysis of variance is carried out as described in chapters 4 and 5. The number of degrees of freedom for the regression is one less than the number of coefficients, p - 1. Thus for 3 components and a first-order model, there are 2 degrees of freedom. Results are given in table 9.5.

We conclude that the regression is highly significant. Using the reduced cubic model does not improve the regression and the second-order model could be used for response-surface analysis. The first-order model is seen to be inadequate.

Table 9.5 ANOVA of the Regression on the Reduced Cubic Model: Simplex Centroid Design with Test Points

	Degrees of freedom	Sum of squares	Mean square	F	Probability > F
Total	9	11764.9	-	-	
Regression	6	11733.7	1955.6	187.9	***
Linear	2	10501.3	5250.7	29.09	***
Quadratic	3	1219.8	406.6	36.96	**
Reduced cubic	1	12.6	12.6	1.21	0.35
Residual	3	31.2	10.4	-	

$R^2 = 0.997$ $\qquad$ $R_{adj}^2 = 0.992$ $\qquad$ $s = \sqrt{10.4} = 3.2$

Since no experiments were replicated we cannot determine lack of fit. Nonetheless the adequacy of the second-order model and the good agreement at the test points between prediction and experiment suggests that the model will be reliable. The response surface can be plotted and analysed by the methods of chapters 5 and 6.

7. Optimization of the placebo tablets

Analysis of the disintegration time data
The disintegration times of the tablets were also measured. Mean values (5 tablets) for each experiment were given in table 9.2, along with the previously discussed hardness data. The disintegration times were also analysed according to the special cubic model. It was noted that there were (a) moderate differences between calculated and measured values in the case of the test points, and (b) negative disintegration times were predicted within part of the domain.

We cannot really know the significance of the differences between calculated and measured values as none of the experiments was duplicated. The considerable variation in the disintegration times within the domain leads one to study a possible transformation on *y*. A Box-Cox analysis (chapter 7) indicates a logarithmic transformation (λ=0), although the improvement on the untransformed response, λ = 1, is at the limit of significance (figure 9.4). This transformation has the effect of making all predictions of $\hat{y}$ positive. However the larger values are predicted with a much lower precision, although precision of prediction of the lower disintegration times is improved.

The authors addressed the problem of negative dissolution times by a *logit* transformation of the data (see chapter 7). The simple logarithmic transformation, although not giving quite as good a fit as the logit, is used for the optimization below.

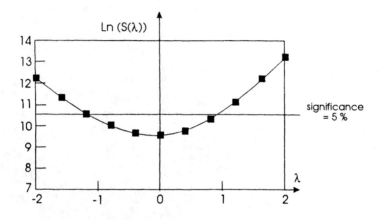

Figure 9.4 Box-Cox transformation study on the tablet disintegration time data.

Graphical optimization

The method described in section II of chapter 6 for independent variables may also be used for mixtures. The parameter estimates for crushing strength and disintegration time may be used to calculate response surfaces over the whole of the domain (figure 9.5a). An optimum (minimum) disintegration time may thus be identified. Disintegration is rapid for tablets containing only small amounts of anhydrous lactose. However, crushing strengths are maximized at high proportions of anhydrous lactose.

Other functions of the composition may be calculated. One of these is the price of the tablet. The authors, taking somewhat arbitrary figures for each of the diluents, used them to calculate the price of the excipients used at each part of the experimental domain. These 3 responses were then combined to determine an over-all optimum region.

For illustration purposes, we assume that the following requirements must be met: crushing strength greater than 100 N, disintegration time less than 60 s, and price less than 400 cts/kg. An acceptable zone may be defined (figure 9.5b).

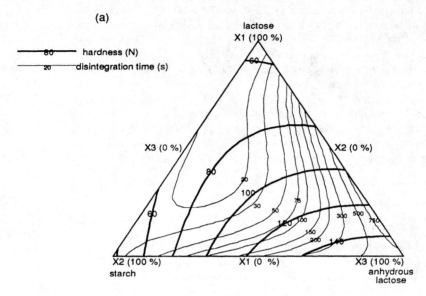

Figure 9.5a   Contour plot of tablet crushing strength and disintegration time.

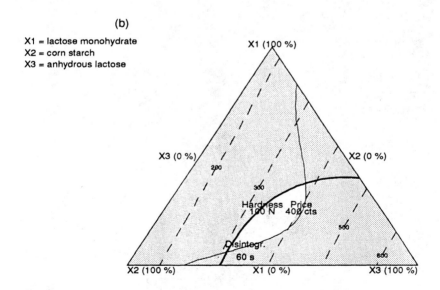

Figure 9.5b   Contour plot of "price" and graphical definition of an optimum region.

Optimization by plotting a desirability function

The desirability method described in section III of chapter 6 may also be used. The minimum hardness allowed is 80 N. The tablets are considered better as the hardness increases to 150N. Similarly, the disintegration time is minimized. The maximum allowable disintegration time is 5 minutes. For times of 1 minute and below, the partial desirability function is equal to 1.

The desirability function for the price is allowed to vary in a non-linear fashion from the maximum price (desirability = 0) to the minimum (desirability = 1), with an exponent $s = 5$, so $d_{price} = [(675 - price)/535]^5$. The overall desirability is the geometric mean of the values of the 3 functions for each point of the design space. It is shown as a contour plot in figure 9.6.

An optimum response is found with desirability 0.37 for 5.5% lactose monohydrate, 63.5% potato starch, and 31% anhydrous lactose:

	value	partial desirability
hardness	99.4 N	0.32
disintegration time	60 s	1.00
price	308 cts/kg	0.15

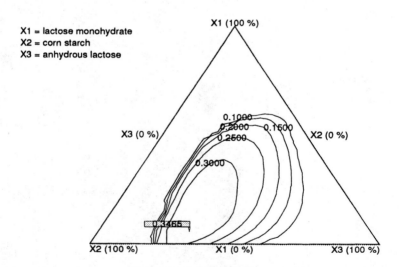

Figure 9.6 Contour plot of desirability function

C. Mixture Models and the Corresponding Designs

The polynomial models, first described by Scheffé (4) and traditionally used for studying mixtures, are of two kinds:

- Reduced models (reduced cubic, reduced quartic etc.) which are first-order with respect to each constituent and all higher order terms are products.
- Complete polynomial models, in their canonical forms (obtained by simplification of the general polynomials, taking into account the restriction that all components sum to unity, as we saw in section I.B.2).

The first group of models corresponds to Scheffé's centroid designs, the second to the Scheffé lattice designs. In fact, the models and the designs are only distinct from the third-order onwards and centroid and lattice designs are identical for the first and second degree.

So far, the only restrictions imposed have been the obvious ones – that the sum of the proportions of all the components equals 1 (or 100%) and that each component can take values from 0 to 1. For pharmaceutical dosage forms we will need to apply further constraints on the level of each component depending on its nature and function. There is little point in considering a system with zero drug substance and even less point in studying one which consists of 50% binder and 50% magnesium stearate. It might be concluded from this that designs where the experimental domain is a simplex are not of much use in pharmaceutical formulation. We treat simplex centroid and lattice designs in some detail nevertheless, for three main reasons:

- They are relatively simple and their analysis will help in our approach of studying the more realistic constrained systems.
- In many constrained systems, especially liquids, the experimental domain is either in itself a simplex or may be adjusted to one by modifying the limits, possible examples being the optimization of diluent mixtures, solvent systems, and coating polymers for sustained release.
- The Scheffé models (whether canonical polynomial or reduced models) can be used for the analysis of both simplex and non-simplex designs.

An experimental matrix or design should ideally have these properties:

- The number of experiments is minimal.
- It is possible to perform them sequentially.
- The mathematical model can be validated with respect to both its analytical form and the coefficient estimates.

1. The first- and second-order canonical equations

The canonical forms of the first- and second-order polynomial models for 3 components were introduced in the previous section (equations 9.1 and 9.2). The first-order model for 4 components can be derived by adding a term $\beta_4 x_4$ to equation 9.1. The general form of the first-order canonical equations, for q components, is:

$$y = \beta_1 x_1 + \beta_2 x_2 + ...+ \beta_i x_i +...+ \beta_q x_q + \varepsilon$$

where x_i is the fraction of the component i in the mixture. In the case of the corresponding predictive model, each coefficient b_i (estimating β_i) is equal to the predicted response for the pure component i. The response surface is planar over the domain.

The second-order model for 4 components is:

$$y = \beta_1 x_1 + \beta_2 x_2 + \beta_3 x_3 + \beta_4 x_4 + \beta_{12} x_1 x_2$$
$$+ \beta_{13} x_1 x_3 + \beta_{14} x_1 x_4 + \beta_{23} x_2 x_3 + \beta_{24} x_2 x_4 + \beta_{34} x_3 x_4 + \varepsilon$$

For reasons explained previously there are no squared terms in the model. The general form of the equations is:

$$y = \beta_1 x_1 + \beta_2 x_2 + ... + \beta_i x_i + ... + \beta_q x_q$$
$$+ \beta_{12} x_1 x_2 + \beta_{13} x_1 x_3 + ... + \beta_{ij} x_i x_j + ... + \beta_{q-1,q} x_{q-1} x_q + \varepsilon$$

2. Interpretation of the coefficients

The meaning of the coefficients in these equations may be illustrated using the previously discussed data for the tableting of placebo mixtures (2), α-lactose monohydrate, potato starch and anhydrous α-lactose. We again consider the crushing strength of the tablets.

Take binary mixtures of lactose monohydrate (proportion x_1) and potato starch (x_2). The first-order model for the response is:

$$y = \beta_1 x_1 + \beta_2 x_2 + \varepsilon$$

with the coefficients β_1, β_2 equal to the crushing strengths of the placebo tablets formulated with pure lactose monohydrate and pure potato starch respectively. We estimate these coefficients therefore as b_1, b_2:

$$b_1 = 55.8 \text{ N} \qquad b_2 = 36.4 \text{ N}$$

and write down a predictive linear model for the response:

$$\hat{y} = b_1 x_1 + b_2 x_2 = 55.8\, x_1 + 36.4\, x_2$$

We can try to predict strengths of mixtures of intermediate compositions by interpolation, shown by the dotted line of figure 9.7. For a mixture of 50% lactose monohydrate, 50% starch, the predicted tablet strength is:

$$55.8 \times 0.5 + 36.4 \times 0.5 = 46.1 \text{ N}$$

The measured tablet strength was 68.8 N. The reproducibility of the experiment is unknown, so there can be no statistical tests. If the difference is assumed to be of "significance" then there is a positive deviation of 22.7 N. The effect of the two components on the hardness is greater than what would be expected from the individual effects. The second-order model may be fitted to these data.

$$\hat{y} = b_1x_1 + b_2x_2 + b_{12}x_1x_2$$

It can easily be shown that the maximum deviation from the predictions of the first order model occurs at the binary point ($\frac{1}{2}$, $\frac{1}{2}$), and that the deviation is $\frac{1}{4}b_{12}$. So the predicted equation is:

$$\hat{y} = 55.8 \ x_1 + 36.4 \ x_2 + 90.8 \ x_1x_2$$

which is plotted in figure 9.7 (solid line).

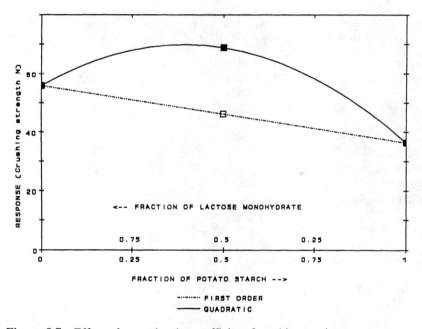

Figure 9.7   Effect of second-order coefficient for a binary mixture.

3. Third-order (and higher) reduced models

In the previous chapters the highest order treated has been 2. It is relatively common to design experiments for mixtures and analyse the data using models that contain third-order terms.

The example of a placebo tablet, described in section I.B of chapter 8, is of a ternary mixture, but the arguments employed there in using the simplex-centroid design may be applied equally well to mixtures of 4 or more components.

For example, in the case of the 3 component mixture we may assume that the response is well or at least adequately modelled by a third-order polynomial. However if we are not sure of the best model, we may prefer a step-wise approach, first building up a second-order design, and then augmenting it as necessary. The possibility of such a sequential approach to experimental design is as valuable for mixtures, as it is for independent variables. Thus the ternary test point 7 is added to the 3 component second-order design (figure 9.3) and the special cubic model determined. Extra test points may be added but these points do not allow fitting of the full cubic model.

In general, the reduced cubic model is obtained by adding the terms $\beta_{ijk}x_ix_jx_k$ to the second-order model. The model for 3 components has already been described. For 4 components it is:

$$
\begin{aligned}
y = {} & \beta_1 x_1 + \beta_2 x_2 + \beta_3 x_3 + \beta_4 x_4 \\
& + \beta_{12} x_1 x_2 + \beta_{13} x_1 x_3 + \beta_{14} x_1 x_4 + \beta_{23} x_2 x_3 + \beta_{24} x_2 x_4 + \beta_{34} x_3 x_4 \\
& + \beta_{123} x_1 x_2 x_3 + \beta_{124} x_1 x_2 x_4 + \beta_{134} x_1 x_3 x_4 + \beta_{234} x_2 x_3 x_4 + \varepsilon
\end{aligned}
\tag{9.5}
$$

The reduced quartic model is obtained by adding the terms $\beta_{ijkl}x_ix_jx_kx_l$ to the reduced cubic model. For 4 components add the term $\beta_{1234}x_1x_2x_3x_4$ to equation 9.5.

4. Significance of the coefficients b_{ijk}

We have already seen that large values of these coefficients do not necessarily indicate statistical significance. They are multiplied by the product $x_ix_jx_k$ and the largest value this can possibly take is 1/27. Some authors prefer to divide the estimations by 27, so that they may be more readily compared with the coefficients of lower order terms.

It is the same for the β_{ijkl} coefficients of the quartic model. The product $x_ix_jx_kx_l$ varies only between zero and 1/256.

5. Scheffé simplex-centroid designs

Construction
Scheffé's simplex centroid designs may be constructed sequentially:

- First-order: q experiments corresponding to the q pure substances (q vertices of the simplex).
- Second-order: the first-order (centroid) design plus all the binary mixtures (½, ½, 0, ...). For the 4-factor system of figure 9.8, add 6 binary mixtures (the mid-points of the 6 edges).
- Third-order: the second-order (centroid) design plus all the ternary mixtures (⅓, ⅓, ⅓, 0,...) mid-points of the faces of the simplex. For the 4-factor system of figure 9.8 add 4 ternary mixtures (⅓, ⅓, ⅓, 0), (⅓, ⅓, 0, ⅓), (⅓, 0,

⅓, ⅓) and (0, ⅓, ⅓, ⅓).
- Quartic: the third-order centroid design plus all the quaternary mixtures (¼, ¼, ¼, ¼, 0, ...), mid-points of the hyperfaces of the simplex.

These designs, which are saturated, allow the corresponding reduced models to be determined directly, and are optimal for the Scheffé reduced models. The points for the 4 component factor space are shown in figure 9.8.

Test points
$q + 1$ test points are selected to be as far as possible from the experimental points of the basic model. The first test point is the centre of gravity of the domain ($1/q$, $1/q$,...). The remaining q points are the points mid-way between the centre of gravity and each of the i vertices. They are thus $(q+1)/2q$ in X_i, and $1/2q$ in the remaining components. Thus we have for:

- 3 components: (⅔, ⅙, ⅙), (⅙, ⅔, ⅙), (⅙, ⅙, ⅔)
- 4 components: (⅝, ⅛, ⅛, ⅛), (⅛, ⅝, ⅛, ⅛), (⅛, ⅛, ⅝, ⅛), (⅛, ⅛, ⅛, ⅝)

These points are also introduced in the axial designs for component screening (section III.B)

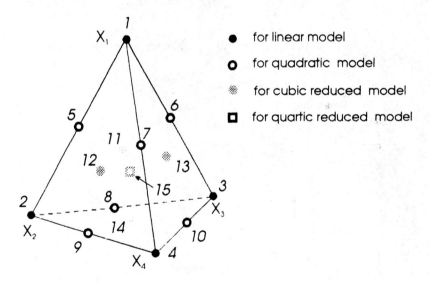

Figure 9.8 Centroid designs for 4 components.

6. Third-order canonical equations

The *full* third-order equation is relatively complex. It is normally written in the form (for 3 components):

$$y = \beta_1 x_1 + \beta_2 x_2 + \beta_3 x_3 + \beta_{12} x_1 x_2 + \beta_{13} x_1 x_3 + \beta_{23} x_2 x_3$$
$$+ \gamma_{12} x_1 x_2 (x_1 - x_2) + \gamma_{13} x_1 x_3 (x_1 - x_3) + \gamma_{23} x_2 x_3 (x_2 - x_3) + \beta_{123} x_1 x_2 x_3 + \varepsilon$$

As in the previous chapters, we use b_i, b_{ij}, b_{ijk} to represent the estimates of the coefficients β_i, β_{ij}, β_{ijk}. Similarly the estimates of γ_{ij} are written as g_{ij}. Equations 9.2-5 may be rewritten using these estimated coefficients. Then we should replace y by $\hat{y}$, the calculated value of the response, and omit the random error term, ε.

7. The Scheffé simplex lattice designs

Like the centroid designs, the Scheffé simplex lattice designs, described below, are easy to construct. Their R-efficiencies are equal to 100%, as the number of experimental points is equal to the number of coefficients in the corresponding mathematical model. They may be resolved directly, that is the coefficients may be calculated directly without using a computer for least squares regression. They can sometimes be built up sequentially. The precision of the calculated response is optimal, that is, the variance over the design space is minimal for the number of experiments.

The validity of the model can be tested by the addition of further points (test points), which in turn can be incorporated into the design for analysis using the same or possibly a more complex model.

The structure of the designs is as follows. The order of the model is defined as $\{q, m\}$ for q constituents and a polynomial of degree m. The constituent fractions take $m + 1$ values, in multiples of $1/m$. For example, the design $\{4, 1\}$ is first-order, with 4 components, and the fractions of the X_i take values 0, 1; that is only the pure substances are tested. It is thus identical to the centroid design, discussed earlier, as is the second degree lattice model $\{4, 2\}$, with the 4 pure substances and 6 binary mixtures.

Table 9.6 Third-Order Simplex Lattice Design for 3 Components $\{3, 3\}$

No.	X_1	X_2	X_3	U_1 %	U_2 %	U_3 %
1	1	0	0	100	0	0
2	0	1	0	0	100	0
3	0	0	1	0	0	100
4	⅔	⅓	0	66.7	33.3	0
5	⅔	0	⅓	66.7	0	33.3
6	⅓	⅔	0	33.3	66.7	0
7	0	⅔	⅓	0	66.7	33.3
8	⅓	0	⅔	33.3	0	66.7
9	0	⅓	⅔	0	33.3	66.7
10	⅓	⅓	⅓	33.3	33.3	33.3

In the case of the third-order design {3, 3}, the X_i for each of the 3 components take values 0, ⅓, ⅔, 1. The design (table 9.6 and figure 9.9) may be used directly and it is optimal for determining the full cubic polynomial. Cornell (1) has given a detailed description of the canonical equations and the simplex lattice designs.

On comparing the simplex centroid design and the third-order simplex lattice design for 3 components (figures 9.3 and 9.9), we see that, for the third-order design, 9 out of the 10 points are binary mixtures and only a single point is a ternary mixture. In the simplex-centroid design with test points, 4 out of the 10 experiments are ternary mixtures, though most of these are test points. These designs where most of the basic points are at the exterior may be compared with the *axial* designs, described later, where more points are placed at the interior of the domain.

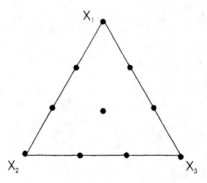

Figure 9.9 Simplex lattice design: 3 components and third-order model {3, 3}.

8. Interpretation of the third-order terms

Higher order models are necessary if the maximum deviation in the binary mixture is other than at the midpoint. Figure 9.10 shows the effect of a cubic term, with the second-order term set at zero:

$$\hat{y} = b_1 x_1 + b_2 x_2 + g_{12}(x_1 - x_2)x_1 x_2$$

and on it is superimposed the synergism of the previous example.

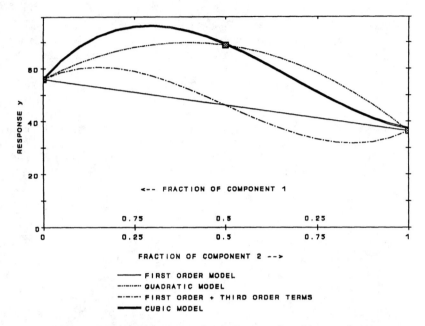

Figure 9.10 Effect of the third-order term in a binary mixture.

The g_{ij} terms therefore describe cubic interactions of *two* components. It is worth looking in detail at this cubic term to understand why it has this particular form. If we took $g_{12}x_1^2x_2$ instead, as a cubic term, this could be transformed on replacing x_1 by $(1 - x_2)$ to:

$$g_{12}x_1x_2 - g_{12}x_1x_2^2$$

thus including part of the second-order coefficient. The advantage of using $g_{12}(x_1 - x_2)x_1x_2$ terms is that they may be added onto the second-order model without altering the estimates of the second-order coefficients when these have been calculated from the points $(1, 0)$, $(\frac{1}{2}, \frac{1}{2})$, and $(0, 1)$.

Thus $g_{12}(x_1 - x_2)x_1x_2$ is equal to zero at the points for the pure components $(x_1 = 1)$ and $(x_2 = 1)$ and for the mid-point of the binary mixture $(x_1 = x_2 = \frac{1}{2})$. The difference from the quadratic model is maximized at $(x_1 = \frac{1}{4}, x_2 = \frac{3}{4})$ with a deviation equal to $3g_{12}/32$, and at $(x_1 = \frac{3}{4}, x_2 = \frac{1}{4})$ with a deviation of $-3g_{12}/32$. The deviations at the design points $(\frac{1}{3}, \frac{2}{3})$ and $(\frac{2}{3}, \frac{1}{3})$ are $2g_{12}/27$ and $-2g_{12}/27$ respectively.

The above argument enables us to select experimental designs for the binary mixture. For a linear model, values for the pure substances are required. For the quadratic model, 3 equally spaced points are needed. For the cubic model, 4 points are required. From the point of view of sequentiality, one can go from a first-order to a second-order design, but not from there to a cubic design. Designs for 3, 4, or more components are analogous.

D. The Solubility of a Hydrophobic Drug in a 3-Component Mixture

1. Outline of the problem and experimental domain

This is an example of the measurement of the solubility of an active substance in water, polyethylene glycol, ethanol, and their binary and ternary mixtures. No restrictions are imposed on the proportions of the components. The experimental domain is therefore like the one given for the placebo tablets in table 9.1. However, this time certain measurements are duplicated.

2. Experimental design and plan

A simplex centroid design with test points was used. Duplicate solubilities were measured in the pure solvents and the binary mixtures. The design, experimentation plan, and results are given in table 9.7.

3. Direct determination of first and second-order models

The first-order model 9.1 can be estimated from data 1, 2 and 3 and used to estimate the binary points 4, 5 and 6. The calculated values for test points 4 and 5 (in column ¶) are much higher than the experimental values. This might lead us to conclude immediately that the first-order model is invalid. If instead we transform the solubility to its logarithm, the predictive model becomes:

$$\log_{10}\hat{y} = b_1x_1 + b_2x_2 + b_3x_3$$

and agreement between the calculated (column §) and experimental values at the test points is improved. However, the first-order model still appears to be insufficient (though this may be confirmed using ANOVA). We try the second-order model, equation 9.2, while maintaining the logarithmic transformation. Using the methods described above we obtain:

$$\log_{10}\hat{y} = -2.602\,x_1 + 0.760\,x_2 + 1.196\,x_3 + 1.27\,x_1x_2 + 3.14\,x_1x_3 + 0.55\,x_2x_3$$

Using this equation, the solubility in a ternary mixture of water, propylene glycol and ethanol ($\frac{1}{3}$, $\frac{1}{3}$, $\frac{1}{3}$) is predicted to be 2.17 mg/mL. This can be compared with the experimental value (point 7) of 2.59. The difference of 15% is large compared with the standard deviation (see analysis of variance, below) and suggests the use of a more complex model. We try a reduced cubic model, which requires addition of the term $b_{123}x_1x_2x_3$.

Calculation of λ by the Box-Cox method (chapter 7), using the 16 data of table 9.7 and the reduced cubic model, leads immediately to the conclusion that the data should be logarithmically transformed. The variance is thus equalized over the domain. It also has the effect of making all predictions positive.

Table 9.7 Solubility of an Active Substance in a Mixture of Water (X_1), Polyethylene Glycol (X_2) and Ethanol (X_3) by a Simplex Centroid Design with Test Points

No.	X_1	X_2	X_3	Water %	PEG %	Ethanol %	Solubility mg/mL			
							Experimental		Calculated	
1	1	0	0	100	0	0	0.0022	0.0028		
2	0	1	0	0	100	0	5.75	5.45		
3	0	0	1	0	0	100	16.4	15.0	¶	§
4	1/2	1/2	0	50	50	0	0.239	0.245	2.8	0.118
5	1/2	0	1/2	50	0	50	1.18	1.24	7.8	0.198
6	0	1/2	1/2	0	50	50	13.2	12.6	10.6	9.37
7	1/3	1/3	1/3	33	33	33	2.59	-	-	-
8	2/3	1/6	1/6	66	17	17	0.046	-		
9	1/6	2/3	1/6	17	66	17	3.77	-		
10	1/6	1/6	2/3	17	17	66	11.5	-		

¶ using first-order equation
§ using first-order equation with logarithmic transformation

4. Determination of reduced cubic model by multi-linear regression

The last datum (point 7) may be used to calculate b_{123}, estimate of β_{123} in the model 9.4. If the data were analysed according to the reduced cubic model the design would be saturated and it would not be possible to test the validity of the model. The solubility was also measured at 3 additional test points, listed in table 9.7.

Instead of calculating the solubility at each of the test points and comparing it with the experimental value we fit the coefficients of the reduced cubic model to the data by least squares multi-linear regression, and investigate the goodness of fit by analysis of variance. The resulting equation is:

$$\log_{10}\hat{y} = -2.662x_1 + 0.759x_2 + 1.222x_3 + 1.16x_1x_2 + 3.09x_1x_3 + 0.63x_2x_3 - 0.35x_1x_2x_3$$

and is plotted as a response surface in figure 9.11. The coefficient of the third-order term is small and probably not significant.

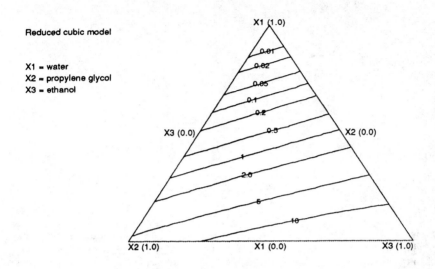

Figure 9.11 Contourplot of calculated solubility (mg.mL⁻¹) of hydrophobic drug as function of solvent composition.

5. Analysis of variance for the mixture model

Not only are the models for mixtures different from those for independent variables, but the analysis of variance also shows certain differences. There is no constant term in the model equation, but the regression sum of squares should be calculated with respect to the mean term. Since we have decided to use a logarithmic transformation of the response, it will be the mean logarithm of the solubility or the logarithm of the geometric mean. We will write the mean of $\log_{10}y$ as $\bar{Y}$. Similarly, we will write Y, $\hat{Y}$ for $\log_{10}y$, $\log_{10}\hat{y}$.

The total adjusted sum of squares is:

$$SS_T = \sum_{i=1}^{N} (Y_i - \bar{Y})^2$$

and this total sum of squares is separated into (a) the contribution explained by the model equation, (b) random experimental variation, and (c) lack of fit. As points are replicated, random experimental error can be estimated. Since there are more individual points in the design than there are terms in the model (7 and 6 respectively), the design is unsaturated and the lack of fit may be estimated and compared with the random error.

The regression SS_{REGR}, residual SS_{RESID}, error SS_{ERR}, and lack of fit SS_{LOF} sums of squares are calculated as described in chapters 4 and 5.

$$SS_{REGR} = \sum_{i=1}^{N} (\hat{Y}_i - \bar{Y})^2$$

$$\text{where } \bar{Y} = \frac{1}{N} \sum_{i=1}^{N} \log y_i$$

$$SS_{RESID} = \sum_{i=1}^{N} (Y_i - \hat{Y}_i)^2$$

$$SS_{TOTAL} = SS_{REGR} + SS_{RESID}$$

$$SS_{ERR} = \sum_{j=1}^{n} \sum_{k=1}^{s_j} (Y_{jk} - Y_j)^2$$

where j is the number of the cell (distinct experiment), n is the number of cells, Y_j is the logarithm of the geometric mean of the replicated readings within the cell (or mean logarithm), and s_j is the number of values within the cell j.

$$SS_{LOF} = SS_{RESID} - SS_{ERR}$$

The analysis of variance table can therefore be set up as described previously.

Table 9.8 Analysis of Variance of Solubility Data for a Hydrophobic Drug

	Degrees of freedom	Sum of squares	Mean square	F	Sign.
Total adjusted	$N-1 = 15$	24.19			
Regression	$p-1 = 6$	24.00	4.00	3429	***
first-order	2	23.00	11.50	9857	***
second-order	3	1.00	0.33	283	***
special cubic	1	0.00	0.00	0	n.s.
Residual	$N-p-1 = 9$	0.194	0.0215		
lack of fit	3	0.187	0.062	53.4	***
pure error	6	0.007	0.001		
$R^2 = 24/24.19 = 0.992$				$s = 0.034$	

The standard deviation is the square root of the pure error mean square. Taking its antilogarithm we obtain 1.081, indicating an overall error of about 8%.

In conclusion, despite the indication of the test point 7, going from a quadratic to a reduced cubic model does not improve the model. There is a substantial and statistically significant lack of fit of the model to the data. The probability that the lack of fit is due to random error is less than 0.1%. Values of the F ratio are therefore calculated using the pure error mean square.

One reason for the significant lack of fit is the considerable variation (more than 3 orders of magnitude) of the solubility over the domain, while the experimental standard deviation is only 8%. It is not surprising that such a simple relationship as a reduced cubic model is insufficient. Examination of the predicted and experimental data shows that almost all the lack of fit is concentrated in the 3 test points. Since they contribute least to the estimations of the model coefficients, it is these points that would normally show up any deviation from the model. In particular, it is seen that the solubility at point 8, with 66.7% water content, is overestimated by a factor of 2.4. For the other test points the error is 33% or less. This is still very high compared to the pure error.

No.	Solubility mg/mL	
	Experimental	Predicted
7	2.59	2.00
8	0.046	0.109
9	3.77	3.51
10	11.5	8.64

The approach to be taken now depends on the use to be made of the data. There are a number of possibilities.

(a) *Decrease the size of the experimental region*, eliminating the region with more than 67% water, for example. This results in a non-simplex experimental domain.

(b) *Replicate the test points* in order to be certain of their validity. A "special quartic" model, with terms $b_{1123}x_1^2x_2x_3$ etc., has sometimes been used (5). The design will be saturated with respect to this model. There is no way of testing for the validity of this model without doing further experiments at different mixture compositions.

(c) Add further binary points to be able to use a *full cubic model*.

(d) Perhaps a *non-polynomial model* may give a better fit (see section III).

(e) *Accept the model*, for the moment, as being sufficiently accurate for our purposes, in spite of the statistically significant lack of fit.

II. CONSTRAINED MIXTURES

A. The Effect of Constraints on the Shape of the Domain

There are limits on the proportions of their constituents of almost all formulations, sometimes very narrow ones. Magnesium stearate, for example, is added as a lubricant, normally between 0.5% and 2%. The ranges for other excipients, especially for diluents, may be wider. We may require at least 20% microcrystalline cellulose in a certain formulation, but with no upper limit, other than that which is implicit in the minimum levels of the remaining components.

We consider here only *absolute* constraints, where the upper and lower limits are constant percentages or fractions and do not depend in any way on the levels of the other components. The limits are therefore parallel to the 0% boundary. However limits may also be *relative*, defined in terms of ratios to the amount of one or more of the other components (see section III and chapter 10). Also, in this section we consider only those cases where individual constraints result in a domain that remains a simplex, though reduced in extent.

1. Components with lower limits only

For simplicity we take a 3-component mixture. A finite concentration L_i of each excipient, i, is required, but no upper limits are imposed.

$$0 \leq L_i \leq x_i \leq 1$$

The explicit lower limits give rise to *implicit* upper limits U_i^* so that

$$0 \leq L_i \leq x_i \leq U_i^* \leq 1$$

If a design space exists within these limits, they are *consistent* (figure 9.12a). However if the sum of the lower limits exceeds 1, it is clear that the experimental domain does not exist (figure 9.12b) and the limits L_i are *inconsistent*. If only lower limits are imposed and the experimental zone exists, the design space is *always* a simplex.

We imagine the case of each excipient present at a minimum level, as follows: 30% for excipient 1, 10% for excipient 2, and 20% for excipient 3. The domain exists as the *L-simplex*, having the same orientation as the total factor space for the mixture. The explicit lower limits give rise to *implicit* upper limits U_i^*, respectively equal to 70%, 50% and 60%, for the components 1, 2, and 3.

Any single point in the factor space may be defined in terms of the constituents of the mixture, X_1, X_2, X_3. All points within the L-simplex may also be defined, in exactly the same way, in terms of the three mixtures X_1', X_2', X_3' at the vertices of the simplex defining the experimental domain. These mixtures X_1', X_2', X_3' are termed *pseudocomponents*. The pure substances 1, 2, and 3 are the *components*. The conversion of system defined in terms of the pure components into one defined in these pseudocomponents is known as the *L-transformation*. The compositions of any point defined in L-pseudocomponents, x_i', or in pure components, x_i, may easily be interconverted (1).

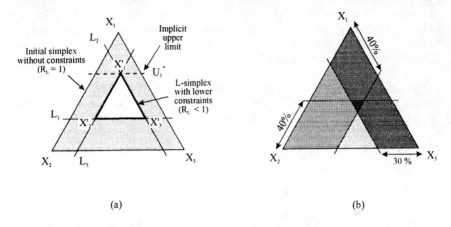

(a)   (b)

Figure 9.12 Mixtures with (a) consistent and (b) inconsistent lower limits.

2. Both upper and lower limits

In addition to these lower constraints there may be an upper limit for one or more of the components. If that *explicit upper limit* is lower than the *implicit* upper limit then the experimental domain is no longer a simplex (figure 9.13). Rules to determine whether or not the experimental domain is a simplex are given by Cornell (1) and Piepel (6).

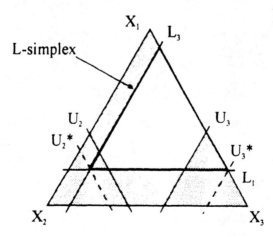

Figure 9.13 Upper and lower limits.

3. Upper limits only: the inverted simplex

If there are upper bounds only, the design space is *sometimes* a simplex, but inverted with respect to the designs discussed previously. This inverted simplex (figure 9.14a) can still be analysed in terms of pseudocomponents and the same reduced Scheffé or simplex lattice designs as discussed previously may be used. Transformation to these pseudocomponents is known as the *U-transformation.*

In the same way as for the lower constraints, the explicit upper constraints give rise to implicit lower limits, as in figure 9.14a, which shows a U-simplex. The upper constants of figure 9.14b result in a non-simplex domain. If the experimental zone does not exist the limits are said to be inconsistent (figure 9.14c).

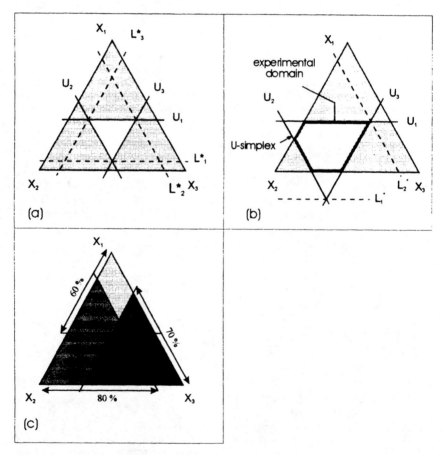

Figure 9.14 Components with upper limits (a) (inverted) U-simplex; (b) non-simplex; and (c) inconsistent limits.

B. Simplex Experimental Designs with Constraints

1. The shape of the experimental domain

We must judge whether the experimental domain space exists within the simplex factor space, whether the upper and lower limits are compatible with one another, and if the domain is a simplex. Rules and equations for the general case are given by Cornell (1) and Piepel (6). However, although they are mathematically simple, in practice we use a computer program. Here we consider the 3-component example shown in figure 9.15:

$$0.1 \leq x_1 \leq 0.2$$
$$0.3 \leq x_2 \leq 0.4$$
$$0.2 \leq x_3 \leq 0.7$$

Existence of the domain
The sum of the lower limits is less than 1 ($R_L = 1 - \Sigma L_i \leq 0$) and that of the upper limits is greater than one. None of the upper limits is set at or below the corresponding lower limit. The domain therefore exists.

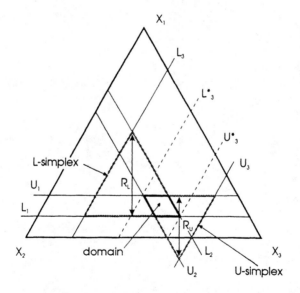

Figure 9.15 Adjustment of inconsistent limits.

Compatibility of the limits
For the limits to be consistent, they must coincide with the true limits of the domain. Thus, the implicit upper limits must not be lower than the explicit upper

limits and the implicit lower limits must not be higher than the explicit lower limits. In figure 9.15, we see that whereas L_1 and U_1 are consistent, as are L_2 and U_2, the limits L_3 and U_3 are inconsistent and the limits for component 3 require readjustment. L_3 is replaced by $L_3^* = 0.4$ and U_3 is replaced by $U_3^* = 0.6$, as shown by the broken lines in figure 9.15.

Shape of the domain

The computer program will give the number of summits of the polyhedron of the constraints. If it is equal to the number of components then it is a simplex. If there are more summits than components then it is not. Here the domain is *not* a simplex. Evidently, for a 3 component mixture, use of the computer or the algebra is fairly pointless. Inspection has rapidly shown the shape of the domain more rapidly. For 4 components (3 dimensions) it is already more difficult, and for more than 4 components such methods are imperative for determining whether the domain is coherent and the limits are compatible.

    If the experimental domain is a simplex, all the strategies of section I (simplex lattice and centroid designs) may be used with the pseudocomponents. If the domain is not a simplex, designs may be set up using the techniques of chapter 8 (exchange algorithms). This will be developed in the next and final chapter. The U- and L-transformations may also be used for non-simplex domains, such as that shown in figures 9.13b and 9.14a.

2. Example of a pellet coating formulation

Experimental domain

A pellet formulation was coated with a mixture of 2 polymers and a plasticizer (polyethylene glycol). Allowed ranges were defined for each component in the coating:

        polymer coating A $x_1 \leq 0.70$
        polymer coating B $0.05 \leq x_2 \leq 0.20$
        polyethylene glycol $0.10 \leq x_3 \leq 0.20$

and coating conditions were maintained constant.

    Simple inspection on a triangular graph shows the domain to be an (inverted) U-simplex. The interior of the simplex can be defined in terms of the following pseudocomponents:

Pseudocomponent	Polymer A	Polymer B	PEG
1	70%	10%	20%
2	70%	20%	10%
3	60%	20%	20%

Model and design

Instead of defining the response in terms of the components X_i, the U-pseudocomponents X'_i were used:

$$y = \beta'_1 x'_1 + \beta'_2 x'_2 + \beta'_3 x'_3 + \beta'_{12} x'_1 x'_2 + \beta'_{13} x'_1 x'_3 + \beta'_{23} x'_2 x'_3 + \epsilon$$

Substituting the expressions connecting the components and pseudocomponents into this equation we obtain an expression equivalent to 9.2, with all the β_i expressed in terms of the β'_i and the constraints U_i – all constant terms. The relationship is a complex one for the first-order coefficients, though simpler for the second-order coefficients (1), and it is not given here. The essential conclusion to note is that the expression for the Scheffé models in terms of pseudocomponents is mathematically equivalent to the model expressed in components. The domain was studied using a 6-point simplex lattice design (figure 9.16). It was thus possible to observe the effects of the different coating polymers on the liberation of drug from the pellets. Results are shown for the dissolution time $t_{50\%}$, the time in hours for 50% of the drug to be released from the pellets.

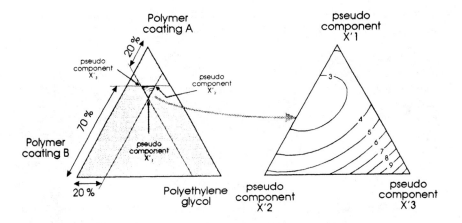

Figure 9.16 Inverted U-simplex factor space for a pellet coating experiment – experimental design – contour plot of the time for 50% dissolution (h).

3. Advantages of a simplex design space

In pharmaceutical formulation, it is very frequent to consider one of the excipients as a diluent or "filler". Thus after fixing the probable levels of the various essential components of the dosage form, drug substance, disintegrant, binder, glidant, lubricant, other diluents, a filler component, usually one of the diluents, is added *q.s.* So that if the experimenter wants to know the effect of increasing a component in the formulation, what he is actually measuring is the effect of increasing the component at the expense of the filler.

When statistical mixture designs are used, this approach is still popular. The formulator will define upper and lower limits for all excipients except one, the filler. This has no explicit boundaries and is allowed to vary *q.s.* The advantage of this approach is that the experimental domain is almost always coherent, as the filler is the constituent present in the greatest proportion. However the design space will not be a simplex. It may be cubic, as in the case of the paracetamol effervescent tablet of chapter 2. Since the proportion of this filler does vary considerably over the experimental domain it is important that its effect be estimated with the greatest possible precision. Designs where the filler is added *q.s.* are not always ideal for this purpose and it is worth while considering fixing limits on the percentages of *all* components in the formulation. If the limits can be adjusted *easily* to give a simplex, then this is to be recommended.

Where all components but one are allowed to vary only within narrow limits and the remaining excipient is a filler, the factor space is most often cubic, and RSM designs in the cubical domain (chapter 5, section VI) may be used.

4. What is the point of using pseudocomponents?

The calculated response within the simplex of figure 9.16 can be expressed in terms of the percentages of the pseudocomponents X_i'.

$$\hat{y} = b_1' x_1' + b_2' x_2' + b_3' x_3' + b_{12}' x_1' x_2' + b_{13}' x_1' x_3' + b_{23}' x_2' x_3'$$

where as before the b_i' coefficients are numerically equal to the responses at the vertices and the b_{ij}' coefficients are estimated directly from the deviations from linearity at the midpoints of each line limiting the simplex (that is at each experimental point representing a 50:50 mixture of 2 pseudocomponents). The response surface can also be defined in terms of the components:

$$\hat{y} = b_1 x_1 + b_2 x_2 + b_3 x_3 + b_{12} x_1 x_2 + b_{23} x_2 x_3 + b_{13} x_1 x_3$$

But here, if the domain is small, the coefficients b_i are unlikely to have a real physical meaning. They are, in actual fact, the calculated responses for the pure components, but these pure components are so far outside the experimental domain that the errors in extrapolation will probably be enormous. This is not in general a problem in the case of response surface analysis, as here we are not concerned with direct interpretation of the coefficients, but only in mapping the form of the response surface within the experimental region.

Even when the design space is not a simplex, it can be expressed in terms of pseudocomponents, and it is sometimes advantageous to do so. In fig. 9.13 we saw a normal *L*-simplex where two of the corners are outside the experimental domain. The coefficients in the model will probably not have any physical significance as the pseudocomponents are outside the domain and cannot be tested. Nevertheless, the design defined in terms of the pseudocomponents has rather better properties than that defined in terms of the pure components.

Figure 9.14b showed a non-simplex domain fitting well inside an inverted U-simplex, but two of the corners of the simplex were not only outside the domain but outside the ternary diagram. They are "imaginary" points with negative amounts of two of the components. This also can give matrices with rather better properties than if the model is defined in terms of the pure components. Here also the coefficients have no direct physical significance and it is only the prediction of the response surface within the domain that is valid.

5. "Irregular" simplexes

The experimental regions we have treated, when in the form of a simplex, have had the same shape as the total factor space. Thus they have been equilateral triangles, regular tetrahedra, etc. However this is not a necessary limitation. Any point in an irregular simplex, like the simplex of figure 9.17a, may be defined in terms of a mixture of the pseudocomponents at each of the vertices. A response may then be analysed within that simplex by means of the Scheffé models and designs. This approach was used by the authors of references (7) and (8).

Relational constraints may also give rise to irregularly shaped simplexes, which can be a basis for using the Scheffé experimental designs. For example, figure 9.17b shows the domain defined by the relationship:

$$0.5 \leq x_1/x_2 \leq 2$$

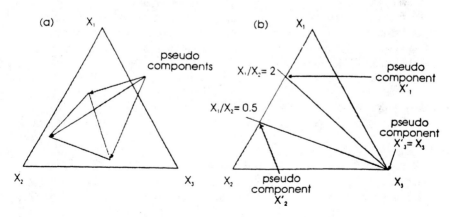

Figure 9.17   (a) Example of an irregular simplex. (b) Simplex defined by a relational constraint.

III. ALTERNATIVE MODELS AND DESIGNS FOR MIXTURES

A. Non-Polynomial Models and Models with Non-Polynomial Terms

The choice of model in the analysis of experimental data is closely tied to the choice of experimental design. We have so far used mainly polynomial models and for mixture experiments we have used the canonical forms of the polynomial equations. These models are usually, but not always, the most appropriate for analysing mixtures. In this section we will briefly describe some others and indicate circumstances where they may be useful.

1. Additive models

Linear blending of components
Consider a ternary mixture, 1-2-3. Estimates of the coefficients in the second-order polynomial model may be obtained experimentally from a simplex lattice design. We will simplify this discussion still further by assuming that we know that the effect of varying component 3 with respect to the other two components is purely first-order, with no synergism between it and either of them. Thus, β_{13} and β_{23} may be omitted and the equation (9.2) reduces to:

$$y = \beta_1 x_1 + \beta_2 x_2 + \beta_3 x_3 + \beta_{12} x_1 x_2 + \varepsilon$$

For example:

$$\hat{y} = 10\, x_1 + 15 x_2 + 5\, x_3 + 80\, x_1 x_2$$

The contour diagrams for this model are shown in figure 9.18a. It is clear that deviations from the first-order model are at their greatest when $x_1 = x_2 = 0.5$ and the level of component X_3 is zero. We continue to look at the calculated deviation from the purely first-order model, along the line representing equal amounts of components 1 and 2 ($x_1 = x_2$), but with steadily increasing amount of component 3. The model predicts a very rapid decrease on the part of the effect $b_{12} x_1 x_2$ as the proportion of X_3 increases.

Is the use of this type of model justified? This depends to a great extent on the physical or other phenomenon studied, but the answer is quite often "yes". For example the physico-chemical models proposed for certain properties of liquid mixtures, such as vapour pressure, are often similar in form to the above equations.

An additive blending model
However, in doing a series of experiments at different levels of the component X_3, along the axis $x_1 = x_2$, we might perhaps find that the form of the synergism is rather different to what was expected, the "extra" effect along this line being proportional to the total amount of the $x_1 + x_2$ components and not to its square. The effect of X_3 is to diminish linearly that of the other two components. If a model is strictly additive with respect to x_3, the response should vary proportionally to the

variation in x_3, which in figure 9.18a is not the case. The following model was suggested by Becker (9) for such a system:

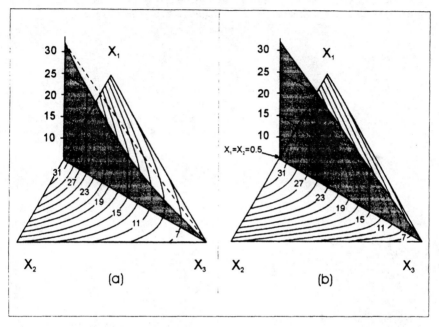

Figure 9.18 Deviations from first-order model by (a) second-order blending and (b) linear blending.

X_3 is acting as a diluent, decreasing linearly the effects of components 1 and 2, enabling linear blending to be described, as in figure 9.18b. The term $x_1 x_2/(x_1 + x_2)$ equals zero when $x_1 = x_2 = 0$. Further terms representing linear blending of X_1 with X_3 (δ_{13}) and X_2 with X_3 (δ_{23}) would normally be added to the equation, replacing the second-order terms in the polynomial. The model may be generalised:

$$y = \beta_1 x_1 + \beta_2 x_2 + \beta_3 x_3 + \delta_{12}\frac{x_1 x_2}{x_1 + x_2} + \delta_{13}\frac{x_1 x_3}{x_1 + x_3} + \delta_{23}\frac{x_2 x_3}{x_2 + x_3} + \varepsilon \qquad (9.6)$$

Designs for the additive blending model

It would not be possible to choose between the second-order polynomial model and the additive blending model of Becker, using only the second-order simplex lattice design of 6 experiments. Measuring the response at a test point, the ⅓, ⅓, ⅓ point, would show the inadequacy of the second-order polynomial and would suggest the use of the special cubic model (addition of a $\beta_{123}x_1x_2x_3$ term to equation 9.2). A larger number of points along the axis would be needed to demonstrate the interest of using such a model rather than the second-order polynomial model.

Becker has proposed other additive models (9).

2. Use of reciprocal terms

If the response changes very rapidly on adding a small amount of one component, the polynomial model may be difficult, or impossible to fit to the experimental data. Draper and St. John (10) suggested adding reciprocal terms, $\beta_{-i}x_i^{-1}$. The 3 factor model becomes:

$$y = \beta_1 x_1 + \beta_2 x_2 + \beta_3 x_3 + \beta_{12} x_1 x_2 + \beta_{13} x_1 x_3 + \beta_{23} x_2 x_3$$
$$+ \beta_{-1} x_1^{-1} + \beta_{-2} x_2^{-1} + \beta_{-3} x_3^{-1} + \varepsilon \qquad (9.7)$$

Such a model cannot be used in a domain where the component proportions, X_i, are allowed to be zero. Here, Draper and St. John suggest adding a small constant value c_i to the x_i, usually between 0.02 and 0.05. The reciprocal terms $\beta_{-i}x_i^{-1}$ become $\beta_{-i}(c_i + x_i)^{-1}$.

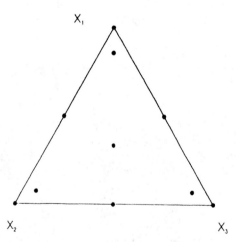

Figure 9.19 Design for a mixture model incorporating reciprocal terms.

They point out that the need to include such terms will probably come to light as the experimenter tries to fit progressively more and more complex models to the experimental data, and they give a kind of "family tree" for sequential model fitting, going from the first-order Scheffé model by stages to the third-order model with reciprocal terms, for treating third-order asymmetric curvature and "edge" effects (1, 6). A suitable design in the unconstrained simplex domain, for such a model, where $c_i = 0.02$, is given in figure 9.19. Here three points (in the case of 3 components) are added to the simplex centroid design (table 9.9). Each represents one main constituent, but with small quantities of the others, 6% in this case. The resulting design is nearly D-optimal.

Table 9.9 Additional Experiments for Reciprocal Terms $\beta_{-i}(0.02 + x_i)^{-1}$

No.	X_1	X_2	X_3
8	0.88	0.06	0.06
9	0.06	0.88	0.06
10	0.06	0.06	0.88

Addition of a reciprocal term to the model might be useful in circumstances such as the addition of an acid to increase the dissolution rate of an insoluble base from a matrix tablet. Ochsner *et al.* (11) have found it useful in solubility models for binary solvents.

3. Ratio models

In setting up the domain the experimenter may define the limits in terms of ratios of one component to another, rather than in terms of proportions (1). These may sometimes be more meaningful as variables for analysis of the data. Unlike all other examples treated in this chapter, the design space is not a simplex, but is generally cubic in form.

As usual we take the three component system. We define r_1, the ratio x_1/x_3 and r_2, the ratio x_2/x_3. Thus any point in the factor space can be defined in terms of r_1 and r_2. Figure 9.20a shows the factor space for upper and lower limits in r_1 and r_2, and a 3^2 design that may be used to determine the second-order model:

$$y = \beta_0 + \beta_1 r_1 + \beta_2 r_2 + \beta_{11} r_1^2 + \beta_{22} r_2^2 + \beta_{12} r_1 r_2 + \varepsilon$$

These may be rescaled as coded variables, taking values of ± 1 at their upper and lower limits. The model is not suitable for treating responses in a domain which is a simplex. It may be extended to 4 or more factors, the model comprising one less ratio term than there are components.

Alternatively, where there are absolute limits imposed (both upper and lower) on the proportion of one of the components, this may replace one of the ratios. For example, r_1 in the above model may be replaced by the proportion of the component X_1. For this model to be useful, the range in X_1 should generally be quite narrow and it should be transformed from a mixture component variable to a coded variable, z_1 (figure 9.20b). This gives the model:

$$y = \beta_0 + \beta_1 z_1 + \beta_2 r_2 + \beta_{11} z_1^2 + \beta_{22} r_2^2 + \beta_{12} z_1 r_2 + \varepsilon$$

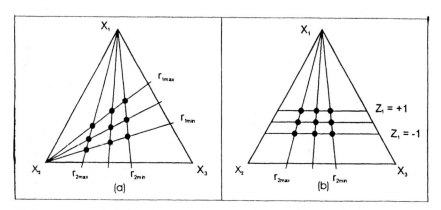

Figure 9.20 (a) Design for ratio model. (b) Design for mixed model, ratio and coded variable.

The variables are independent and so any of the second-order designs of chapter 5 may be employed. For more than 3 ratio variables, use one of the designs of chapter 5, section VI, as the full 3 level factorial design will be too expensive.

Compare this with the similar, but not quite identical treatment of the solubility of a drug in a mixed surfactant system, in chapters 4 and 5. Experimental domains defined in this way, by ratios or by mixed ratios and absolute limits may also be treated by the multicomponent constraints method (see chapter 10).

B. Screening Designs for Mixtures

1. Axial designs in the non-constrained experimental domain

One property that we have observed for the majority of experimental designs, both for mixture models and for independent variables, has been the concentration of experimental points around the exterior of the domain. This is an inevitable consequence of our aim of maximizing the precision, and has therefore been true both for standard designs and D-optimal designs. Although quantitatively this might be an advantage, qualitatively we might sometimes wish to place more experiments at the interior of the factor space. This can be achieved in mixtures by using *axial designs*. Experiments are placed on the *principal axes*, which are the lines joining the pure component *A* to the centre of the opposite face, representing equal amounts of all components except A.

In these designs we study mixtures of a number of components, each having normally the same or a related function and we observe the effect of changing the concentration of each component, *in the presence of all the others*, in order to establish its utility, or perhaps to eliminate it. The method can therefore complement the qualitative screening methods of chapter 2.

The axial design for a non-constrained mixture of q components X_i is set up with the following experiments:

- Each pure component:
  $$x_i = 1$$
- The mixture of all components except one on each opposite "face" of the domain (a hyperface of n-2 dimensions), present in equal amounts:
  $$x_i = 0 \quad \text{and} \quad x_j = 1/(q\text{-}1) \quad \text{for} \quad i \neq j$$
- An intermediate point on each of the q axes midway between the centre point and the pure component points, as suggested by Snee and Marquardt (12). This is the centre of gravity of the simplex consisting of a vertex i and the midpoints of the q - 1 edges coming out from it. These points are also the test points which we have used to test the validity of the reduced cubic model (section I.C.5), as they are the points which are furthest separated from their nearest neighbours.
  $$x_i = (q\text{+}1)/2q \quad \text{and} \quad x_j = 1/2q$$
- The mixture of all components in equal quantities at the centre of the domain:
  $$x_1 = x_2 = \dots = x_i = \dots = x_q = 1/q$$

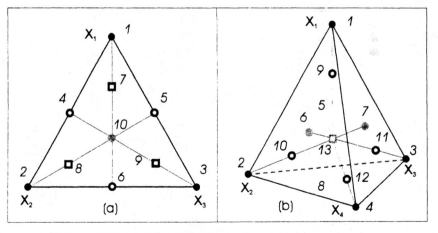

Figure 9.21   Axial designs for (a) 3 and (b) 4 components.

The total number of points is therefore $(3q + 1)$. It is shown in figure 9.21a for 3 components and in 9.21b for a 4 component, unconstrained mixture. As we move along each X_i axis, each component in turn is varied. At the same time, the proportions of all the remaining variable components in the mixture are held equal to one another.

2. Graphical analysis in screening components

A graphical treatment of the results is always necessary, and may be sufficient, with no numerical analysis. Some different possible cases are shown in figures 9.22 a-f. Figure 9.22b suggests that the constituent X_2 has little influence on the response as it is practically the same whatever the proportion. On the contrary, in figure 9.22c we see an example where X_2 has a very strong influence in the region of the pure substance. Here one could perhaps add a term in x_2^{-1} to the model. The curvature of the response surface appears clearly in figure 9.22d. In 9.22e the variations along the axes for the constituents X_1 and X_2 are close (figure 9.22f) and leads one to suppose that the two components have similar effects. Thus the mathematical model might be simplified, by combining them as (x_1+x_2). We might also eliminate one of them from the formulation.

3. Example of a placebo tablet

Experimental design and plan
The use of axial designs is illustrated in an experiment by Chariot, Lewis and Mathieu (13), who used it to investigate a conventional tablet formulation, containing a mixture of up to 5 diluents, disintegrant, binder, lubricant, binder (see table 9.10). The resulting axial design is shown in table 9.11.

Table 9.10 Formulation for a Placebo Tablet

Function	Excipient	
Binder	polyvinylpyrrolidone	5%
Disintegrant	sodium starch glycolate	3%
Lubricant	magnesium stearate	1%
Diluent	lactose microcrystalline cellulose (M.C.C.) corn starch calcium phosphate dihydrate mannitol (see experimental design, table 9.10)	91%

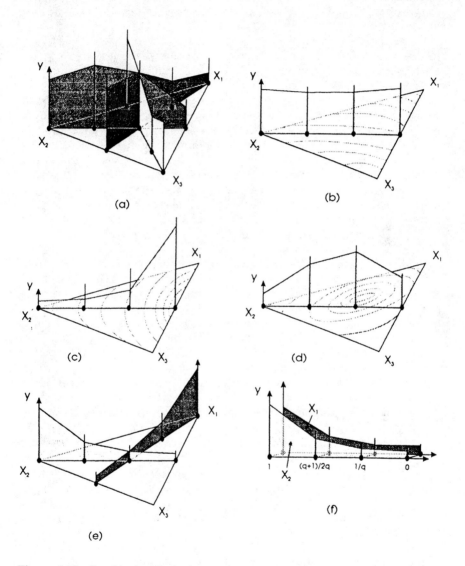

Figure 9.22 Graphical analysis for screening designs (see text).

Results and discussion

The mixtures were granulated and tableted, and the usual properties of granulates and tablets were measured – flow, friability, hardness, disintegration. They are listed in table 9.11.

A peculiarity of this type of design is that the data are analysed graphically, almost without any mathematical or statistical analysis. The response is plotted against the amount of component, for each component axis. Thus, the data for a given response could be plotted on the same graph (figure 9.23a-b).

One possible criticism of this design is that the centre point enters in consideration for each component. If there is an error in the experiment the effects on the analysis could be serious. The centre point should therefore preferably be replicated 2 or 3 times. Also the data points for the pure diluents are rather less useful than the data for mixtures. Those five points could have been omitted, cutting down very considerably on the resources needed.

The properties of the complex mixture (5 diluents) were surprisingly good, and although analysis of the graphs gives some indications for the direction to go in optimizing the formulation it is quite difficult to simplify the formulation.

The results are summarised as follows:

- Increasing the amount of lactose gives improved hardness but slower disintegration and slightly poorer flow. Effect on the friability is slight.
- The effect of microcrystalline cellulose is more complex. Adding up to 60% improves the friability and shortens disintegration time with little effect on the flow, but increasing the percentage from 20% to 60% leads to decreased hardness. A high concentration of microcrystalline cellulose gives poor flow.
- Increasing the proportion of corn starch to 20% leads to more rapid disintegration, and apparently lower friability. However, increasing it further, to 60% and above, leads to soft tablets and poor flow.
- Amounts of calcium phosphate up to 60% have little apparent effect on the properties. Increasing the concentration further gives soft and friable tablets with long disintegration times. These results suggest that calcium phosphate could be removed without a deleterious effect.
- Addition of mannitol improves flow and slightly improves friability. Increased levels also lead to hard tablets, comparatively slow to disintegrate.

It is interesting to compare this design with the Scheffé design of 15 experiments (quadratic model). The two designs have only the 5 pure diluents in common. Otherwise the Scheffé design consists entirely of binary mixtures whereas 9 of the experiments of the screening design are performed on 4 or 5 component mixtures. The screening design cannot be used to estimate a second-order model. The data indicate that the first-order model is inadequate.

Table 9.11 Screening Design for a Placebo Tablet and Measured Responses (13)

Excipient (varied)	X_1	X_2	X_3	X_4	X_5	flow (s)	friability (%)	hardness (kP)	disintegration (min)
[Centre point]	0.20	0.20	0.20	0.20	0.20	4.7	0.03	10.3	5.0
Lactose monohydrate	1.00	0.00	0.00	0.00	0.00	9	0.3	13.3	11
	0.60	0.10	0.10	0.10	0.10	4	0.3	13.3	9
	0.00	0.25	0.25	0.25	0.25	5	0.3	6.6	3
Micro-crystalline cellulose	0.00	1.00	0.00	0.00	0.00	35	4.0	3.0	0.5
	0.10	0.60	0.10	0.10	0.10	6	0.3	5.7	3
	0.25	0.00	0.25	0.25	0.25	4	8.9	7.6	7
Corn starch	0.00	0.00	1.00	0.00	0.00	36	100.0	0.0	0
	0.10	0.10	0.60	0.10	0.10	9	0.6	5.0	6
	0.25	0.25	0.00	0.25	0.25	4	0.3	12.7	12
Calcium phosphate	0.00	0.00	0.00	1.00	0.00	6	78	1.7	20
	0.10	0.10	0.10	0.60	0.10	5	0.2	10.9	6
	0.25	0.25	0.25	0.00	0.25	8	0.03	11.9	7
Mannitol	0.00	0.00	0.00	0.00	1.00	7	0.3	17.9	12
	0.10	0.10	0.10	0.10	0.60	4	1.0	18.5	14
	0.25	0.25	0.25	0.25	0.00	10	1.1	8.0	3

X_1 = lactose
X_2 = microcrystalline cellulose
X_3 = corn starch
X_4 = calcium phosphate
X_5 = mannitol

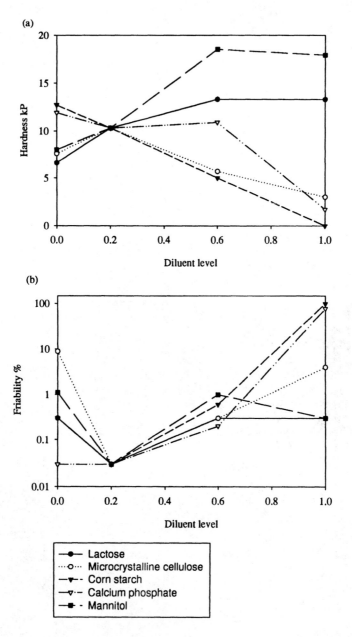

Figure 9.23 Screening design: effect of diluents on (a) hardness (b) friability.

3. Component effects

First-order model
If the first-order model holds, the effect of component i on the response may be considered as the difference in response between the base of the simplex, where component i is absent and all other components have equal proportions $(q-1)^{-1}$ and its vertex where $x_i = 1$ (pure i). It is a measure of the slope along the principle axis for this component.

Thus the "component effect" E_i of constituent i is given by the difference between the coefficient b_i and the mean of the coefficients b_j of the remaining components:

$$E_i = b_i - \frac{1}{q-1}\sum_{j \neq i}^{q} b_j$$

Even if the behaviour is not first-order, the concept may still be useful. In the above example we see that the responses for formulations containing equal amounts of all but one components are not equal to the mean values of the responses for those 4 pure components. It may also be more interesting to compare the response at $x_i = 0.6$ and at $x_i = 0$, and divide by the factor of the change in x_i, that is 0.6. The results of this calculation are shown in table 9.12.

Note that this method of defining component effects, along the principle axes, is not valid for constrained systems.

Table 9.12 Modified Component Effects (0 to 60%)

	Flow	Friability	Hardness	Disintegration
lactose	-1.7	0.0	11.2	10.0
cellulose	3.3	-14.3	-3.2	-6.7
starch	8.3	0.5	-12.8	-10.0
phosphate	-5.0	0.3	-1.7	-1.7
mannitol	-10.0	-0.2	17.5	18.3

Transformation of a second-order model
In the experiment we have described, the component effects are measured and plotted directly. Similar plots can be obtained from the fitted models, where the second-order model may be determined, calculating the response along the X_i principle axis, with the proportions of the other components held equal to one another. That is:

$$x_j = (1 - x_i)/(q-1) \qquad \text{for all } i \neq j$$

The calculated response along the X_i principle axis for the second-order model is:

$$\hat{y}(x_i) = b_i x_i + \frac{1-x_i}{q-1}\sum_{j\neq i} b_j + \frac{x_i(1-x_i)}{q-1}\sum_{j\neq i}^{q} b_{ij} + \left(\frac{(1-x_i)}{q-1}\right)^2 \sum_{i\neq j\neq k}^{q} b_{jk}$$

Response values along each axis may be plotted using this equation. These component plots may easily be compared on the same graph, even for a large number of components, allowing the effects of the components to be assessed. They are shown in figure 9.24 for the 3-component placebo tablet described in section I of this chapter. Note that all the curves pass by the same point, the centre of gravity of the factor space. The method may be extended to constrained systems, although the axes will require redefinition.

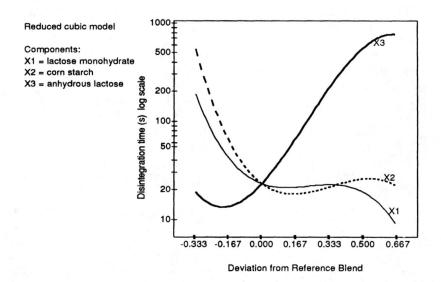

Figure 9.24 Calculated disintegration times along the principle axes for the placebo tablet of table 9.2 (2).

IV. PROBLEMS INVOLVING BOTH MIXTURES AND INDEPENDENT FACTORS

A. Combining Mixture and Process Variables

1. Separate or joint formulation and process study

Up until now we have dealt with two kinds of factors and associated variables:

- In this chapter, mixture or formulation variables, the sum of whose values is restricted to 1 or some lesser fraction.
- In the previous chapters, factors whose levels can be varied independently of one another. These independent factors are often ones that are to do with a pharmaceutical *process*, and though this is not always the case, in what follows we will most often refer to the variables as "process variables".

Each experimental problem treated, therefore, contains factors belonging to one or the other of these classes of factor, or sometimes to both. Most often we have dealt with the classes separately. Only where the mixture variables have been so constrained that they may themselves be treated as independent variables have we been able to combine the two (see for example the paracetamol tablet formulation experiment in chapter 3). Otherwise it is more usual to simplify the problem by separating them, the formulation being optimized first and then the process. This is taking a risk, because the optimum formulation under non-optimum processing conditions may not be the best one for the optimized process. However, the approximation is often acceptable, because the essential properties of the system often depend less on the processing conditions than on the formulation. Also, there may be more than one combination of formulation and processing conditions that will give a satisfactory product.

For example, it may be possible to obtain the desired drug release profile from a coated pellet either by adjusting the coating conditions or by modifying the coating formulation. Formulation and manufacturing conditions may therefore be optimized separately, though the result may not be the best possible, not the most robust system. If the satisfactory zone is relatively small then it may be necessary to investigate both process and formulation factors at the same time, or in several stages. The dependence on both processing and formulation factors may also be so great, and the interactions between two kinds of variables so important, that it becomes necessary to investigate the joint effects of the component fractions and manufacturing conditions on the formulation (1).

When process and mixture variables are combined, the models have a special form, because of the restrictions on the formulation components.

2. Three components and one independent variable

Solubility as function of composition and temperature: experimental domain
This experiment was done to characterise the solubility of a poorly soluble drug at

ambient temperature (22°C) and at 4°C in a mixture of water, ethanol, and propylene glycol. The only limit on the components was that the proportion of water should be not less than 50% ($x_1 \geq 0.5$). The design space for the *mixture* is a simplex, with vertices at 100% water, 50:50 water/ethanol, and 50:50 water/propylene glycol. If we include the temperature variation, within its limits of 4 - 22°C, we obtain the domain of table 9.13.

Table 9.13 Experimental Domain for Drug Solubility Measurement

Factor		Associated variable	Lower limit	Upper limit
water	%	X_1	50	100
ethanol	%	X_2	0	50*
propylene glycol	%	X_3	0	50*
temperature	°C	Z_4	4	22

* Implicit upper limits resulting from imposed lower limit on water

The total design space is a *triangular prism* (figure 9.25). Each point, on a triangular face representing a given temperature, can be defined in terms of its composition as a mixture of water, ethanol and propylene glycol, or alternatively as a mixture of the 3 pseudocomponents which have the mixture composition for each vertex of the simplex. We wish to obtain a predictive model for the solubility of the drug within this space.

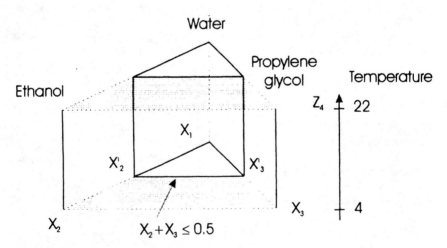

Figure 9.25 Experimental design for solubility as function of composition and temperature.

We will therefore express the factor space and the model in terms of 3 pseudocomponents, water X_1', ethanol/water 50:50 X_2' and propylene glycol/water 50:50 X_3'. The corresponding variables in the model, corresponding to the pseudocomponents (and not to the components), will be written as x_i.

Mathematical models

Let us assume first of all that dependence on the temperature (Z_4) is first-order:

$$y = \beta_0 + \beta_4 z_4 + \varepsilon \tag{9.8}$$

but that the value of β_0 is only constant for a given composition and that it depends on the proportions of the constituents: $\beta_0 = \beta_0(x_1, x_2, x_3)$. If this dependence, in its turn, is assumed to be first-order, we have:

$$\beta_0 = \beta_1 x_1 + \beta_2 x_2 + \beta_3 x_3$$

and so equation 9.8 may be rewritten as:

$$y = \beta_1 x_1 + \beta_2 x_2 + \beta_3 x_3 + \beta_4 z_4 + \varepsilon \tag{9.9}$$

In addition, we may assume that the coefficient β_4 depends on the composition in the same way as β_0, that is $\beta_4 = \beta_4(x_1, x_2, x_3)$. Again assuming first-order dependence, we may write:

$$\beta_4 = \beta_{14} x_1 + \beta_{24} x_2 + \beta_{34} x_3$$

and so the model 9.9 becomes:

$$y = \beta_1 x_1 + \beta_2 x_2 + \beta_3 x_3 + \beta_{14} x_1 z_4 + \beta_{24} x_2 z_4 + \beta_{34} x_3 z_4 + \varepsilon \tag{9.10}$$

Experimental design and plan

Equation 9.8 suggests the use of a 2^1 factorial design to study the effect of the temperature. Equation 9.9 would require a first-order Scheffé design at each temperature (simplex vertices). In fact two independent measurements of solubility were carried out at each point. Also unreplicated test points were set up at the midpoints of the binary mixtures (points 7-12) that would allow use of a more complex model, if necessary. The resulting design is given in table 9.14.

Experimental results and model determination

The experimental results are also listed in table 9.14. The data were analysed according to the two models, equations 9.9 and 9.10. Since some measurements of solubility were in duplicate, we can estimate the reproducibility of the experimental technique. It is possible to estimate the model with the data for the duplicated points 1-6, and then validate it with the test points 7-12. Instead we estimate it by multilinear regression over all the data and then test by analysis of variance.

Table 9.14 Design for Measuring the Solubility of drug X in Mixture of Water, Ethanol, and Propylene Glycol at 10°C and 25°C

No.	X_1	X_2	X_3	Z_4	Volume fraction of solvent			Temperature	Solubility			
					Water %	Ethanol %	Propylene glycol %	°C	Measured y		Calculated $\hat{y}$	
											Equation 9.9	Equation 9.10
1	1	0	0	-1	100	0	0	4	0.005	0.006	0.005	0.0055
2	0	1	0	-1	50	50	0	4	1.25	1.28	1.48	1.30
3	0	0	1	-1	50	0	50	4	1.30	1.34	1.27	1.31
4	1	0	0	+1	100	0	0	25	0.008	0.007	0.009	0.08
5	0	1	0	+1	50	50	0	25	3.12	2.90	2.62	2.93
6	0	0	1	+1	50	0	50	25	2.25	2.40	2.24	2.34
7	½	½	0	-1	75	25	0	4	0.102		0.09	0.10
8	½	0	½	-1	75	0	25	4	0.073		0.08	0.08
9	0	½	½	-1	50	25	25	4	1.24		1.37	1.17
10	½	½	0	+1	75	25	0	25	0.165		0.15	0.17
11	½	0	½	+1	75	0	25	25	0.135		0.14	0.12
12	0	½	½	+1	50	25	25	25	2.2		2.43	2.34

Using the first-order model, equation 9.9, Box-Cox analysis showed a better fit if the calculation were done with the solubility transformed to its logarithm (this is a common result for solubility of hydrophobic substances in water-organic mixtures). The transformation was retained for the remainder of the analysis.

Estimates of the coefficients for the first-order model are listed in table 9.15 and predicted values may be compared with the experimental measurements in table 9.14. These differences are considerable when compared with the experimental variability (by ANOVA). In other words, the effect of the temperature appears to depend upon the composition of the solvent. The effect of temperature on the solubility is relatively higher in ethanol-water mixtures as the proportion of ethanol increases. ANOVA indicates significant lack-of-fit (sign. = 0.026). We therefore analyse the data using the model of equation 9.10 (coefficients in table 9.15).

Table 9.15 Estimations of the Coefficients of the Different Models (significant terms are in bold type)

			Model			
	9.9	9.10	9.11	9.12	9.13	
b_1	**-5.031**	**-5.031**	**-5.051**	**-5.051**	b_1	**-5.051**
b_2	**0.677**	**0.677**	**0.668**	**0.668**	b_{2+3}	**0.592**
b_3	**0.522**	**0.522**	**0.560**	**0.560**		
b_{12}			**0.597**	**0.597**	$b_{1(2+3)}$	0.214
b_{13}			-0.258	-0.258		
b_{23}			-0.450	-0.450		
b_4	**0.286**					
b_{14}		**0.165**	**0.165**	**0.156**	b_{14}	**0.156**
b_{24}		**0.407**	**0.407**	**0.433**	$b_{(2+3)4}$	**0.344**
b_{34}		**0.288**	**0.288**	**0.283**		
b_{124}				-0.216	$b_{1(2+3)4}$	0.096
b_{134}				0.352		
b_{234}				-0.285		
LOF	*	0.095	0.259			*

Is the model a good one? It describes the data significantly better than does equation 9.9 (see last column of table 9.14), but ANOVA still shows some evidence of lack of fit. The fact that 6 test points were introduced allowed us to extend model 9.10. We thus assume second-order dependence of $\beta_0(x_1, x_2, x_3)$, giving a model which is second-order in the mixture variables:

$$y = \beta_1 x_1 + \beta_2 x_2 + \beta_3 x_3 + \beta_{12} x_1 x_2 + \beta_{13} x_1 x_3 + \beta_{23} x_2 x_3$$
$$+ \beta_{14} x_1 z_4 + \beta_{24} x_2 z_4 + \beta_{34} x_3 z_4 + \varepsilon \qquad (9.11)$$

We may also assume second-order dependence of $\beta_1(x_1, x_2, x_3)$, which gives:

$$y = \beta_1 x_1 + \beta_2 x_2 + \beta_3 x_3 + \beta_{12} x_1 x_2 + \beta_{13} x_1 x_3 + \beta_{23} x_2 x_3$$
$$+ [\beta_{14} x_1 + \beta_{24} x_2 + \beta_{34} x_3 + \beta_{124} x_1 x_2 + \beta_{134} x_1 x_3 + \beta_{234} x_2 x_3] \times z_4 + \varepsilon \quad (9.12)$$

for which the design is perfectly adapted but where the number of distinct points is equal to the number of coefficients. Thus, it is not possible to test the validity of the model.

The final row (LOF) is the significance of the lack of fit sum of squares in the analysis of variance.

The use of a model such as 9.11 where we assume different orders of dependence of $\beta_0(x_1, x_2, x_3)$ and $\beta_1(x_1, x_2, x_3)$ may be justified by the fact that all terms in the model are first- or second-order, and that third-order terms are excluded. However there are enough experiments in the design to analyse the data according to model 9.12 and estimate the significance of the coefficients in the model (table 9.15). We find that the coefficients β_{124}, β_{134}, β_{234} are not significant, suggesting simplification of the model to 9.11.

The validity of model 9.11 may be tested by ANOVA. No significant lack of fit is found. It appears that the model may be accepted. As we have already seen further simplification to model 9.10, with elimination of the second-order mixture terms leads to a worse fit.

The predictions of the model (coefficients in table 9.15) are shown as a contour plot in figure 9.26. More test points, ($\frac{1}{3}$, $\frac{1}{3}$, $\frac{1}{3}$) at each temperature may be added, if desired, to validate the model further.

Further simplification of the model
It may be seen in table 9.15 that the coefficients of terms in x_2 (proportion of the "ethanol" pseudocomponent) are not significantly different from the corresponding terms in x_3 (the propylene glycol pseudocomponent). Also b_{23} is not significant. This suggests that we may simplify the model by combining the two pseudocomponents. Since they are both equivalent in their composition, the former being 50% ethanol/50% water, the second 50% propylene glycol/50% water, the combination is easy to interpret.

The model 9.12 becomes:

$$y = \beta_1 x_1 + \beta_{2+3}(x_2 + x_3) + \beta_{1(2+3)} x_1(x_2 + x_3) + \beta_{14} x_1 z_4 + \beta_{(2+3)4}(x_2 + x_3) z_4$$
$$+ \beta_{1(2+3)4} x_1(x_2 + x_3) z_4 + \varepsilon \qquad (9.13)$$

indicating that ethanol and propylene glycol behave very similarly. Coefficients are given in the last column of table 9.15. Such analysis may be used to justify simplification of a formulation, especially where there are large numbers of components. See the discussion on screening designs in section III of this chapter.

However analysis of variance of the regression of equation 9.13 shows significant lack of fit. The model 9.13 is rejected, and we retain the model 9.11.

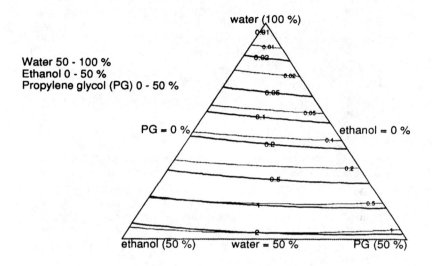

Figure 9.26 Contour plots for solubility at 4°C (thin lines) and 22 °C (thick lines).

3. Generalisation

Mathematical models (product models)
We may use a similar argument to that of the above example. Any polynomial model may be postulated for describing the response as a function of the independent variable(s). For example:

$$y = \beta_0 + \beta_4 z_4 + \beta_{44} z_4^2 + \varepsilon \tag{9.14}$$

or

$$y = \beta_0 + \beta_4 z_4 + \beta_5 z_5 + \beta_{45} z_4 z_5 + \varepsilon \tag{9.15}$$

Each of the coefficients of these models may be a function of composition of the mixture, the dependence being linear, quadratic, or other. Suppose there is one process variable, as in model 9.14, for which the coefficients have a quadratic dependence on the composition, so that:

$$\beta_0 = \beta_{0,1} x_1 + \beta_{0,2} x_2 + \beta_{0,3} x_3 + \beta_{0,12} x_1 x_2 + \beta_{0,13} x_1 x_3 + \beta_{0,23} x_2 x_3$$

Model 9.14 becomes:

$$y = \beta_{0,1}x_1 + \beta_{0,2}x_2 + \beta_{0,3}x_3 + \beta_{0,12}x_1x_2 + \beta_{0,13}x_1x_3 + \beta_{0,23}x_2x_3 + \beta_{4,1}x_1z_4$$
$$+ \beta_{4,2}x_2z_4 + \beta_{4,3}x_3z_4 + \beta_{4,12}x_1x_2z_4 + \beta_{4,13}x_1x_3z_4 + \beta_{4,23}x_2x_3z_4$$
$$+ \beta_{44,1}x_1z_4^2 + \beta_{44,2}x_2z_4^2 + \beta_{44,3}x_3z_4^2$$
$$+ \beta_{44,12}x_1x_2z_4^2 + \beta_{44,13}x_1x_3z_4^2 + \beta_{44,23}x_2x_3z_4^2 + \varepsilon$$

It has 18 coefficients.

If we choose the same type of mixture model for each coefficient in the process model, then the overall model we obtain is called a product model. The total number of coefficients p is obtained by multiplying the number of coefficients in the mixture model, p_1, by the number in the process model, p_2.

$$p = p_1 \times p_2$$

Experimental designs: "product designs"
The designs are called *product* designs. A design is chosen for the mixture variables (for example, a Scheffé design) and another for the process variables (factorial, RSM design, etc.). The N_2 experiments of the process design are carried out for each of the N_1 distinct compositions of the mixture design, giving $N_1 \times N_2$ experiments in all. If each of the designs is optimal for its model, mixture, and process, then the product design is also optimal.

Examples of the use of these designs are to be found in the pharmaceutical literature. A combined mixture-factorial product design has been used to optimize tablets for tropical countries (14), studying the proportions of the 3 diluents in the mixture and 2 process variables (lubrification time and the compression force).

We take the example of a formulation of a drug with 3 excipients whose proportions are allowed to vary. The formulation is a granulate produced by wet granulation and the effects of 2 process components, impeller speed and granulation time, is to be studied along with those of the mixture variables. The amount of binder solution is that normally thought optimal for the formulation and is adjusted according to the fractions of the different diluents.

If the mixture factor space is a simplex then the formulation may be studied by a simplex centroid design of 7 experiments (figure 9.3). The effect of the process variables may be investigated by means of a suitable first- or second-order design, according to which model is thought appropriate. For the first-order model a 2^2 factorial design would be selected. Otherwise we might take any suitable second-order design, such as a uniform shell (Doehlert) design. If the Doehlert design (7 experiments) is replicated for each mixture in the simplex centroid design, the result is a design of $7 \times 7 = 49$ runs (figure 9.27).

This design has been used to study a formulation for treating textiles (15). Vojnovic and co-workers have taken a very similar design for the optimization of a granulate produced in a high-speed mixer. They combined a Doehlert design with a mixture design, based on 3 pseudocomponents forming an irregular simplex (7).

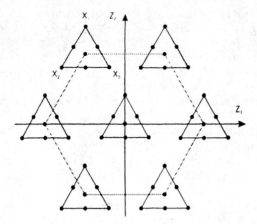

Figure 9.27   Product design of a 2 factor Doehlert design and a 3 factor simplex centroid design

We see that these full designs, obtained by multiplying a mixture design (Scheffé or simplex centroid) by a classical design for independent variables (factorial, Doehlert, or hybrid), contain a very large number of experiments. Methods for reducing this to a smaller number would be useful (16). If the "process" design is a 2-level factorial design, it may sometimes be replaced by a fractional factorial design at each point in the mixture factor space. The number of experiments may thus be halved. We treated another method of reducing the number of experiments, that of D-optimal designs, in chapter 8. In the following chapter we will cover its extension to mixture design, and also to these mixture-process studies.

B. Treatment of a Mixture Component with Narrow Variation Limits

We saw that where there are narrow constraints on the proportions of formulation components, except for a "filler" whose proportions may vary freely, these constrained components may be treated as independent variables. It is not uncommon to treat problems where some components vary within narrow limits, but several components vary freely and so must be treated as mixture components. However the constrained components may still be treated as independent variables.

Consider a formulation study where all components are allowed to vary considerably except for the level of lubricant. We want to know the effect of changing the level of the lubricant, magnesium stearate, from 0.5 to 2%, along with the lubrification time and at the same time changing the levels of certain other excipients (diluents for example). Although the levels of these excipients are not totally independent of the amount of lubricant added, the dependence is extremely slight and can be neglected. The level of the lubricant can be treated as a coded independent variable, level -1 for 0.5% magnesium stearate, and +1 for 2%. The lubrification time is also a process variable. The other excipient levels, those which vary considerably in the design, are treated as mixture components (fractions of the amount remaining after subtraction of the stearate). So if there are 3 variable components, the second-order Scheffé design of 6 experiments may be set up 4 times, once at each combination of the upper and lower levels of the independent variables, magnesium stearate level and lubrification time, making a total of 24 experiments.

Another application, to return to the solubility example, might be the introduction of a low level of a salt, or of a surface active substance to the ternary system. This will have very little effect on the other components and so can be treated as a (coded) independent variable, Z_i in one of the above "mixed" models.

What are the advantages of expressing one or more mixture components as independent variables, in this way? If we test the effect of stearate over a small range we may conclude that it has an effect, perhaps comparable to more considerable changes in the levels of the other components. If it were treated as a mixture component the calculated coefficient would be the extrapolated value (for the response being treated) of the formulation of pure magnesium stearate (or if we were using pseudocomponents, of a formulation very rich in stearate). Whereas if we treat it as a coded variable the coefficient is equal to half the mean effect of increasing the amount of stearate from 0.5% (level -1) to 2% (level +1). Analysis and predictions will be identical but the interpretation of the coefficient is evidently simpler when the excipient level which changes only slightly is treated as a process variable, and co-linearity of the coefficients in the model is much reduced. Chu *et al.* use such a design and model, where the Scheffé design is repeated 4 times at different levels of a constituent present at a low level (5).

Another possible advantage is the flexibility of the model. In the case of a pure mixture model we are almost obliged to have a model that is of the same order in all components. Here there are a number of possible solutions. The model may be entirely first-order (equation 9.9) or second-order in all mixture and "process" components (equation 9.12). Alternatively we can use various "mixed" models, such as that of equation 9.10 that is linear in the mixture terms with interactions between mixture and "process" terms.

C. Mixture-Amount Designs

1. The mixture-amount problem in pharmaceutical development

One important independent factor that can vary alongside the mixture variables is the *total amount* of material in the mixture. If the properties of the mixture are thought to vary according to the total amount as well as with changes in the mixture's composition, we need to use a mixture-amount model (1). The properties of a "bulk" mixture might be thought to be little affected by the amount. This is no longer true once we have formed unit doses, such as tablets, capsules or suppositories. The total amount must be considered as a variable factor, and it may normally be treated as an independent variable, using equations like the mixed models of the previous section. Another possible approach would be to consider the *amount* of each mixture component as an independent variable. This can be only be done when there are no explicit restrictions on the unit mass of the pharmaceutical dosage form.

Consider a tablet containing a drug substance X and a diluent lactose. Several dosages are to be formulated and the amount of drug substance per tablet can vary from 10 to 50 mg. The amount of lactose is allowed to vary from 90 to 150 mg. There are no explicit restrictions on the tablet weight, other than those implied by the restrictions on the lactose and drug substance. The resulting domain is shown in figure 9.28a, and it can be studied by a factorial design, or composite design, or any other of the designs suitable for independent variables that we have discussed.

2. Factor space and model for the mixture-amount problem

Consider a case of a tablet formulation where the factor space is a simplex. An example might be that of the tablet formulation treated earlier, with a fixed proportion (10%) of active substance added to it. The mixture could be compressed to give tablets between 100 mg (dosed at 10 mg) and 200 mg (dosed at 20 mg). The tablet mass may thus be treated as an independent variable and the model equation 9.14 may be used (figure 9.28b)

3. Restriction of the unit dose

This constraint, in the case of mixture-amount problems where the proportion of active substance is allowed to vary, is one which is particularly relevant to pharmaceutical formulation. We take a formulation where the percentage of drug substance is between 5 and 50%. This variation is compensated for by a change in the percentage of diluent (for simplicity we assume a single diluent). The tablet mass is varied from 100 to 200 mg, so the design space is square as shown in figure 9.28c (broken lines).

Now assume that several dosage strengths are to be formulated, from 10 to 50 mg. This adds a further constraint to the factor space, as the 5% mixture will only give a 5 mg tablet at 100 mg, and the 50% mixture is too concentrated for a tablet of mass 200 mg. This leads to the limits shown on figure 9.28c. A product model may be used as described in the previous section.

But, even if the model is simple, the design space is not. The choice of experiments to be carried out is not straightforward and again it normally requires the use of the exchange algorithm method, described in chapter 8, to give a D-optimal design (see chapter 10).

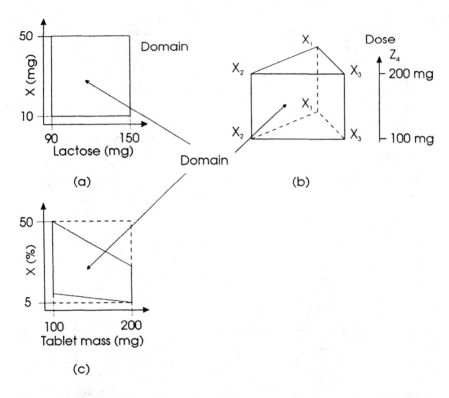

Figure 9.28 Factor space for a 2 component mixture containing 5-50% drug substance. (a) Limits on amount of drug in tablet. (b) and (c) Limits on total tablet weight.

   Sometimes the total amount of material per unit (unit mass) is allowed to vary, but the unit dose is fixed. The tablet or capsule weight is optimized as well as its composition for a given unit dose. So the tablet or capsule fill weight will have to vary in inverse proportion to the proportion of drug substance in the mixture. The proportion of drug substance, and the dosage unit weight are therefore confounded, and the standard mixture model may be used.

References

1. J. A. Cornell, Experiments with Mixtures, 2nd edition, Wiley, N. Y. (1990).
2. R. Huisman, H. V. van Kamp, J. W. Weyland, D. A. Doornbos, G. K. Bolhuis, and C. F. Lerk, Development and optimization of pharmaceutical formulations using a simplex lattice design, *Pharm. Weekbl. [Sci.]*, **6**, 185-194 (1984).
3. S. E. Frisbee and J. W. McGinity, Influence of nonionic surfactants on the physical and chemical properties of a biodegradable pseudolatex, *Eur. J. Pharm. Biopharm.*, **40**, 355-363 (1994).
4. H. Scheffé, Experiments with mixtures, *J. Roy. Stat. Soc., Ser. B.*, **20**, 344-360 (1958).
5. J. S. Chu, G. L. Amidon, N. D. Weiner, and A. H. Goldberg, Mixture experimental design in the development of a mucoadhesive gel formulation, *Pharm. Res.*, **8**, 1401-1407 (1991).
6. G. F. Piepel, Defining consistent constaint regions in mixture experiments, *Technometrics*, **25**, 97-101 (1983).
7. D. Vojnovic, M. Moneghini, and F. Rubessa, Optimization of granulates in a high shear mixer by mixture design, *Drug Dev. Ind. Pharm.*, **20**, 1035-1047 (1994).
8. M. L. Wells, W. Q. Tong, J. W. Campbell, E. O. McSorley, and M. R. Emptage, A four-component study for estimating solubilities of a poorly soluble compound in multisolvent systems using a Scheffé-type model, *Drug Dev. Ind. Pharm.*, **22**, 881-889 (1996).
9. N. G. Becker, Models for the response of a mixture, *J. Roy. Stat. Soc, Ser. B*, **30**, 349-358 (1968).
10. N. R. Draper and R. C. St. John, A mixtures model with inverse terms, *Technometrics*, **19**, 37-46 (1977).
11. A. B. Ochsner, R. J. J. Belotto, and T. D. Sokoloski, Prediction of xanthine solubilities using statistical techniques. *J. Pharm. Sci.*, **74**, 132-135 (1985).
12. R. D. Snee and D. W. Marquardt, Screening concepts and designs for experiments with mixtures, *Technometrics*, **18**, 19-20 (1976).
13. M. Chariot, G. A. Lewis, and D. Mathieu, Use of experimental design in formulation: constituent screening, 2nd Int. Forum Phys. Chem. Formulation & Applic., Toulouse, 1990.
14. C. E. Bos, G. K. Bolhuis, C. F. Lerk, J. H. de Boer, C. A. A. Duineveld, A. K. Smilde, and D. A. Doornbos, Optimization of direct compression tablet formulations for use in tropical countries, *Drug Dev. Ind. Pharm.*, **17**, 2477-2496 (1991).
15. J. Chardon, J. Nony, M. Sergent, D. Mathieu, and R. Phan-Tan-Luu, Experimental research methodology applied to the development of a formulation for use with textiles, *Chemom. Intell. Lab. Syst.*, **6**, 313-321 (1989).
16. C. A. A. Duineveld, A. K. Smilde, and D. A. Doornbos, Designs for mixture and process variables applied in tablet formulations, *Analytica Chimica Acta*, **277**, 455-465 (1993).

10

MIXTURES IN A CONSTRAINED REGION OF INTEREST

Screening, Defining the Domain, and Optimizing Formulations

I. IRREGULAR (NON-SIMPLEX) DOMAINS IN MIXTURES

A. Constraints on the Components in Pharmaceutical Formulations

We have now examined what distinguishes experimental design problems involving mixtures and applied the simplex lattice and centroid designs to explore responses over the whole of the possible factor space. As we said earlier, it is hardly realistic to expect all components to be allowed to vary from zero to one hundred percent in a real system. That this should be so for solid pharmaceutical dosage formulation is especially evident – each component of a solid dosage form has its own function, and it will only fulfill that function within a limited concentration range. It is pointless attempting to study all possible combinations of the proportions of the components when we already know that many of these, probably the vast majority, are excluded.

Not only solids, but many liquid systems also, have discontinuities in their properties over the factor space. Figure 10.1 shows a ternary mixture of water, an oil, and a surfactant. According to the proportions of the components, completely different phases are obtained and the methods of experimental design that we have seen so far, along with the polynomial or related models developed for their analysis, are not suitable for analysis of the total system. For the experimental domain, it would be necessary to select a portion of the factor space with the same phase or phases over all of it.

$$C_7F_{15}CO_2Na/C_3F_7CH_2OH$$

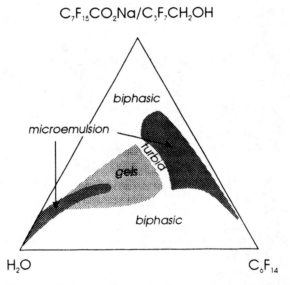

Figure 10.1 Ternary diagram of a water-fluorocarbon-surfactant/cosurfactant mixture showing different phases.

Now, provided the selected portion of the factor space has the same shape as the whole of the factor space (i.e. it is a simplex), it is possible to use the same mathematical models and designs as we used previously to investigate it. We have seen that imposing only lower limits on the domain results only in a decrease in the extent of the experimental region and it still remains a regular simplex.

In chapter 3, we considered the case of narrow constraints for all components but one, this remaining excipient being the filler (making up the 100%). The experimental space can then be considered as cubic. Such systems can still be analysed using the Scheffé models, but we may also treat all variables except the "filler" as independent and use factorial and composite designs.

In practice, many systems are intermediate, certain components with narrow bands, others allowed to vary considerably, and no single filler component. Neither the simplex nor the "factorial" approaches can be used to construct an experimental design.

When confronted with such a system, our problem is as follows:

- To define the experimental space.
- To select the best experiments in this space.

We start with a very simple example, that could have been carried out almost without any statistical analysis, yet which illustrates the advantages of the multifactor, statistical approach to formulation (1,2).

B. Example of the Formulation of a Modified Release Tablet

1. Definition of the problem and the initial domain

The drug concerned was the salt of a weak base, with solubility in water about 2%. The base would precipitate out of solution at pH values above 5 or 6. The requirement was for a tablet formulation releasing the drug substance over about 3 hours. To ensure an adequate release rate in the neutral pH conditions of the small intestine it was decided to introduce an acidic ingredient, that would maintain a low pH within the dosage form and prevent formation of the insoluble base. Carbomer 934 was chosen as a matrix-forming polymer which would delay release but would also maintain the acidic environment within the tablet.

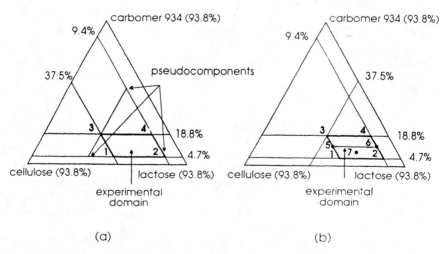

Figure 10.2 Factor and design space for formulation of a modified release tablet. (a) Initial and second design. (b) Final design.

The tablet mass was fixed at 200 mg. Each unit contained 10 mg (5%) drug and 1.2% lubricant. It was considered that the lubricant level could be adjusted and optimized later as part of a process study. Thus, although there were 5 components in the mixture, 2 of them were fixed. The factor space in the remaining components, microcrystalline cellulose, lactose, and carbomer (totalling 93.8%) can be represented as a ternary diagram. The tablets were manufactured by direct compression of the powder mixture and the dissolution profiles were measured at pH 2 (0.01 M hydrochloric acid) and pH 7 (phosphate buffer). The time for 50% dissolution was measured in each case.

The initial domain was defined as in table 10.1. Limits were fixed for carbomer and for the cellulose, with the lactose allowed to vary $q.s.$ The factor space and the domain are shown in figure 10.2a.

Table 10.1 Initial Experimental Domain for Formulation of a Sustained Release Tablet (1)

Factor (component in mixture)	Associated variable	Lower limit	Upper limit
carbomer 934 (%)	X_1	4.7	18.8
microcrystalline cellulose (%)	X_2	9.4	37.5
lactose (%)	X_3	$q.s.$ 93.8	
active substance (%)	fixed	5.0%	
lubricant (%)	fixed	1.2%	

2. Experimental design and results

Initial design and model
Experiments were carried out at each vertex of the space (table 10.2, experiments 1-4). It thus resembles a 2^2 factorial design. Analysis would normally be carried out using one of two models. There is the first-order Scheffé model (see chapter 9):

$$y = \beta_1 x_1 + \beta_2 x_2 + \beta_3 x_3 + \varepsilon$$

where x_i are the proportions of each component, expressed either directly or in terms of pseudocomponents. Even though the design is not a simplex, we can define the mixtures in terms of the pseudo-components (figure 10.2a), which are the mixtures defining the corners of the smallest simplex that encloses the experimental factor space. It might have been preferable to have chosen a lower limit for lactose (37.5%) instead of the upper limit for microcrystalline cellulose – the design would have "fitted better" within the simplex.

Alternatively, the synergistic model, used for factorial designs, may be employed to analyse the influence of the factors:

$$\hat{y} = \alpha_0 + \alpha_1 z_1 + \alpha_2 z_2 + \alpha_{12} z_1 z_2 + \varepsilon$$

where z_1 represents the replacement of lactose in the mixture by carbomer and z_2 its replacement by microcrystalline cellulose. The variables may be transformed, or coded, in which case $2\alpha_1$ would be the increase of the response y, on increasing the level of carbomer from 4.7% (level -1) to 18.8% (level +1), the proportion of lactose being accordingly decreased by 14.1%. A disadvantage of this model is that, although the lactose has been eliminated from the model, its proportion varies very considerably over the experimental space and is dependent both on the amount of cellulose and of carbomer.

The dissolution profiles for experiments 1 and 2 were sufficiently encouraging for the project to be continued. However, the tablets of formulations 3 and 4 were not tested for dissolution, as both showed unacceptable variation in tablet weight and hardness and the powder of experiment 3 would hardly flow.

Table 10.2 Formulation of a Slow Release Tablet: Plan and Responses (1)

No.		carbomer (%)	microcrystalline cellulose (%)	lactose (%)	$t_{50\%}$ (h)	
					pH 2	pH 7
1	*	4.7	37.5	51.6	0.80	1.69
2		4.7	9.4	79.7	0.81	1.37
3		18.8	37.5	37.5	***	***
4		18.8	9.4	65.6	***	***
5	*	9.4	37.5	46.9	1.86	2.07
6		9.4	9.4	75.0	0.95	2.46
7	*	7.0	23.5	63.3	0.67	1.95
8	*	4.7	25.8	63.3	0.70	1.49
9	*	9.4	25.8	58.6	0.82	1.85
10	*	7.0	30.5	56.3	0.72	1.54

*   Experiments comprising the final design (step 3)

*** Tablets not tested for dissolution as flow and compression properties were very poor

Step 2

Owing to the failure of experiments 3 and 4, a reduced upper limit for carbomer of 9.4% was established (that is, 10% of the variable components). The new design is shown in figure 10.2b and table 10.2 (experiments 1, 2, 5, and 6 and a centre point, 7). All 3 new experiments were feasible and it appeared that there was an interesting region within the experimental space. The dissolution data are shown in table 10.2. The data could be analysed using either of the models given above. Because of the narrowness of the carbomer limits and the relatively large amount of lactose, the effect on the amount of lactose of going from one carbomer limit to the other is minimal. So in the case of the canonical equation, X_1 is almost independent of X_3 (lactose). On the other hand, X_2 and X_3 are almost totally correlated. The "factorial" approach appears to be quite suitable for this domain and this design. Comparison of the results of experiment 7 with the other 4 experiments indicates that neither the first-order model nor the synergistic could be fitted. At pH 2, the dissolution rate of experiment 7 is much more rapid than those of the tablets of the 4 "factorial" experiments.

Step 3

The scientist developing this formulation wished to optimize it using RSM. The results of the experiment at the centre indicated that the first-order model was not valid within the domain. We have seen that in such circumstances we may either choose to use a more complex model, augmenting the design, or may reduce the extent of the domain. Here she decided to reduce the domain, eliminating low levels of microcrystalline cellulose, as shown in figures 10.2b and 10.3, but also to use a second-order model. Thus, 3 additional experiments (nos. 8-10) were carried out, selected to allow determination of a quadratic model for predicting the dissolution profile over a limited region of the total factor space, and following on from that, optimization of the formulation.

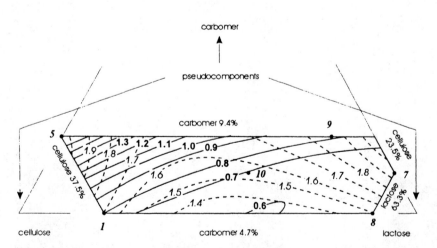

Figure 10.3 Formulation of a sustained release tablet: final domain and contour plots for 50% dissolution time at pH 2 (solid lines) and pH 7 (broken lines).

3. Analysis of the final model

The design and response data marked by asterisks in table 10.2 were analysed according to the second-order Scheffé model:

$$y = \beta_1 x_1 + \beta_2 x_2 + \beta_3 x_3 + \beta_{12} x_1 x_2 + \beta_{13} x_1 x_3 + \beta_{23} x_2 x_3 + e$$

The x_i are defined in terms of the pseudocomponents.

Contour lines for the dissolution time at pH 2 and pH 7 are given in figure 10.3. A formulation for further study could be selected from the region of the top right on the basis of the desired release rate and a reduced pH dependance.

C. Screening Mixture Components

The above example demonstrates a number of changes in size and shape of the domain. With only 3 variable components this was easily done graphically, but where there are more components, systematic methods are necessary, both for selecting components and defining their limits. Screening, or axial designs, described previously for the simplex domain, provide a possible means. Here we extend the treatment of section III.B of chapter 9 to the more usual case of a constrained mixture.

1. Principle of the method

In an axial design in the unconstrained mixture, where the domain is the whole of the factor space for the variable components, a reference mixture is chosen where the q components are present in equal proportions q^{-1}. Then, for each component, we may do experiments along an axis between the vertex which is the pure component and the mid-point of the face opposite the vertex, where the component we are studying is absent but still all others are present in equal proportions, $(q - 1)^{-1}$. All of these principal axes pass through the reference mixture.

When there are constraints on the mixture, even if the domain remains a regular simplex, we may no longer do this. Firstly, the mixture with all components set at q^{-1} will quite possibly not be available. Secondly, if instead we choose the centre of gravity of the simplex as the reference mixture and do experiments along the pseudocomponent principle axes, the relative proportions of some or all of the components will change. We see this in figure 10.4a for the ternary mixture, water/propylene glycol/ethanol, where a lower limit of 50% is imposed for water. Taking the water axis, X_1, and moving up from the base of the domain to the vertex, the proportions of the other two components remain equal to one another. If, on the other hand, we take the ethanol principal axis X_3, the ratio of the other *pseudocomponents* remains constant, but that of the (*pure*) components varies considerably (table 10.3 and figure 10.4a).

Table 10.3 Points Along the "Ethanol" Pseudocomponent Principal Axis: Variation of the Proportion Ratio of the Remaining Components

No.	Point	Water (%)	Propylene glycol (%)	Ethanol (%)	Ratio $X_2{:}X_1$
1	base	75.0	25.0	0.0	0.333
2	centre of gravity	66.7	16.7	16.7	0.250
3	test point	58.3	8.3	33.3	0.143
4	vertex	50.0	0.0	50.0	0.000

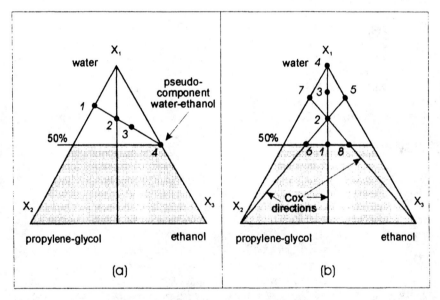

Figure 10.4 Ternary mixture with one constraint: (a) points along a principle axis, table 10.3 and (b) the Cox directions.

As a solution to this problem, we may still select the centre of gravity, or similar convenient point, as a *reference mixture* and draw lines for each component through the point of reference mixture, allowing it its full extent of variation, but keeping the ratios of the other component constant. These lines are called the "Cox directions" (3). These are shown in figure 10.4b, the centre of gravity being the reference mixture.

If the water concentration is allowed to vary, the ratio of the remaining components is unity, as the reference mixture contains 16.7% of each. Thus the Cox direction line passes though the water vertex, as before. The line for ethanol, if extended outside the domain, would pass through the ethanol pure component vertex, the ratio propylene glycol to water remaining constant, at 0.250. The design points are given in table 10.4. This design allows the effect of any one of the given components to be defined, while keeping the relative proportions of all other variable components constant. Clearly, there is no longer any particular advantage in using the simplex for treating this kind of problem.

Table 10.4 Points on Cox Directions for the Constrained Simplex

No.	Point	Water (%)	Propylene glycol (%)	Ethanol (%)	Ratio $X_2{:}X_3$	$X_3{:}X_1$	$X_3{:}X_1$
1	Base	50.0	25.0	25.0	1.00		
2	Centre of gravity	66.7	16.7	16.7	1.00		
3	Intermediate	83.3	8.3	8.3	1.00		
4	Vertex	100.0	0.0	0.0	1.00		
5	Minimum	80	0.0	20		0.25	
[2]	Centre of gravity	66.7	16.7	16.7		0.25	
6	Maximum	50	37.5	12.5		0.25	
7	Minimum	80	20	0.0			0.25
[2]	Centre of gravity	66.7	16.7	16.7			0.25
8	Maximum	50	12.5	37.5			0.25

It is therefore usual practice to determine the effect of component *i* by taking lines from a selected *reference mixture* in the domain (usually its centre of gravity), allowing component *i* to vary within its allowed limits, while keeping the ratios of the remaining components constant. The proportions of the remaining components are no longer equal to one another as they were in the unconstrained simplex example of chapter 9, but rather keep the same relative proportions as in the reference mixture (figure 10.4b).

We now examine the example of the formulation of a delayed release tablet by the dry-coating method.

2. Dry-coated tablet

Definition of the problem: experimental domain
A tablet with delayed drug release may be formulated by dry-coating a conventional immediate release tablet, containing a soluble drug, with a mixture containing a wax matrix forming excipient (4). Dissolution begins very slowly and then after a given time the drug starts to be released, in some cases rapidly, as a pulse. It is possible to modulate both the lag-time before release starts and the rate of release. A

formulation of this type, based on hydrogenated castor oil, was studied by Hoosewys *et al.* (5), who wished to screen the effects of the various excipients on the formulation, and particularly on its release profile, within the limits of table 10.5. A reference mixture was selected within the domain central to the ranges of all components, except magnesium stearate.

Table 10.5 Delayed Release Dry-Coated Tablet – Domain

		Minimum	Maximum	Reference mixture
hydrogenated castor oil (HCO)	(%)	10	75	39
lactose	(%)	0	40	19.5
microcrystalline cellulose (μCC)	(%)	0	40	19.5
polyethylene glycol 600 (PEG)	(%)	0	50	20
magnesium stearate	(%)	0	2	2

Screening design
In this design each component is allowed to vary from its upper to its lower limit, whilst maintaining the relative proportions of the remaining components equal to their ratios in the reference mixture. For example, in the case of hydrogenated castor oil, the remaining major components are at about 20%, so, as the concentration of HCO was decreased from 39% to 10%, they were each increased to about 30%. The magnesium stearate, at about a tenth of their level, was adjusted in proportion. It is not, however, allowed to increase above 2%. The resulting design is shown in table 10.6. The similarity between this and the axial design for a placebo tablet (table 9.11) is obvious.

There are thus 5 series of experiments – one for each component, each consisting of 3 experiments, the component at its upper and its lower extreme value, plus the reference mixture, which therefore enters into each series. It would have been preferable to have replicated the reference experiment, but this was not done because of restrictions of time and material.

Hydrogenated castor oil being varied over a wider range than the other components, an intermediate experiment was carried out at 60% HCO. Magnesium stearate was only decreased from its reference state level of 2% to zero, so an intermediate experiment, at 1% stearate, was also inserted here.

Results and discussion
The experiments were carried out and responses were measured. These are also listed in table 10.6. The main response measured was the dissolution profile, which was characterised by the delay before the beginning of release, which was several hours, and the time over which the release took place, which could be relatively short. However in some cases the release was so slow and incomplete that this latter quantity had to be estimated by extrapolation. The formulation containing

75% hydrogenated castor oil gave such slow release that the result is not given in the table.

Table 10.6 Design for the Screening of a Drycoated Tablet Formulation: Results

No.		HCO (%)	lactose (%)	μCC (%)	PEG 600 (%)	Mg stearate (%)	lag-time (h)	dissolution time (h)
1	Reference	39	19.5	19.5	20.0	2.0	4.2	4.4
2	HCO	**10.0**	29.0	29.0	30.0	2.0	3.1	2.0
3		**60.0**	12.8	12.8	13.0	1.4	8.6	10.0
4		**75.0**	8.0	8.0	8.0	1.0	>14	
5	lactose	48.7	**0.0**	24.3	25.0	2.0	7.8	5.5
6		29.4	**40.0**	14.7	15.1	1.5	4.3	3.1
7	μCC	48.7	24.3	**0.0**	25.0	1.5	6.6	5.1
8		29.4	14.7	**40.0**	15.1	1.5	2.5	4.0
9	PEG	49.0	24.5	24.5	**0.0**	2.0	3.0	9.7
10		24.5	12.25	12.25	**50.0**	1.0	4.0	1.8
11	Mg	39.6	19.7	19.7	20.2	**1.0**	5.3	3.7
12	stearate	39.8	19.9	19.9	20.4	**0.0**	5.3	4.3

The effects of each of the five components on lag-time and the dissolution time are given in figures 10.5a-b. The independent axis represents the change in percentage of each excipient, with respect to the reference mixture.

- The percentage of *hydrogenated castor oil* (HCO) had a major effect on the release, lengthening the time before release begins and then slowing the release.
- The effects of *lactose* and *microcrystalline cellulose* (μCC) on the release rate were similar to one another, causing release to be more rapid (this being possibly an effect of the corresponding decrease in hydrogenated castor oil) and decreasing the lag-time.
- Increasing the percentage of *polyethylene glycol* (PEG) had a similar effect on the dissolution rate. Omitting it gave tablets with very slow release. Changing the proportion of polyethylene glycol had very little effect on the lag-time.
- The effect of *magnesium stearate* is unclear, though removing it seems to increase the lag-time and make the subsequent dissolution more rapid.

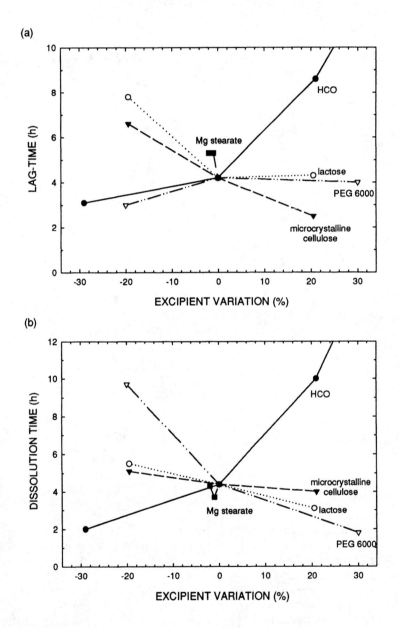

Figure 10.5 Results of screening design for dry-coated tablet (5). (a) Lag-time. (b) Dissolution time.

3. Other screening designs

Since hydrogenated castor oil has the greatest effect on the dissolution profile, a modified approach for testing the effects of the diluents (in which class we include polyethylene glycol) might be to maintain the hydrogenated castor oil (HCO) level constant at 39% and to vary one component, keeping the ratios of the remaining excipients constant.

Thus for lactose we would have:

	HCO (%)	lactose (%)	µCC (%)	PEG 600 (%)	Mg stearate (%)
(low)	39.0	**0.0**	29.1	29.9	2.0
reference	39.0	19.5	19.5	20.0	2.0
(high)	39.0	**40.0**	9.6	9.9	1.5

4. Continuation

The 11 experiments allow many of the important trends to be identified by a direct graphical analysis and tablets with different profiles to be prepared. Mixture screening designs may therefore reveal major effects of the proposed constituents, allowing levels of each of them to be chosen for further study. Such a design may allow the choice of a satisfactory formulation, but it is insufficient for optimization of the kind described in chapter 5, even if the selected limits for screening were found to enclose an interesting region for study. A good design for *estimating* a model will have most of its experiments on the edge of the domain.

However, following the choice of a profile and new limits for the excipients, a D-optimal design could be set up in a reduced domain and the formulation optimized (perhaps taking into account other properties besides the dissolution profile). In the following two sections we will see how good designs for RSM may be obtained for an irregular domain.

II. DEFINING THE MIXTURE DESIGN SPACE

Once the individual limits of each component are known, the design space for the experiment may be defined. For 3 components it may be determined graphically. For more than 3 or 4 components the *McLean-Anderson algorithm* (6) is used. Experiments must then be selected within the design space. As examples we take the solubility of a drug in a ternary mixture where the constraints give rise to a non-simplex design space and the formulation of a sustained release tablet, with 4 variable components.

A. The McLean-Anderson Algorithm

1. Principle

To describe the domain it is sufficient in the first instance to calculate the coordinates of the corners of the q-1 dimensional polyhedron, defined by the upper and lower limits on the formulation. The algorithm used for this purpose is remarkably simple and does not require a computer program.

Let the formulation consist of q components each associated with an upper and a lower limit. For each component, calculate the range (that is the difference between upper and lower levels) and order the components 1 to q in order of increasing allowed range.

Set up a full 2^{q-1} factorial design for all the components except the last, which will normally be the diluent or one of the diluents. The lower limit is equivalent to the -1 level in the factorial designs of chapter 3, the upper limit is the +1 level. Then for each experiment, calculate the concentration of the last component needed for a total of 100% (or whatever is the total of the variable components).

Finally, examine each of these experiments to see whether the qth component's concentration is within the predefined limits for this component. If it is above the upper limit, set it equal to that limit. If it is below the minimum allowed value, set it equal to that value. Then adjust the percentage of each remaining component in turn, so that their total adds up to 100%. Adjustment of each component will give rise to a separate experiment, although not all of them will be feasible, as the proportion of the component will sometimes be outside its limit. In this case the experiment is simply deleted.

2. First example: solubility of phenobarbital

Defining the problem
Belotto *et al.* (7) have reported the determination of the response surface for the solubility of phenobarbital in a mixture of water, ethanol and glycerol. The domain was defined by the following limits (see figure 10.6):

0%	$\leq$	propylene glycol	$\leq$	50%
0%	$\leq$	glycerol	$\leq$	80%
10%	$\leq$	water	$\leq$	100%

A large number of experiments had been previously reported within this domain (8) and this allows us to use this example to demonstrate the advantages of reducing the numbers of experimental runs, using the McLean-Anderson method (this section) and the exchange algorithm method of chapter 8 (in the following section).

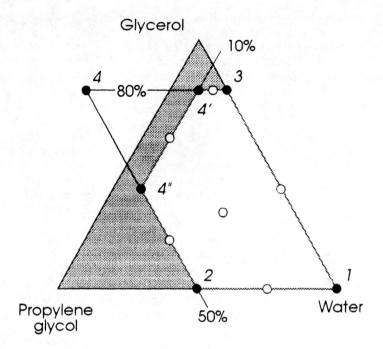

Figure 10.6 Experimental domain for solubility of phenobarbital (see table 10.7).

The McLean-Anderson algorithm

The least extensive range is that for propylene glycol, followed by glycerol. The 2^2 factorial design is therefore set out for those 2 components, with the percentage of water as the slack variable, calculated for each of the 4 experiments (table 10.7, lines 1-4).

Table 10.7 Vertices of Factor Space for Solubility of Phenobarbital (7)

No.	propylene glycol	glycerol	water	
1	0.0	0.0	1.0	
2	0.5	0.0	0.5	
3	0.0	0.8	0.2	
4	0.5	0.8	-0.3	Outside limits
4'	0.1	0.8	0.1	
4"	0.5	0.4	0.1	

In the case of experiment 4, the concentration of water is below its lower limit, which is 0.1, or 10%. The difference is 0.4 (40%), which must be subtracted in turn from the percentage of propylene glycol and of glycerol. So experiment 4 is replaced by experiments 4' and 4", where water is increased to 10%. To compensate for this, propylene glycol is adjusted from 50% to 10% (point 4') or glycerol, initially at 80% is set to 40% in point 4" (table 10.7). Thus, the 5 vertices of figure 10.6 are defined.

Extreme vertices design
If the solubility is assumed to be linearly dependent on the concentrations of the components, then a reasonable choice of experimental points would be the 5 points at the vertices of the design space, shown in figure 10.6. Such a design, proposed by McLean and Anderson (6), is known as the *extreme vertices design*. It would be expected to be entirely adequate for use with the first-order model, equation 9.1, which consists of 3 terms. However, it is clear from figure 10.6 that two of the experiments, 3 and 4', are very close to one another and we will see later that one of them might be omitted without compromising the quality of the design.

Note that this is a case of a Scheffé equation being used for a non-simplex experimental domain. The example is evidently a trivial one, as far as defining the domain goes, as the problem of the design space is so easily solved graphically. With 4 or more components it becomes more difficult.

3. Second example: a hydrophilic matrix tablet of 4 variable components

Defining the problem
Mixture design was used to help in formulation of a sustained release tablet, based on a hydrophilic cellulose polymer, which swells in the presence of water, and so impedes the release of the soluble active substance. Drug release is by a combination of diffusion and erosion. The formulators wished to examine the effect of changing the proportions of polymer, and of the different diluents (9). The constraints on the formulation are given in table 10.8. Other components (drug substance, lubricant) are to be considered as fixed, at least for the time being.

We test the limits to check that they are consistent, as described in chapter 9, section II. The experimental space is evidently not a simplex, so we use the McLean-Anderson method to determine the extent of the domain.

Table 10.8 Domain for Formulating a Hydrophilic Matrix Tablet

Factor		Associated variable	Lower limit	Upper limit
hydrophilic polymer	%	X_1	17	25
lactose	%	X_2	5	42
calcium phosphate	%	X_3	5	47
microcrystalline cellulose	%	X_4	5	52

Table 10.9 Initial 2^3 "Factorial" Design

	Polymer	Lactose	Phosphate	Cellulose	
1	17	5	5	73	outside limits
2	25	5	5	65	outside limits
3	17	42	5	36	*
4	25	42	5	28	*
5	17	5	47	31	*
6	25	5	47	23	*
7	17	42	47	-6	outside limits
8	25	42	47	-14	outside limits

McLean-Anderson algorithm

Microcrystalline cellulose, variable X_4, has the greatest range, so the initial 2^3 "factorial" design is set up using the other 3 variables, as in table 10.9. Of the 8 "factorial" experiments, 4 require correction – those indicated as being outside the limits. Each such point is replaced by 2 new ones (the third is outside the domain in all cases), as shown in table 10.10. The final vertices are marked "▲". The experimental domain polyhedron has 12 vertices. As it is 3-dimensional it may be represented diagrammatically (figure 10.7).

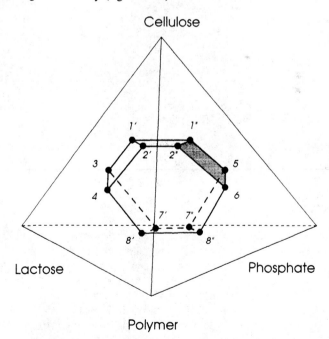

Figure 10.7 Design space for hydrophilic tablet formulation.

Table 10.10   Vertices of the Design Space for Formulation of a Hydrophilic Matrix Tablet

	Polymer	Lactose	Phosphate	Cellulose	
1	17	5	5	*73*	replaced
1'	17	5	24	52	▲
1"	17	24	5	52	▲
1'''	*36*	5	5	52	outside limits
2	25	5	5	*65*	replaced
2'	25	5	18	52	▲
2"	25	18	5	52	▲
2'''	*38*	5	5	52	outside limits
3	17	42	5	36	▲
4	25	42	5	28	▲
5	17	5	47	31	▲
6	25	5	47	23	▲
7	17	42	47	*-6*	replaced
7'	17	42	36	5	▲
7"	17	31	47	5	▲
7'''	*6*	42	47	5	outside limits
8	25	42	47	*-14*	replaced
8'	25	42	28	5	▲
8"	25	23	47	5	▲
8'''	*6*	42	47	5	outside limits

B. Designs Using the Algorithm

The domain is defined by the vertices, whose positions we have determined. Experiments can then be selected in the domain, according to the postulated model.

1. First-order models

The extreme vertices design consisting of the points at each of the vertices, is usually an adequate design for determining the first-order Scheffé model. Quite often it provides rather more experiments than are required. We saw that in the case of the 5 vertices of the solubility experiment, above (figure 10.6), two of these experiments were very close (3 and 4') and one of them could be omitted (4' for example). In the example of the hydrophilic matrix tablet, the domain in factor space (figure 10.7) has 12 vertices, but there are only 4 coefficients to be determined in the first-order model. A choice may accordingly be made between the 12 points.

On the other hand, it may be that the first-order model is inadequate for describing the system within the domain or that it was decided from the beginning to postulate a more complex model. If only the vertices are taken, the number of experiments will probably be insufficient or the design will be of poor quality for determining the second-order or the reduced cubic model. In these cases, it is usual to add further experiments.

2. Experiments for second-order models

From the coordinates of the vertices we calculate:

- Mid-points of the edges, joining adjacent experiments.
- Centres of gravity of hyperfaces of 2, ... k-2 dimensions.
- The overall centre of gravity of the domain.

In the case of the solubility problem, figure 10.6, it can be seen that 5 mid-points of edges have been added, as well as a point close to the overall centre of gravity. There are 11 experiments to determine a second-order Scheffé model (R_{eff} = 55%).

In the case of the hydrophilic matrix (figure 10.7) we may add the centroids of the faces of the polyhedron to the 12 vertices, 18 mid-points of edges, and centre of gravity. The total number of experiments would be 39 in order to determine 10 coefficients in the second-order model (R_{eff} = 26%).

Note that most of the experiments are on the edge of the domain.

3. Choosing experiments for the second-order model

The vertices are generally selected, except when two or more are very close to one another. After that, the midpoints of the edges are chosen. The centre of gravity should be selected, if only for validation of the model. If some points are very close to one another, as are 3 experiments in the top edge of figure 10.6, all but one *may* perhaps be eliminated.

It is thus easy to determine the extreme vertices and related designs and they have been used quite widely (10-13). These designs enable the models to be determined with adequate precision, but they often require a far larger number of experiments than the number of coefficients. We therefore need a method to select those experiments which carry the most information. In chapter 8 we described how the exchange algorithm may be used to reliably select experiments, often in irregularly shaped domains, of independent variables according to the D-optimal criterion. The method is equally applicable to the mixture problem and will be demonstrated for the same examples as the extreme vertices. Other examples are to be found in the pharmaceutical literature (14-17).

III. D-OPTIMAL DESIGNS FOR MIXTURE AND MIXTURE-PROCESS EXPERIMENTS

The optimal design criteria for mixtures are the same as those described in chapter 8: $|X'X|$, $|M|$, d_{max}, and $tr(X'X)^{-1}$. The same exchange algorithms are used to determine the D-optimal designs from a candidate design. For these, there may be advantages in using a model defined in pseudocomponents, especially where the design space is small compared with the total factor space.

Cornell (3) has compared mixture designs optimal according to different criteria (A-, G-, and D-optimality, see chapter 8, section II.C). The designs were of very similar quality. As in chapter 8, we recommend the D-optimal criterion for selecting designs initially, and using the remaining parameters to aid the final choice. Exchange algorithms are also useful for selecting a reduced number of experiments from a product design.

Note that the exchange algorithm selects all or most of the experiments at the exterior of the domain. The worker may wish to add a number of (replicated) experiments at the centre of gravity. This may often improve the precision uniformity of the design.

A. Solubility of Phenobarbital

1. Problem: experimental domain and mathematical model

The response surface of solubility was to be determined as a function of the composition of a ternary mixture, as described in section II.A.2. The domain is defined in table 10.7 and figure 10.6. A large number of experiments, 42 in all, had previously been carried out within this domain (8). These are shown by the diamond shaped points of figure 10.8. The reduced cubic model is proposed:

$$y = \beta_1 x_1 + \beta_2 x_2 + \beta_3 x_3 + \beta_{12} x_1 x_2 + \beta_{13} x_1 x_3 + \beta_{23} x_2 x_3 + \beta_{123} x_1 x_2 x_3 + \varepsilon$$

2. Calculation of the coefficients

Analysis of the data of reference 8 by the Box-Cox method shows that the experimental variance cannot be considered constant within the domain and the most adequate value of λ is zero. This corresponds to a logarithmic transformation of the data, which was the one used by the authors. The coefficients of the model, (with the logarithmic transformation) are given in table 10.11 (column A). The response surface predicted by the model is shown in figure 10.8 (dotted line contours).

3. Reduction of the number of experiments

The authors remarked on the fact that 42 experiments is a large number for estimating only 7 coefficients. We have seen above how a smaller number, 11 experiments, may be selected using the McLean-Anderson method (7). In this paragraph we simulate the selection of experiments from the total data-set of 42. Imagine that they are the only possible experiments, but they have not actually been carried out. We therefore try to construct the best experimental design, the one that allows the model to be determining with the best precision, for as few experiments as possible. We do not take into account in any way the *responses* for these 42 experiments, as these are assumed not to be available.

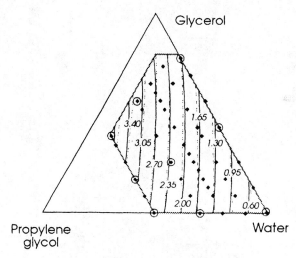

Figure 10.8   Design space and previous experiments for phenobarbital: contour plots.

There are 7 coefficients in the model, so at least 7 experiments are needed. We have therefore determined the statistical properties of D-optimal designs for 7 to 16 experiments. We carried out 30 different runs for each value of N, the initial design being selected randomly in each case, so as to be reasonably sure of having obtained a true optimum on taking the best result. The properties of the optimal designs, as functions of the number of experiments, are shown in figures 10.9a-d. The determinant of the information matrix increases over the whole range, from $N = 9$ to 16, but the other properties show that the best design consists of 9 experiments. These are shown enclosed by circles, in figure 10.8.

The coefficients estimated from these 9 experiments are given in column B of table 10.11, where they may be compared with the values estimated from the entire data set of 42. The solid contour lines on figure 10.8 represent the predictions of the reduced design, dotted lines the predictions of the full data set. The two response surfaces are very close to one another.

Table 10.11 Estimations of the Reduced Cubic Model for Phenobarbital Solubility in a Ternary Mixture from 2 Experimental Data-Sets

	A	B
b_1	2.60	2.56
b_2	5.47	5.43
b_3	0.33	0.25
b_{12}	1.45	0.72
b_{13}	-0.66	-0.16
b_{23}	-1.32	-1.06
b_{123}	-2.06	0.26

A = data-set of 42 experiments (from reference 8)

B = reduced data-set of 9 experiments selected from A by an exchange algorithm

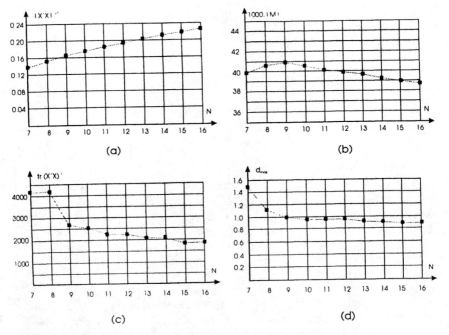

(a)

(b)

(c)

(d)

Figure 10.9 Properties of D-optimal designs for phenobarbital solubility (candidate design of 42 experiments).

One may also compare the *predictive quality* of the designs. The maximum variance function d_{max} from the 9 experiment D-optimal design is 0.97. The value obtained with all the experiments is 0.561. It appears that if we were to do each of the 9 experiments *twice* we would have a design of the same quality, from the point of view of prediction, as the complete 42 point design. One can also show that the determinant of $X'X$ will be larger and the replication of each experiment would also provide information on the reproducibility and allow the validity of the model to be tested.

4. Reduction of the McLean-Anderson design

The vertices of the constrained polyhedron, the midpoints of the edges and the centre of gravity of the domain together make up the 11 experiment design of figure 10.6. Taking this as the candidate design we may extract experiments and determine the "best" designs for 7 to 11 experimental runs. The characteristics of the resulting designs are listed in table 10.12.

Table 10.12 Properties of D-optimal Designs from the McLean-Anderson Candidate Design

N	$\|X'X\|^{1/7}$	$\|M\|^{1/7}$	$tr(X'X)^{-1}$	d_{max}
7	0.139	0.0404	6216	1.00
8	0.157	0.0417	4136	1.00
9	0.171	0.0423	2514	0.97
10	0.183	0.0422	2474	0.96
11	0.190	0.0413	2424	0.96
9'	0.166	0.0409	2679	0.98

9' is the D-optimal design from the 42 experiment data-set of reference (8)

As before, we find a design of 9 experiments to be the most efficient. The remaining points, 4 and 9, add little information. It can be seen in figure 10.6 that they are very close to point 3. The individual points in the design are very close to those of the D-optimal design extracted from the total data-set of 42 experiments, as just described in section 3. The statistical properties of the two designs, 9 and 9' in table 10.12, are also very close.

5. Conclusions

This example has demonstrated a number of important points:

- If the experimental domain is limited to a certain number of possible experiments, a reduced experimental design may give satisfactory results.
- If extra experiments are able to be carried out, it is often better to repeat some of them (the "best"ones) rather than to choose to carry out new ones.
- The exchange algorithm allows us to analyse the set of experiments generated by the McLean-Anderson algorithm and to choose those which provide most information, with respect to the postulated model.

B. Formulation of a Hydrophilic Matrix Tablet

We continue the discussion of the formulation (9) of sustained release tablets of a soluble drug, using a hydrophilic matrix-forming polymer. The polymer absorbs water, forming a gel through which the drug may slowly diffuse and which is gradually eroded away. The total amount of polymer, a mixture of two hydrophilic polymers in a fixed ratio, was allowed to vary, as were the proportions of excipients. The experimental domain was derived above by the McLean-Anderson algorithm (tables 10.9 and 10.10) and is shown in 3 dimensions in figure 10.7.

The dosage was fixed, but it was decided to investigate the effect of tablet *size*, and therefore drug substance concentration.

1. Formulation, process, and experimental domain

The objective was to formulate tablets, with a dissolution rate as constant as possible, a profile suitable for once daily dosing, and satisfactory physical properties (hardness, friability).

The main details of the formulation and the limits of the experimental domain were given in table 10.8. Other components not noted in the table were the drug substance (1 - 2%) and a lubricant (magnesium stearate). Tablets at 6 mg were made by direct compression of the mixture. The concentration of drug substance (a soluble salt) were allowed to vary between 1% and 2%. Thus, the domain shown in figure 10.7 may be considered as being repeated for each concentration of drug substance tested.

2. Mathematical models

Since the variation in the level of drug substance is so small in relation to that of the other variable constituents, it may be treated as an independent variable Z_5 (see chapter 9, section IV) and set at coded levels -1 to +1, corresponding to 1% and 2%. Two possible models are proposed, a first-order model:

$$y = \beta_1 x_1 + \beta_2 x_2 + \beta_3 x_3 + \beta_4 x_4 + \beta_5 z_5 + \varepsilon \qquad (10.1)$$

and then a second-order model:

$$y = \beta_1 x_1 + \beta_2 x_2 + \beta_3 x_3 + \beta_4 x_4 + \beta_{12} x_1 x_2 + \beta_{13} x_1 x_3 + \beta_{14} x_1 x_4 + \beta_{23} x_2 x_3$$
$$+ \beta_{24} x_2 x_4 + \beta_{34} x_3 x_4 + \beta_{15} x_1 z_5 + \beta_{25} x_2 z_5 + \beta_{35} x_3 z_5 + \beta_{45} x_4 z_5 + \varepsilon \qquad (10.2)$$

Note the absence of a *separate* $\beta_5 z_5$ term, in the model 10.2. It is included implicitly in the $\beta_{i5} x_i z_5$ terms. The $\beta_{55} z_5^2$ term, part of the full second-order model, was left out, though this omission is not necessarily to be recommended.

3. Candidate experimental design

The vertices of the domain in the factor space were obtained using the McLean-Anderson algorithm, as described previously. A candidate experimental design, suitable for the second-order model 10.3 was derived from these. The midpoints of edges, centres of faces, and the centre of gravity of the domain were determined and added to the vertices in the design. The result was duplicated for the 2 levels of drug substance (since there was no $\beta_{55} z_5^2$ term in the model, the intermediate level $z_5 = 0$ could be omitted). Z_5 is thus treated the same as a process variable at 2 levels. The candidate design was a product design, with a total of $39 \times 2 = 78$ experiments. Candidate designs for mixture-process problems are derived and treated in the same way, with the models described in chapter 9.

4. D-optimal designs

Second-order D-optimal design
Federov's exchange algorithm was first of all used to obtain D-optimal designs of 14 to 22 experiments for the second-order model 10.2. On the basis of their det(**M**) values, a design of 17 experiments was selected as having a maximized D-efficiency and a fairly constant variance function over the experimental region, requiring a relatively small number of experiments. These experiment points are shown in table 10.13.

First-order D-optimal design
So that the design could be carried out in two stages, the 17 experiments of the second-order design were themselves used as *candidate points* from which experiments could be extracted to give a D-optimal design for the first-order model 10.1. Eight experiments were thus chosen. The points for the first-order design are shown in the first part of table 10.13.

The D-optimal design approach thus allows sequentiality. The first-order design of 8 experiments can be compared with the D-optimal design selected using the complete candidate design of 78 experiments, which has det(**M**) = 1.01×10^{-5}. The former is nearly as efficient, with det(**M**) = 0.95×10^{-5}. This approach, of selecting the best final design and then doing a selection of the experiments chosen according to a simpler model but ready to complete the design in a second stage, was preferred to the alternative method, of determining the best first-order design and then augmenting it as described in chapter 8.

Table 10.13 Experimental Designs and Results for Sustained Release Hydrophilic Matrix Tablet Experiment

No.	Polymer %	Lactose %	Phosphate %	Cellulose %	Drug substance %	$t_{50\%}$ h	shape factor	hardness kP	friability %
								Responses	
First-order design									
1	17	5	26	52	2	7.65	1.45	8.0	0.00
2	25	5	18	52	1	12.05	1.00	1.0	0.03
3	17	42	5	36	2	4.28	1.29	9.8	0.00
4	25	42	5	28	1	8.09	1.28	8.6	0.01
5	17	5	47	31	1	8.76	1.50	10.1	0.02
6	25	5	47	23	2	9.60	1.50	10.6	0.00
7	17	42	36	5	1	6.75	1.42	6.9	3.00
8	25	42	28	5	2	6.65	1.39	8.0	0.50
Additional experiments for second-order design									
9	17	26	5	52	1	6.08	1.35	8.5	0.03
10	25	18	5	52	2	7.80	1.41	8.1	0.06
11	17	31	47	5	2	4.27	1.58	4.0	3.0
12	25	23	47	5	1	11.84	1.60	6.3	2.0
13	21	5	22	52	2	6.60	1.45	9.0	0.00
14	21	42	5	32	2	5.05	1.18	10.4	0.00
15	21	27	47	5	2	5.48	1.50	9.0	6.0
16	21	32	5	42	1	6.57	1.17	8.9	0.00
17	21	24	26	29	2	6.92	1.31	8.7	0.00

Adequacy of the design
The design is D-optimal for the proposed model. There were no repeated points, so it was not possible to estimate lack of fit in the statistical analysis. Possible improvements in the design, but requiring further experimentation, would be to replicate some experiments and also to carry out additional experiments far from all other points for testing the validity of the model.

5. Results and analysis

Responses
The dissolution profile was described by the time for 50% dissolution $t_{50\%}$, and the shape factor, S. This shape factor was defined by: $S = (t_{75\%} - t_{25\%})/t_{50\%}$. Thus:

- for zero order release $S = 1$
- for a sigmoid profile $S < 1$
- for a "first-order" or "matrix" release profile $S > 1$

For true first-order release, $S = 1.5$, and if the amount released is strictly proportional to the square root of time, $S = 3.75$.

Experimental results for the mean dissolution time, the shape factor of the dissolution curve, and hardness and friability of the tablets, are also given in table 10.13.

Results of first-order design
The 8 experiments of the first-order design were carried out first. The resulting tablets all gave sustained release over the domain, though one mixture (point 8) was tableted only with great difficulty. The results for the 8 experiments were sufficiently close to the desired profile for it to be worth continuing and doing the remaining experiments for the second-order design.

Results of second-order design
The data were analysed according to the second-order equation 10.2. The model was estimated and contour lines were drawn (figure 10.10). Subsequent treatment depends on the exact properties required, particularly those of dissolution profile. The formulation may be optimized by the methods of chapter 6, graphical analysis or desirability, to identify regions of (for example) good flow and friability properties, a median dissolution time of about 9 hours, and a shape factor as near to 1 as possible.

Graphs are given for 1% drug substance and 17%, 21%, and 25% polymer. Release rates from formulations with 2% drug substance are rather more rapid. Formulations with very low levels of cellulose turn out to be very friable and should be avoided. A region where the $t_{50\%}$ value is from 9 to 10 hours and where the shape factor is less than 1.2 are shown in figures 10.10b-c. However, tablets with a variety of profiles may be formulated using these results.

A complex system may thus be characterised efficiently, choosing experiments by an exchange algorithm. The method also allows the problem to be tackled in stages.

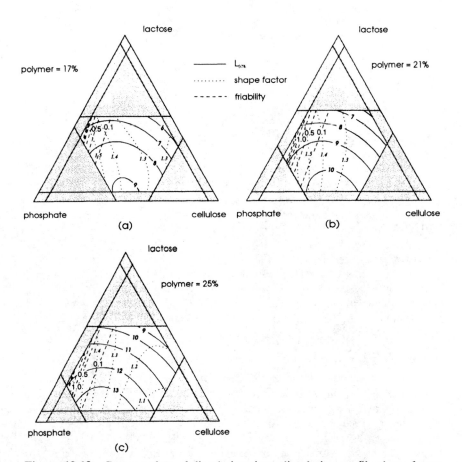

Figure 10.10 Contour plots of dissolution time, dissolution profile shape factor, and friability, for 1% drug substance and (a) 17%; (b) 21%; and (c) 25% polymer content.

IV. RELATIONAL CONSTRAINTS ON MIXTURES

A. Mixtures of Mixtures

1. Component classification: major and minor components

When defining the experimental domain of a mixture up to now, each variable component has been treated simply as one of a number of mixture factors. Where there were restrictions on the proportions of each excipient, these were individual and independent of one another. We may however wish to classify the excipients according to type or function. For example, in the placebo tablet screening study of chapter 9, the excipients could have been divided into two classes, one being the diluents, consisting of up to 5 components, and the other being the remaining functional excipients, binder, lubricant, and disintegrant. The total proportion of diluents and the proportions of the remaining excipients were held constant for that particular experiment, but if they had been allowed to vary, then it would have had a profound effect on the shape of the domain.

We call the class of component (e.g. diluent) the *major*, or *M-component*. Each member of that class (lactose, cellulose, calcium phosphate) is a *minor component* (*m-component*). Instead of there being individual restrictions on each excipient, $L_i \leq x_i \leq U_i$, let there be an overall restriction on the class. So if there are three polymers in a sustained release formulation, X_1, X_2, $X_{and\ 3}$, we may, in optimizing the composition, fix the total amount of polymer between 10 and 25%, but allow the individual polymers to vary freely within that range. Therefore:

$$0.10 \leq x_1 + x_2 + x_3 \leq 0.25$$

This is an example of a *multiple constraint* (also called *relational constraint*). It is the simplest and also the most usual form for such a constraint. The number of ingredients of an *M*-component is normally not large, generally two or three. Additional limits may be set on the individual components' proportions, within the *M*-component.

There must be at least two major components, at least one of them with two or more minor components, for the method to be applicable, or the domain becomes that given by the individual constraints.

2. Major components in fixed proportions

Shape of the experimental domain
A typical example is one of a matrix tablet, containing a mixture of 2 polymers (X_1, X_2) and 2 diluents (X_3, X_4). The total proportion of polymer is allowed to vary between 10% and 50%. The total level of diluents varies *q. s.* The shape of the domain is quite different from that of the previous examples, where there were only individual constraints.

There are 4 components, and therefore the domain may be defined in 3-dimensional space within the tetrahedron of the 4-variable mixture factor space.

The domain is given in figure 10.11. Note that if the total level of polymer is a fixed proportion, the domain becomes a plane within the factor space.

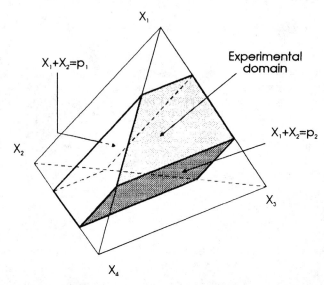

Figure 10.11 Domains for mixture of mixtures: 2 polymers totalling 10 - 50% and 2 diluents.

Models
First-order, second-order, or reduced third-order Scheffé models may be used. Alternatively, it may be postulated that certain components have more pronounced effects than others on the curvature of the response. "Interaction terms" between certain components might thus be omitted. In the model which follows, it is assumed that neither the 2 polymers, nor the 2 diluents exhibit synergism. The second-order terms $\beta_{12}x_1x_2$ and $\beta_{34}x_3x_4$ are left out.

$$y = \beta_1 x_1 + \beta_2 x_2 + \beta_3 x_3 + \beta_4 x_4 + \beta_{13} x_1 x_3 + \beta_{14} x_1 x_4 + \beta_{23} x_2 x_3 + \beta_{24} x_2 x_4 + \varepsilon$$

Experimental design
The approach is close to that used for the simple mixture constraints. A candidate design consists of the 8 vertices, the 12 mid-points of the edges, the 6 centres of the faces, the centre of gravity, 27 points in all. If the full model were postulated at least 10 experiments would be required. The number of experiments seems excessive. The exchange algorithm can be used to eliminate some of these.

It is more difficult to determine the domain and the candidate design for a mixture with relational constraints than for one with only simple upper and lower limits. A computer program designed for treating relational constraints is necessary, both for obtaining the design, but also for showing the limits of the domain graphically, on contourplots.

We will now look at a similar example, but with 5 components.

B. Formulation of an Inert Matrix Tablet Using Component Classification

1. Components and experimental domain

Objectives, formulation and limits
The formulation of an inert matrix tablet for a soluble drug substance (dosed at 5 mg per tablet) was investigated (18). It was based on 2 acrylic polymers and 3 diluents, as shown in table 10.14. Polymer A is of low permeability, polymer B relatively permeable.

The formulators wanted to vary the composition of the polymer within the formulation, from consisting entirely of polymer A to containing only polymer B. The total polymer content (A + B) was allowed to vary between 10% and 50% of the mixture excluding the drug substance and the lubricant. These are given as percentages of the total tablet mass in column 2 of table 10.14, after correction for the proportions of the remaining components (drug substance and lubricant).

Table 10.14   Domain for Inert Polymer Matrix Tablet Formulation (18)

Fixed and major components		Minor components		
drug substance	2.4%			
polymer	9.5 - 47.8%	X_1	acrylic polymer A	0 - 100%
		X_2	acrylic polymer B	0 - 100%
diluent	*q.s.*	X_3	calcium phosphate dihydrate	0 - 100%
		X_4	lactose	0 - 100%
		X_5	μ-crystalline cellulose	0 - 100%
lubricant	2.0%			

It is thus a sustained release formulation, containing a number of diluents and of polymers with restrictions on the total percentage of each class, but where the percentages of individual polymers or diluents are allowed to vary freely within their class. The figure in the last column is the range of minor component with respect to the total within the class. Clearly, this is a "mixture of mixtures" problem.

The major components (*M*-components) are the polymers and diluents. The individual excipients which make up each of these are the minor or *m*-components. Here the composition as well as the proportion of each *M*-component is variable.

So that we may best illustrate the treatment of this kind of problem, potentially a common one, but one that has not, to our knowledge, been described in the pharmaceutical literature, we will take certain liberties in describing the order of experimentation. We demonstrate experimentation in stages, whereas in fact the data were obtained in one step. We also add a number of simulated experimental data (experiments 31 to 36 in table 10.16), to allow treatment of a more complex model, whereas in actual fact the project was halted before these data could be obtained.

Domain and candidate design
In terms of relational multiple constraints we have:

$$0.1 \leq x_1 + x_2 \leq 0.5$$

This relation defines a domain with 12 vertices. It is not anticipated that a model of order higher than 2 will be determined, therefore only vertices and midpoints of edges are required in the design. A candidate design of 37 points, given in table 10.15, was derived. The experiments, the groups separated by dotted lines so that the structure of the design is easier to distinguish, are set out in the following order:

- 12 vertices.
- 24 midpoints of edges, consisting of:
  - mixtures containing both polymers but only one diluent,
  - mixtures containing one polymer and two diluents, and
  - only one polymer, and one diluent at intermediate levels.
- Centre of gravity of the domain.

Responses
The normal tablet properties were measured, but of these only the *mean dissolution time*, equal to the first statistical moment of the release profile, is analysed here. This parameter is often close to the time for 50% dissolution (the median), but all the data of the dissolution profile are used in its determination.

2. An initial study

Mathematical model
A first-order model was proposed for the response. In addition, it was expected that because the polymers had such different permeabilities, their relative proportions would have an important effect on the tablet dissolution properties. The second-order term $\beta_{12}x_1x_2$ was thus also included. The model becomes:

$$y = \beta_1 x_1 + \beta_2 x_2 + \beta_3 x_3 + \beta_4 x_4 + \beta_5 x_5 + \beta_{12} x_1 x_2 + \varepsilon \tag{10.3}$$

Experimental design
A possible design for determining the above model would consist of the vertices and the points with mixtures of the two polymers (the first 18 rows of table 10.15). This is too large for the determination of only 6 coefficients. From 6 to 9 experiments may be selected by an exchange algorithm to give a D-optimal design. A design of 9 experiments appears to be best, according to the criteria described in chapter 8: determinant of **M**, trace of $(\mathbf{X'X})^{-1}$, d_{max}. The best design, extracted from

the candidate design of table 10.15, is indicated by the sign "D9" in the second column of the table. This design should be used if no further experimentation is anticipated.

However, a different design was proposed, as the use of a more complex model was anticipated (see the next section). We shall see that the best design for this model consisted of 21 experiments. These 21 points were taken as a candidate design and a D-optimal design for the model 10.3 was extracted from these. Its properties were only slightly inferior to those of the best design and it could be augmented to the higher order design. The design is indicated by the starred experiments in table 10.15.

Results and analysis

Analysis of the data for the 9 starred experiments in table 10.15 gives estimations for the model coefficients and would enable contour graphs to be plotted. The results showed a non-significant regression. Little apparent difference between the coefficients for the two polymers was seen and the interaction term was not large. However it appears that the influence of microcrystalline cellulose on the formulation is complex, giving rapid release, especially at higher levels of polymer. It was decided to go on to the next stage, postulating a more complete model with cross terms between polymer and diluents.

3. Continuation

A higher order mathematical model

The new model was to include cross terms between the polymers and diluents. The rectangular terms between diluents would still be omitted, and the design would therefore not include mixtures of diluents (experiments 19 - 30 in table 10.15). It would thus become necessary to add 6 "interaction" terms to the model. The model's Scheffé form is:

$$y = \sum_{i=1}^{5} \beta_i x_i + \beta_{12} x_1 x_2 + \sum_{i=1}^{2} \sum_{j=3}^{5} \beta_{ij} x_i x_j + \varepsilon \qquad (10.4)$$

and its 12 terms are:

- first-order in polymer components (2 terms) and diluents (3 terms),
- second-order in polymers (1 term), and
- rectangular terms between polymer and diluent (6 terms).

Experimental design

Using an exchange algorithm, D-optimal designs of 12 to 24 experiments were extracted from the candidate design of table 10.15. A design of 21 experiments was selected on the basis of maximized $|M|$ and a low value of the maximum variance function d_{max} (figure 10.12). It was these 21 experiments that made up the candidate design for deriving the 9 experiment initial design. Mean dissolution times are shown for these experiments in table 10.15.

Table 10.15 Candidate Design for Inert Matrix Tablet in a Domain Defined by Relational Constraints (18)

Major components		1 Polymer		2 Diluents			MDT (h)
No.		X_1	X_2	X_3	X_4	X_5	y
1*	D9	0.1	0	0.9	0	0	2.42
2		0	0.1	0.9	0	0	1.83
3		0.5	0	0.5	0	0	3.45
4*	D9	0	0.5	0.5	0	0	2.13
5		0.1	0	0	0.9	0	0.69
6*		0	0.1	0	0.9	0	0.61
7*	D9	0.5	0	0	0.5	0	1.72
8	D9	0	0.5	0	0.5	0	0.84
9		0.1	0	0	0	0.9	1.47
10*	D9	0	0.1	0	0	0.9	0.94
11*	D9	0.5	0	0	0	0.5	0.60
12*		0	0.5	0	0	0.5	0.57
13		0.05	0.05	0.9	0	0	
14*	D9	0.25	0.25	0.5	0	0	2.47
15	D9	0.05	0.05	0	0.9	0	
16*		0.25	0.25	0	0.5	0	1.64
17		0.05	0.05	0	0	0.9	
18	D9	0.25	0.25	0	0	0.5	0.56
19		0.1	0	0.45	0.45	0	
20		0.1	0	0.45	0	0.45	
21		0.1	0	0	0.45	0.45	
22		0	0.1	0.45	0.45	0	
23		0	0.1	0.45	0	0.45	
24		0	0.1	0	0.45	0.45	
25		0.5	0	0.25	0.25	0	
26		0.5	0	0.25	0	0.25	
27		0.5	0	0	0.25	0.25	
28		0	0.5	0.25	0.25	0	
29		0	0.5	0.25	0	0.25	
30		0	0.5	0	0.25	0.25	
31		0.3	0	0.7	0	0	2.8
32		0.3	0	0	0.7	0	1.9
33		0.3	0	0	0	0.7	1.0
34		0	0.3	0.7	0	0	0.8
35		0	0.3	0	0.7	0	1.2
36		0	0.3	0	0	0.7	0.8
37		0.15	0.15	0.233	0.233	0.233	

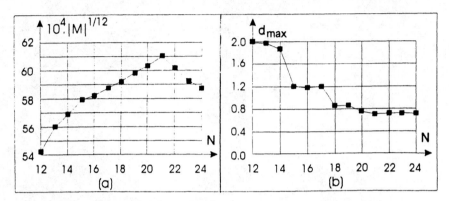

Figure 10.12 Derivation of a D-optimal design for the model 10.4: $|M|^{1/12}$ and d_{max} plotted against the number of experiments.

Analysis and conclusions

Model 10.4 was found to be highly significant. Ternary diagrams for response surfaces of the effects on the mean dissolution time of varying polymer A, polymer B, and each diluent are given in figure 10.13. Diagrams for mixtures of diluents are not given. The original model was probably adequate to describe the behaviour of the polymer-lactose-phosphate mixtures, but cellulose apparently has a very different effect from the other two diluents on the dissolution profile. In mixtures containing only microcrystalline cellulose as diluent, increasing proportions of polymer, even of the less permeable polymer A, lead to accelerated release from the tablet.

In all cases the effect of polymer A, relative to that of the more permeable polymer B, was to slow down the release rate, but the maximum delay in dissolution, expressed as mean dissolution time, was still only about 3 hours.

4. Alternative approaches for the above experiment

Axial screening of excipients

It would have been possible to have begun with the screening design described in the first section of this chapter, taking as reference a blend at the centre of the design space. An alternative is to use the same excipient classification system in a screening design and screen the major components separately. This leads to the design of table 10.16.

The design would have allowed the effects of changing the polymer composition, varying the composition of the diluent, as well as varying the amount of polymer while keeping the composition of individual polymers and diluents constant, to be identified in a minimum of experiments. Note how the total proportion of polymer and the proportions of the individual polymers remain constant as the diluent composition is varied.

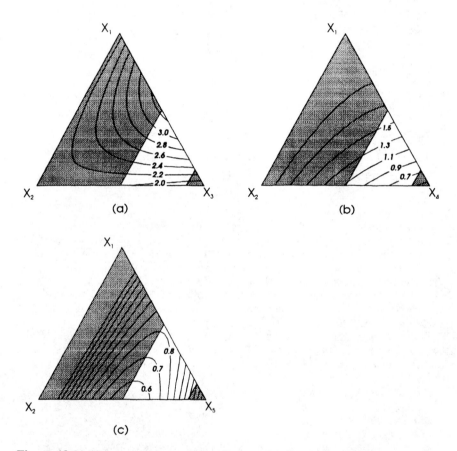

Figure 10.13 Ternary diagram of the effect on the dissolution profile (mean dissolution time) of varying proportions of polymer A, polymer B, and diluent.

Full quadratic model and design
A D-optimal design for the complete Scheffé model of 15 terms may be extracted from the candidate design of table 10.15, using the methods previously described.

Polymer composition as an independent variable
Instead of the 12 element model of equation 10.4 or the full second-order Scheffé model, we may postulate a model where the total polymer is to be considered as one component, X_1, but where the proportion of permeable acrylic polymer is taken as an independent coded variable Z_2. This approach is possible because there are only 2 components in the class. The *full* second-order model is:

$$y = \beta_1 x_1 + \beta_3 x_3 + \beta_4 x_4 + \beta_5 x_5 + \beta_{12} x_1 z_2 + \beta_{13} x_1 x_3 + \beta_{14} x_1 x_4 + \beta_{15} x_1 x_5 + \beta_{23} z_2 x_3 \\ + \beta_{24} z_2 x_4 + \beta_{25} z_2 x_5 + \beta_{22} z_2^2 + \beta_{34} x_3 x_4 + \beta_{35} x_3 x_5 + \beta_{45} x_4 x_5 + \varepsilon$$

The candidate design is obtained by the McLean-Anderson method, multiplied by a 3 level design in Z_2. Experiments are then selected by the exchange algorithm.

Table 10.16 Screening Design with Component Classification for Matrix Tablet

Major components	1 Polymer		2 Diluents			Response[*] MDT
No. [minor components]	X_1	X_2	X_3	X_4	X_5	y
Reference	0.15	0.15	0.233	0.233	0.233	1.4
2 [major]	0.25	0.25	0.167	0.167	0.167	1.6
3	0.05	0.05	0.30	0.30	0.30	1.5
4 [polymers]	0.30	0.00	0.233	0.233	0.233	1.6
5	0.00	0.30	0.233	0.233	0.233	1.2
6 [lactose]	0.15	0.15	0.00	0.35	0.35	0.92
7	0.15	0.15	0.70	0.00	0.00	2.4
8 [cellulose]	0.15	0.15	0.35	0.00	0.35	1.6
9	0.15	0.15	0.00	0.70	0.00	0.23
10 [phosphate]	0.15	0.15	0.35	0.35	0.00	1.8
11	0.15	0.15	0.00	0.00	0.70	0.74

[*] Mean dissolution times simulated using data of table 10.15

C. Relational Constraints

1. Mixtures of mixtures

Relational constraints are those where the limits of one component depend on the levels of other components. Mixtures of mixtures, such as the example described in the previous section, are simple cases of these. Where there are Q major components there are $Q - 1$ independent relational constraints. For the inert tablet above, with 2 major components, we saw the the single relational constraint was:

$$0.1 \leq x_1 + x_2 \leq 0.5$$

Relational constraints for mixtures of mixtures have the general form:

$$L_j \leq x_1 + x_2 + x_3 + \dots \leq U_j$$

where x_1, x_2, x_3, etc., are the proportions of minor components.

2. General form for relational constraints

More generally (19) they have the form:

$$L_j \leq w_1 x_1 + w_2 x_2 + w_3 x_3 + \dots \leq U_j$$

where the weighting factors w_i are between 0 and 1. Vertices are calculated by an algorithm similar to that of McLean and Anderson, but, of necessity, far more complex. A suitable computer program is required to determine the domain and to plot response surfaces. For example, because the polymer B is more permeable than the polymer A in the above example, we might wish to compensate for this in the constraint, setting it at:

$$0.2 \leq x_1 + 0.67 x_2 \leq 0.4$$

The resulting design space (with a single diluent) is shown in figure 10.14a.

3. Ratio models

Another type of relational constraint that we have already seen (chapter 9) is that used with a ratio model. For example, we may rewrite:

$$1 \leq x_2/x_1 \leq 3 \qquad \text{and} \qquad 0.2 \leq x_3 \leq 0.5$$

as a set of relational constraints (see figure 10.14b):

$$0 \leq x_2 - x_1 \qquad\qquad 0 \leq 3x_1 - x_2 \qquad\qquad 0.2 \leq x_3 \leq 0.5$$

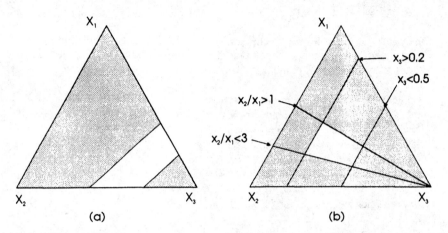

Figure 10.14 Domains of 3 components with relational constraints (see text).

4. Major components in fixed proportions

Shape of the experimental domain

A typical example is one of a matrix tablet containing a mixture of 2 polymers (X_1, X_2) and 2 diluents (X_3, X_4), where the total level of polymer is fixed at a fraction P. There are 4 components, and therefore the domain may be defined in 3 dimensional space within the tetrahedron of the 4 variable mixture factor space. There is a second restriction, however, of $x_1 + x_2 = P$ and so the domain must lie on a plane within that space. For fixed major components, the number of independent variables is equal to the number of minor components, minus the number of major components. In the following discussion we will restrict ourselves to 2 major components.

The possible models and designs depend on the number of minor components associated with each major component.

Product models and designs

In this special case where the proportions of the major components are fixed, the Scheffé models are not applicable, but we may use product models (3). Consider a mixture of 2 polymers (X_1, X_2) their sum being fixed, and 3 diluents $(X_3, X_4$ and $X_5)$. We assume that the response is described by a second-order Scheffé model, in terms of the polymers:

$$y = \alpha_1 x_1 + \alpha_2 x_2 + \alpha_{12} x_1 x_2 + \varepsilon$$

but each coefficient depends on the composition of the diluent, according to a first-order Scheffé model. Thus:

$$y = (\beta_{13}x_3 + \beta_{14}x_4 + \beta_{15}x_5)x_1 + (\beta_{23}x_3 + \beta_{24}x_4 + \beta_{25}x_5)x_2 + (\beta_{123}x_3 + \beta_{124}x_4 + \beta_{125}x_5)x_1x_2 + \varepsilon$$

The optimal design of 9 experiments for this model is obtained by multiplying the second-order Scheffé design for the polymers (3 points) by the first-order Scheffé design for the diluents (3 points). It is saturated.

A major component with 2 minor components

Another way of treating the above problem, where the major component "polymer" is fixed and has only 2 minor components, is to define a Scheffé model for the remaining components and treat the relative proportions of polymers A and B as an independent "process" variable. It may be transformed to a coded variable Z_1. Thus when all the polymer is "A", $z_1 = -1$, when all of it is "B", $z_1 = +1$, and for a 50:50 mixture, $z_1 = 0$. The models and designs are those of mixture/process models and designs. For example, the second-order model is:

$$y = \beta_1 z_1 + \beta_3 x_3 + \beta_4 x_4 + \beta_5 x_5 + \beta_{11} z_1^2 + \beta_{13} z_1 x_3 + \beta_{14} z_1 x_4 + \beta_{15} z_1 x_5$$
$$+ \beta_{34} x_3 x_4 + \beta_{35} x_3 x_5 + \beta_{45} x_4 x_5 + \varepsilon$$

All (fixed) major components each with 2 minor components

The state of each major component may be described by a coded independent variable, as above. Standard first- or second-order models for RSM in a cubic experimental domain may be used and the designs described in chapter 5, section V, are suitable.

D. Mixture-Amount Designs and Unit Doses

Mixture-amount designs (20) were introduced at the end of chapter 9. Let us now consider a 3 component mixture of a drug substance with 2 diluents, with lower limits of 5% for the drug and 25% for each excipient (lower limits only have been chosen so that the design space is a simplex). Figures in brackets are the implicit upper limits:

X_1 (drug substance)	5% - [50%]
X_2 (lactose)	25% - [70%]
X_3 (cellulose)	25% - [70%]

The unit mass U_4 is allowed to vary between 100 and 250 mg. This is associated with the coded variable Z_4. The resulting 3-dimensional allowed experimental region, a triangular prism, is shown in figure 10.15a. It is almost identical to the 2 component case, except that the mixture is represented by a triangular factor space instead of a line.

The unit dose D is the product of the unit mass U_4 and the fraction of drug substance x_1 (pseudocomponents are not used here). In terms of the coded variable z_4, this becomes:

$$D = U_4 x_1 = (75\ z_4 + 175)x_1$$

We impose limits of 10 mg $\leq$ D $\leq$ 50 mg and obtain the rather curious boundary conditions:

$$0.4\ \leq\ 3\ z_4 x_1 + 7\ x_1\ \leq\ 2 \tag{10.5}$$

shown in figure 10.15b. The problem may be solved in the usual way, by setting up a suitable mixture-amount model. Experiments may then be selected within the domain by an exchange algorithm. The main problem will be that of imposing the boundary conditions of the inequality 10.5 on the candidate design and afterwards on the contour plots and possible optimization.

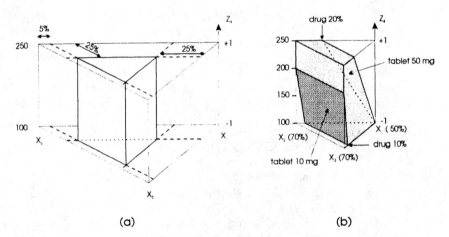

(a)   (b)

Figure 10.15 Domain for mixture-amount experiment: 3 component mixture with limits on (a) tablet mass and (b) tablet mass and dosage.

References

1. M. Chariot, unpublished data.
2. G. A. Lewis, Pharmaceutical development: control and rigour in formulation: optimization of the process, *S.T.P. Pharma Pratiques*, **4**, 455-460 (1995).
3. J. A. Cornell, Experiments with Mixtures, 2nd edition, J. Wiley, N. Y., 1990.

4. R. Ishino, H. Yoshino, Y. Hirakawa and K. Noda, Design and preparation of pulsatile release tablet as a new oral drug delivery system, *Chem. Pharm. Bull.*, **40**, 3036-3041 (1992).
5. E. Hoosewys, V. Andrieu, and G. A. Lewis, unpublished data.
6. R. A. McLean and V. L. Anderson, Extreme vertices design of mixture experiments, *Technometrics*, **8**, 447-454 (1966).
7. R. J. Belotto, Jr., A. M. Dean, M. A. Moustafa, A. M. Molokhia, M. W. Gouda, and T. D. Sokoloski, Statistical techniques applied to solubility predictions and pharmaceutical formulations: an approach to problem solving using mixture response surface methodology, *Int. J. Pharm.*, **23**, 195 (1985).
8. M. A. Moustafa, A. M. Molokhia, and M. W. Gouda, Phenobarbital solubility in propylene glycol-glycerol-water systems, *J. Pharm. Sci.*, **70**, 1172-1174 (1981).
9. V. Andrieu and G. Lewis, unpublished data.
10. A. T. Anik and L. Sukumer, Extreme vertices design in formulation development: solubility of butoconazole nitrate in a multicomponent system, *J. Pharm. Sci.* **70**, 897-900 (1981).
11. A. D. Johnson, V. L. Anderson, and G. E. Peck, A statistical approach for the development of an oral controlled release matrix tablet. *Pharm. Res.*, **7**, 1092-1097 (1990)
12. M. Hirata, K. Takayama, and T. Nagai, Formulation optimization of sustained release tablet of chlorpheniramine maleate by means of exteme vertices design and simultaneous optimization technique. *Chem. Pharm. Bull.*, **40**, 741-746 (1992).
13. O. Reer and B. W. Müller, Investigation of the influence of cosolvents and surfactants on the complexation of dexamethasone with hydroxypropyl-beta-cyclodextrin by use of a simplex lattice design. *Eur. J. Pharm. Biopharm.*, **39**, 105-111 (1993).
14. M. Chariot and G. A. Lewis, Optimisation of a modified release formulation by non-classical experimental design. *Drug Dev. Ind. Pharm.* **17**, 1551-1570 (1991).
15 D. Vojnovic, M. Moneghini, and D. Chicco, Nonclassical experimental design applied in the optimization of a placebo granulate formulation in high-shear mixer, *Drug Develop. Ind. Pharm.*, **22**, 997-1004 (1996).
16. D. Vojnovic and D. Chicco, Mixture experimental design applied to solubility predictions, *Drug Dev. Ind. Pharm.*, **23**, 639-645 (1997).
17. A. Bodea and S. E. Leucuta, Optimization of hydrophilic matrix tablets using a D-optimal design, *Int. J. Pharm.*, **153**, 247-255 (1997).
18. V. Andrieu and G. Lewis, unpublished data.
19. R. Snee, Experimental designs for mixture systems with multicomponent restraints, *Communications in Statistics, Theory and Methods*, A8, 303-326 (1979).
20. G. F. Piepel and J. A. Cornell, Models for mixture experiments when the response depends on the total amount, *Technometrics* 27, 219-227 (1985).

APPENDIX I

A SUMMARY OF MATRIX ALGEBRA

    A. Some preliminary definitions
    B. Operations on matrices
    C. Special types of matrices and their properties

Only those results needed for understanding the structure of the information and variance-covariance matrices and linear regression will be summarised. No proofs are given. The interested reader is directed to introductory and standard works on matrix algebra (1).

A. Some Preliminary Definitions

A matrix is a rectangular array of elements. These may be of many different kinds, but, as far as this work is concerned, all the elements are real numbers. Thus a matrix can represent an experimental design (the rows normally referring to the experiments and the columns to the factor variables). Each element in the experimental design matrix thus identifies the experimental condition for the corresponding experiment and factor.

    A matrix with a single row or a single column is called a *vector*. For example, the series of results for the yield in an extrusion-spheronization experimental design can be expressed as a row vector. Each element of the vector represents the result of the corresponding experiment.

    The *order* of a matrix is defined as the number of rows and columns. If the order is 4×5 then it has 4 rows and 5 columns. A *square matrix* has equal numbers of rows and columns.

B. Operations on Matrices

1. Addition and subtraction

Matrices can only be added or subtracted if they are of the same order, that is, with identical numbers of rows and of columns. In such a case, each element in the second matrix is added to or subtracted from the corresponding element in the first matrix.

2. Multiplication of matrices

A matrix A may be multiplied by a matrix B provided the number of columns of the first matrix is equal to the number of rows of the second matrix. The resulting matrix AB will have the same number of rows as A and the same number of columns as B.

If the matrix A has r_A rows and c_A columns and the matrix B has r_B rows and c_B columns then A can be multiplied by B only if $c_A = r_B$. The matrix AB will have r_A rows and c_B columns.

Multiplication is carried out as follows: an element c^{ij} in the matrix $C = AB$ is obtained from line i of A and column j of B. We have already seen that each line of A and each column of B has the same number of elements k. We multiply each of the corresponding elements together and add them all together. Thus:

$$c^{ij} = a^{i1}b^{1j} + a^{i2}b^{2j} + \dots + a^{ip}b^{pj} + \dots + a^{ik}b^{kj}$$

The multiplication of matrices is *associative*:

$$(AB)C = A(BC)$$

but it is *not commutative*. We have seen that if the necessary conditions of numbers of columns and of rows in A and B are fulfilled, so that A may be multiplied by B to give AB, it is not necessarily the case that we can also multiply the matrix B by A to obtain BA. This can only be done if the number of rows of A equals the number of columns of B, and also *vice versa*. And in such a case we would generally find:

$$AB \neq BA$$

C. Special Types of Matrices and their Properties

1. The transpose matrix

The position of each element in a matrix is defined by its row and column number. So in row i and column j of the matrix X, there is the element x^{ij}. On exchanging rows and columns, we obtain the transpose matrix X'. Thus, the element x'^{ji}, found in column i and row j of the transpose matrix, is equal to the element x^{ij} of the original matrix. An example is as follows:

If $\mathbf{A} = \begin{pmatrix} 4 & 2 & 1 \\ 6 & -3 & 4 \end{pmatrix}$ then its transpose is $\mathbf{A}' = \begin{pmatrix} 4 & 6 \\ 2 & -3 \\ 1 & 4 \end{pmatrix}$

The following important results should be noted:

- If A is a matrix of k rows and m columns, its transpose A′ is a matrix of m rows and k columns. The two matrices may therefore be multiplied together, in both senses. A′A and AA′ can both be obtained, although the resulting matrices are not, in general, equal to one another.

- In the theory of experimental design, where X is the model matrix, we use the X′X matrix, called the *information matrix*.

- X′X is a square matrix. If X has N rows and p columns (N is the number of experiments and p the number of terms in the mathematical model), X′X has p rows and p columns.

- A′A has another important property, that can be shown by inspection using the rules of matrix multiplication outlined above – it is symmetrical about its principal diagonal. The terms in the top right half of the matrix are identical to the terms in the bottom left.

2. Determinant of a matrix

The determinant of a matrix exists only for a square matrix and is a number. The determinant of a matrix is written as $|A|$. For a 2×2 matrix it is simple to define:

$$\mathbf{A} = \begin{pmatrix} a_{11} & a_{12} \\ a_{21} & a_{22} \end{pmatrix}$$

$$|\mathbf{A}| = \begin{vmatrix} a_{11} & a_{12} \\ a_{21} & a_{22} \end{vmatrix} = a_{11}a_{22} - a_{12}a_{21}$$

For a 3×3 matrix it can be defined in terms of the determinants of three 2×2 matrices:

$$A = \begin{pmatrix} a_{11} & a_{12} & a_{13} \\ a_{21} & a_{22} & a_{23} \\ a_{31} & a_{32} & a_{33} \end{pmatrix}$$

$$|A| = a_{11} \begin{vmatrix} a_{22} & a_{23} \\ a_{32} & a_{33} \end{vmatrix} - a_{12} \begin{vmatrix} a_{21} & a_{23} \\ a_{31} & a_{33} \end{vmatrix} + a_{13} \begin{vmatrix} a_{21} & a_{22} \\ a_{31} & a_{32} \end{vmatrix}$$

It is interesting to note that for a square matrix with all non-diagonal elements zero, the determinant of the matrix is equal to the product of the diagonal elements.

3. The null matrix

This is a square matrix with all elements zero. The order of the matrix is sometimes indicated as a subscript. For example:

$$O = O_3 = \begin{pmatrix} 0 & 0 & 0 \\ 0 & 0 & 0 \\ 0 & 0 & 0 \end{pmatrix}$$

4. The identity matrix

The identity matrix, I, is a square matrix with each diagonal term unity and each non-diagonal term zero. In the same way as for the null matrix, the order may be indicated as a subscript:

$$I = I_3 = \begin{pmatrix} 1 & 0 & 0 \\ 0 & 1 & 0 \\ 0 & 0 & 1 \end{pmatrix}$$

5. Inverse matrix

The matrix B is the inverse of the matrix A if their product is a unit matrix. If:

$$AB = I$$

then B may be written as A^{-1}.

The conditions for inverting a matrix **A** are:

- **A** is a *square* matrix.

- The determinant of **A** is non-zero. If it equals zero, then the matrix is termed *singular* and cannot be inverted. To estimate the model coefficients from the results of an experimental design, it is necessary to invert the information matrix **X'X**. If the matrix is singular, then the design is insufficient for use with that model.

I is of the same order as **A** and **B**.

The transpose of the inverse of the square matrix **A** is the same as the inverse of its transpose matrix. That is:

$$(\mathbf{A}')^{-1} = (\mathbf{A}^{-1})'$$

6. Orthogonal matrices

An orthogonal matrix is one where the transpose of the matrix is equal to its inverse:

$$\mathbf{A}' = \mathbf{A}^{-1}$$

7. Trace

The trace of a matrix is the sum of its diagonal elements:

$$\text{Trace} = \sum c^{ii}$$

It is defined only for a square matrix.

Reference

1. H. T. Brown, Introduction to the Theory of Determinants and Matrices, University of North Carolina Press, Chapel Hill, N.C., 1958

APPENDIX II

EXPERIMENTAL DESIGNS AND MODELS FOR SCREENING

    A. Hadamard or Plackett-Burman designs at 2 levels
    B. Symmetrical fractional factorial designs at 2, 3, 4, 5, and 7 levels
    C. Asymmetrical designs
        Designs obtained from the symmetrical designs by collapsing
        Other orthogonal asymmetrical designs
    D. First-order models for qualitative factors
        The presence-absence model
        Conversion to the reference state model

A. Hadamard or Plackett-Burman Designs at 2 Levels

These are designs where each factor is present at 2 levels and where the number of experiments N is a multiple of 4. The variables may be either qualitative or quantitative. The levels may be represented as 0,1 or -1, +1. Here, we write them simply as "-" and "+".

They are very easily constructed. In most cases the first line is given and the remaining lines are obtained by permutation, except for the last line, which consists entirely of minus signs (1).

1. Designs of 4, 8, ... 24 experiments

Table A2.1 shows the first lines of the first 6 designs. For the 28 experiment design, see below. Larger designs are for the most part given in reference (1).

    Take the case of $N = 12$ for 11 factors. Construction of the design by permutation of the first line is shown in table A2.2. To make the second line we move the minus sign at the far right to the beginning of the next line and slide the rest of the signs one place. We then do the same thing for the next line, and so on. The final (12th) row is one of negatives.

Table A2.1 First Lines of Plackett-Burman Designs of 4 to 24 Experiments

N	First line of design
4	+ + -
8	+ + + - + - -
12	+ + - + + + - - - + -
16	+ + + + - + - + + - - + - - -
20	+ + - - + + - + - + - + - - - - + + -
24	+ + + + + - + - + + - - + + - - + - + - - - -

Table A2.2 Plackett-Burman Design for 11 Factors

	X_1	X_2	X_3	X_4	X_5	X_6	X_7	X_8	X_9	X_{10}	X_{11}
1	+	+	-	+	+	+	-	-	-	+	-
2	-	+	+	-	+	+	+	-	-	-	+
3	+	-	+	+	-	+	+	+	-	-	-
4	-	+	-	+	+	-	+	+	+	-	-
5	-	-	+	-	+	+	-	+	+	+	-
6	-	-	-	+	-	+	+	-	+	+	+
7	+	-	-	-	+	-	+	+	-	+	+
8	+	+	-	-	-	+	-	+	+	-	+
9	+	+	+	-	-	-	+	-	+	+	-
10	-	+	+	+	-	-	-	+	-	+	+
11	+	-	+	+	+	-	-	-	+	-	+
12	-	-	-	-	-	-	-	-	-	-	-

We have a design of 12 experiments to calculate 11 parameters (effects) plus a constant term. There is a unique solution to the problem and the results themselves do not give any evidence of the precision of the results. Of course, the experimental data are not exact so the calculated parameters are not exact either, but are *estimates* of the true values β_i. If we define the imprecision of each experimental result by a standard deviation σ, then the corresponding standard deviation of the estimate of β_i can then be calculated as

$$\sigma_b = \sigma/\sqrt{12}.$$

This is the value given earlier for the standard deviation of an optimum matrix, with each factor at two levels. To estimate the standard error of each coefficient estimate b_i, we therefore need an estimate of the experimental standard deviation. Note that this equation applies to coded levels set at ±1. For levels set at 0, 1 the equivalent standard deviation is obtained by multiplying by $\sqrt{2}$.

2. Design for 28 experiments

In the Plackett-Burman designs of 28 experiments for up to 27 factors, a set of three square blocks each of 9 rows and columns is permutated cyclically (table A2.3a). These blocks, labelled A, B, and C, are given in table A2.3b.

Table A2.3 Plackett-Burman Design of 28 Experiments

(a) Permutation of 9×9 Blocks as a Latin Square

Rows 1 - 9	A	B	C
Rows 10 - 18	B	C	A
Rows 19 - 27	C	A	B
Row 28	A row of negative signs		

(b) Structure of the Blocks

A	B	C
+ - + + + + - - -	- + - - - + - - +	+ + - + - + + - +
+ + - + + + - - -	- - + + - - + - -	- + + + + - + + -
- + + + + + - - -	+ - - - + - - + -	+ - + - + + - + +
- - - + - + + + +	- - + - + - - - +	+ - + + + - + - +
- - - + + - + + +	+ - - - - + + - -	+ + - - + + + + -
- - - - + + + + +	- + - + - - - + -	- + + + - + - + +
+ + + - - - + - +	- - + - - + - + -	+ - + + - + + + -
+ + + - - - + + -	+ - - + - - - - +	+ + - + + - - + +
+ + + - - - - + +	- + - - + - + - -	- + + - + + + - +

3. Plackett-Burman designs of 2^k (4, 8, 16, 32, ...) runs

Designs where N is a power of 2 are termed *geometric*. They may be derived in the same way as those with $N = 12, 20, 24$ experiments, and are part of the set of Hadamard designs. The Plackett-Burman design of 8 experiments and 7 factors was used in chapter 2. There is however an alternative derivation, as they are also symmetrically reduced 2^{k-r} factorial designs.

4. Confounding in Plackett-Burman designs

It was shown in chapter 3 how, in 2^{k-r} fractional factorial designs, the interactions may be confounded with the main effects or with each other. In the case of the geometric designs (2^{k-r} fractional factorial with 4, 8, 16, 32, ... experiments), each interaction is *totally* confounded with one of the main effects.

Aliasing in the non-geometric 12 and 20 experiment designs is rather different. Take, for example, the interaction X_1X_2. Its effect is partially confounded with *all* of the main effects except those of X_1 and X_2. Therefore with these designs there is a reduced likelihood of drawing a wrong conclusion on the main effects where there is one important interaction. Analysis taking interactions into account is quite complex (2). This effect is seen to a lesser extent for the 24-experiment design, where the confounding is shared over fewer main effects. In the 28-experiment design each interaction is roughly 50% confounded with a single main effect and 14% confounded with each of the other main effects (other than the main effects of the variables of the interaction in question).

B. Symmetrical fractional factorial designs at 2, 3, 4, 5, and 7 levels

1. $2^k//2^{k'}$ where $k \leq 2^{k'} - 1$

These are identical to the corresponding Plackett-Burman designs described in the previous section. They may be also be derived from the corresponding full factorial designs (resulting in a different standard order of experimentation).

- $2^3//2^2$ (or 2^{3-1}; $N = 4$)
- $2^7//2^3$ (or 2^{7-4}; $N = 8$)
- $2^{15}//2^4$ (or 2^{15-11}; $N = 16$)

2. Designs at ≥ 3 levels

The $3^4//3^2$ design with $N = 9$ experiments is shown in table 2.13 of chapter 2. The other potentially useful symmetric screening design at 3 levels is the $3^{13}//3^3$ design of $N = 27$ experiments, shown in table A2.4.
   The only really useful design at 4 levels is the $4^5//4^2$ design ($N = 16$), shown in table 2.16. It was demonstrated that this also can be a starting point for a large variety of asymmetrical designs, also of 16 experiments, derived by collapsing. In the same way we consider only one design at 5 levels, the $5^6//5^2$, for up to 6 factors and requiring $N = 25$ experiments (table A2.5). Plackett and Burman (1) give cyclic derivations of these designs in the same way as for the 2 level designs.
   There is no reduced symmetrical design of 36 experiments with the factors at 6 levels. There is a $7^8//7^2$ design for 8 variables taking 7 levels and requiring N = 49 experiments. This design is unlikely to be required, but if needed it may be found in reference (1).

Table A2.4 $3^{13}/\!/3^3$ Design for Screening

	X_1	X_2	X_3	X_4	X_5	X_6	X_7	X_8	X_9	X_{10}	X_{11}	X_{12}	X_{13}
1	0	0	0	0	0	0	0	0	0	0	0	0	0
2	0	0	0	0	1	1	1	1	1	1	1	1	1
3	0	0	0	0	2	2	2	2	2	2	2	2	2
4	0	1	1	1	0	0	0	1	1	1	2	2	2
5	0	1	1	1	1	1	1	2	2	2	0	0	0
6	0	1	1	1	2	2	2	0	0	0	1	1	1
7	0	2	2	2	0	0	0	2	2	2	1	1	1
8	0	2	2	2	1	1	1	0	0	0	2	2	2
9	0	2	2	2	2	2	2	1	1	1	0	0	0
10	1	0	1	2	0	1	2	0	1	2	0	1	2
11	1	0	1	2	1	2	0	1	2	0	1	2	0
12	1	0	1	2	2	0	1	2	0	1	2	0	1
13	1	1	2	0	0	1	2	1	2	0	2	0	1
14	1	1	2	0	1	2	0	2	0	1	0	1	2
15	1	1	2	0	2	0	1	0	1	2	1	2	0
16	1	2	0	1	0	1	2	2	0	1	1	2	0
17	1	2	0	1	1	2	0	0	1	2	2	0	1
18	1	2	0	1	2	0	1	1	2	0	0	1	2
19	2	0	2	1	0	2	1	0	2	1	0	2	1
20	2	0	2	1	1	0	2	1	0	2	1	0	2
21	2	0	2	1	2	1	0	2	1	0	2	1	0
22	2	1	0	2	0	2	1	1	0	2	2	1	0
23	2	1	0	2	1	0	2	2	1	0	0	2	1
24	2	1	0	2	2	1	0	0	2	1	1	0	2
25	2	2	1	0	0	2	1	2	1	0	1	0	2
26	2	2	1	0	1	0	2	0	2	1	2	1	0
27	2	2	1	0	2	1	0	1	0	2	0	2	1

Table A2.5 $5^6 /\!/ 5^2$ Design for Screening

	X_1	X_2	X_3	X_4	X_5	X_6			X_1	X_2	X_3	X_4	X_5	X_6
1	0	0	0	0	0	0		16	3	0	3	1	4	2
2	0	1	1	1	1	1		17	3	1	4	2	0	3
3	0	2	2	2	2	2		18	3	2	0	3	1	4
4	0	3	3	3	3	3		19	3	3	1	4	2	0
5	0	4	4	4	4	4		20	3	4	2	0	3	1
6	1	0	1	2	3	4		21	4	0	4	3	2	1
7	1	1	2	3	4	0		22	4	1	0	4	3	2
8	1	2	3	4	0	1		23	4	2	1	0	4	3
9	1	3	4	0	1	2		24	4	3	2	1	0	4
10	1	4	0	1	2	3		25	4	4	3	2	1	0
11	2	0	2	4	1	3								
12	2	1	3	0	2	4								
13	2	2	4	1	3	0								
14	2	3	0	2	4	1								
15	2	4	1	3	0	2								

3. Statistical properties of the symmetrical qualitative designs

Taking the $3^4 /\!/ 3^2$ design as example, and using the reference state model, its dispersion matrix is:

$$(\mathbf{X'X})^{-1} = \begin{pmatrix}
& 0 & 1 & 2 & & 3 & & 4 & \\
1 & -\frac{1}{3} & -\frac{1}{3} & -\frac{1}{3} & -\frac{1}{3} & -\frac{1}{3} & -\frac{1}{3} & -\frac{1}{3} & -\frac{1}{3} \\
-\frac{1}{3} & +\frac{2}{3} & +\frac{1}{3} & \cdot & \cdot & \cdot & \cdot & \cdot & \cdot \\
-\frac{1}{3} & +\frac{1}{3} & +\frac{2}{3} & \cdot & \cdot & \cdot & \cdot & \cdot & \cdot \\
-\frac{1}{3} & \cdot & \cdot & +\frac{2}{3} & +\frac{1}{3} & \cdot & \cdot & \cdot & \cdot \\
-\frac{1}{3} & \cdot & \cdot & +\frac{1}{3} & +\frac{2}{3} & \cdot & \cdot & \cdot & \cdot \\
-\frac{1}{3} & \cdot & \cdot & \cdot & \cdot & +\frac{2}{3} & +\frac{1}{3} & \cdot & \cdot \\
-\frac{1}{3} & \cdot & \cdot & \cdot & \cdot & +\frac{1}{3} & +\frac{2}{3} & \cdot & \cdot \\
-\frac{1}{3} & \cdot & \cdot & \cdot & \cdot & \cdot & \cdot & +\frac{2}{3} & +\frac{1}{3} \\
-\frac{1}{3} & \cdot & \cdot & \cdot & \cdot & \cdot & \cdot & +\frac{1}{3} & +\frac{2}{3}
\end{pmatrix}$$

There are 2 rows and 2 columns corresponding to each (qualitative) variable. Since each of these takes 3 levels, it has 2 independent coefficients associated with it. We can see from the matrix that the design is orthogonal – that the estimates of the coefficients of the different variables are uncorrelated.

However, the estimates for the effects of the different levels of the *same* variable *are* correlated.

The coefficient estimates and the constant term are also correlated. In the presence-absence model, the estimation of the constant term is unbiased. However the reference state is not the (hypothetical) mean but is in fact one of the levels (selected at random), and this gives rise to the negative correlation between the constant term and the coefficients in the reference state model.

Summaries of the properties of the various designs are given in table A2.6. The standard deviations and variance functions apply to the coefficients in the *reference state model* – that is, the effect of changing the level of a variable from the selected reference state (e.g. diluent = lactose) to another level (e.g. diluent = cellulose). They are defined in terms of the standard deviation for a single experiment, σ.

Table A2.6 Statistical Properties of Some Symmetrical Screening Designs

N	Levels	Variance function	Standard deviation	VIF (inflation factor)
9	3	0.667	0.816 σ	1.33
27	3	0.049	0.222 σ	1.33
16	4	0.5	0.707 σ	1.5
25	5	0.4	0.632 σ	1.6
49	7	0.286	0.535 σ	1.71

The standard deviations can be compared with the precision achieved by testing one factor at a time: 1.414σ. Precision is clearly much improved over that obtained with the "one factor at a time" technique. (The standard deviations for the "presence-absence" forms of the coefficients – see chapter 2, section II.B – defined with respect to a hypothetical "mean" state, are smaller by a factor of $\sqrt{2}$.)

All the designs are saturated ($R_{eff} = 100\%$) where the full number of factors is studied.

C. Asymmetrical Designs

1. Designs obtained from the symmetrical designs by collapsing

Properties of the $3^3 2^1$ design derived from $3^4//3^2$
Collapsing of the $3^4//3^2$ and $4^5//4^2$ designs, the most useful, is discussed in detail in chapter 2, section V.C. Statistical properties and the dispersion matrices are similar to those for the saturated designs. We take the $3^3 2^1$ design derived by collapsing A from the $3^4//3^2$ design as an example. The design is unsaturated ($R_{eff} = 89\%$). It remains orthogonal. Notice the slight co-linearity between estimates of the constant

term in the model and the coefficients for the unbalanced variable at 2 levels.

$$
(X'X)^{-1} =
\begin{pmatrix}
1 & -\tfrac{1}{3} & -\tfrac{1}{3} & -\tfrac{1}{3} & -\tfrac{1}{3} & -\tfrac{1}{3} & -\tfrac{1}{3} & -\tfrac{1}{6} \\
-\tfrac{1}{3} & +\tfrac{2}{3} & +\tfrac{1}{3} & . & . & . & . & . \\
-\tfrac{1}{3} & +\tfrac{1}{3} & +\tfrac{2}{3} & . & . & . & . & . \\
-\tfrac{1}{3} & . & . & +\tfrac{2}{3} & +\tfrac{1}{3} & . & . & . \\
-\tfrac{1}{3} & . & . & +\tfrac{1}{3} & +\tfrac{2}{3} & . & . & . \\
-\tfrac{1}{3} & . & . & . & . & +\tfrac{2}{3} & +\tfrac{1}{3} & . \\
-\tfrac{1}{3} & . & . & . & . & +\tfrac{1}{3} & +\tfrac{2}{3} & . \\
-\tfrac{1}{6} & . & . & . & . & . & . & +\tfrac{1}{2}
\end{pmatrix}
$$

where columns are labelled $0 \quad 1 \quad 2 \quad 3 \quad 4$.

Standard deviations for the coefficient estimates remain at 0.816σ, except for the variable at 2 levels, where it is slightly reduced to 0.707σ ($\sqrt{0.5}\,\sigma$). Variance inflation factors are 1.333 for variables 1-3 and 1.00 for the variable 4 at 2 levels.

Collapsing the $5^6//5^2$ design (25 experiments)
Any of the 6 variables, each at 5 levels, can be reduced to 4 or 3 levels by collapsing A. It is interesting to note that once one of the columns is at 4 levels, it may be collapsed in turn to 3 variables at 2 levels (collapsing B). So for example $5^3.4^2.3^1 \rightarrow 5^3.4^1.3^1.2^3$. Therefore this design matrix may be used to derive an even wider range of designs than the 16 experiment symmetrical design. Not all of them are useful. For example, a $2^{18}//25$ design could be derived, but the Plackett-Burman design of 24 experiments would be superior in terms of precision. However, if the problem involves one or more variables at 5 levels, a 25 experiment design derived from the $5^6//5^2$ design matrix should be considered, possibly comparing it with D-optimal designs.

Collapsing the $3^{13}//3^3$ design (27 experiments)
This can be transformed to a $9^1 3^9$ design by collapsing C, by taking combinations of columns 1 and 2, fusing them to give a single column of 9 levels, and then deleting columns 3 and 4 of the $3^{13}//3^3$ design (because these are derived by combinations of columns 1 and 2).

The $9^1 3^9$ design, in turn, may be transformed to $8^1 3^9$ by collapsing A and then to $4^1 3^9 2^4$ by collapsing B.

2. Other orthogonal asymmetrical designs

The $2^1 3^7 //18$ design

This design (table A2.7), sometimes called L_{18}, is nearly saturated ($R_{eff} = 89\%$). Whereas the $3^4 //9$ design allows 4 variables to be tested at 3 levels, and the $3^{13} //27$ allows us to go up to 13 variables, this design allows an intermediate number of factors to be investigated efficiently, with the additional possibility of investigating one factor at 2 levels only. In addition, any of the 3 level factors may be reduced to 2 (collapsing A).

Table A2.7 The $2^1 3^7 //18$ design

	X_1	X_2	X_3	X_4	X_5	X_6	X_7	X_8
1	0	0	0	0	0	0	0	0
2	0	0	1	1	1	1	1	1
3	0	0	2	2	2	2	2	2
4	0	1	0	0	1	1	2	2
5	0	1	1	1	2	2	0	0
6	0	1	2	2	0	0	1	1
7	0	2	0	1	0	2	1	2
8	0	2	1	2	1	0	2	0
9	0	2	2	0	2	1	0	1
10	1	0	0	2	2	1	1	0
11	1	0	1	0	0	2	2	1
12	1	0	2	1	1	0	0	2
13	1	1	0	1	2	0	2	1
14	1	1	1	2	0	1	0	2
15	1	1	2	0	1	2	1	0
16	1	2	0	2	1	2	0	1
17	1	2	1	0	2	0	1	2
18	1	2	2	1	0	1	2	0

The columns for the variable at 2 levels and one of those at 3 levels may be combined to allow one variable to be tested at 6 levels, as follows.

The $3^6 6^1 //18$ design

This design (table A2.8) is saturated ($R_{eff} = 100 \%$). Collapsing A is possible on it, allowing for example a factor to be studied at 5 levels. Thus it is useful, if we have one factor at 5 or 6 levels, and no other factor at more than 3 levels.

Table A2.8 The $3^6 6^1 // 18$ design

	X_1	X_2	X_3	X_4	X_5	X_6	X_7
1	0	0	0	0	0	0	0
2	0	1	1	1	1	1	1
3	0	2	2	2	2	2	2
4	1	0	0	1	1	2	2
5	1	1	1	2	2	0	0
6	1	2	2	0	0	1	1
7	2	0	1	0	2	1	2
8	2	1	2	1	0	2	0
9	2	2	0	2	1	0	1
10	3	0	2	2	1	1	0
11	3	1	0	0	2	2	1
12	3	2	1	1	0	0	2
13	4	0	1	2	0	2	1
14	4	1	2	0	1	0	2
15	4	2	0	1	2	1	0
16	5	0	2	1	2	0	1
17	5	1	0	2	0	1	2
18	5	2	1	0	1	2	0

Other asymmetrical designs
Many other asymmetrical orthogonal designs exist, examples being the $4^1 2^{20}$ design (24 experiments), and the $6^1 3^{11} 2^3$ and $3^{11} 2^{12}$ designs of 36 experiments. They consist of too many experiments to be generally useful for our purposes in pharmaceutical development.

D. First-Order Models for Qualitative Factors

1. The presence-absence model

A single factor at multiple levels
If the factor F_1 is allowed to take s_1 distinct levels (which may be either qualitative or, more usually, quantitative) the screening model may be written as:

$$y = \beta_0 + \beta_{1,A} x_{1,A} + \beta_{1,B} x_{1,B} + \dots + \beta_{1,S_1} x_{1,S_1} + \varepsilon \qquad \text{(A2.1a)}$$

where S_1 is the final s_1^{th} state. This model has $s_1 + 1$ coefficients but only s_1 of these are independent because of the relationship between the variables:

$$x_{1,A} + x_{1,B} + \dots + x_{1,S_1} = 1 \qquad\qquad\qquad\qquad (A2.1b)$$

which gives rise to the following constraint:

$$\beta_{1,A} + \beta_{1,B} + \dots + \beta_{1,S_1} = 0 \qquad\qquad\qquad\qquad (A2.1c)$$

Generalisation: several factors with various numbers of levels

Consider k factors, the first of which takes s_1 distinct levels, the second s_2 etc, up to the kth with s_k levels. The levels for the ith factor go from A to S_i. By analogy with the previous model (A2.1a) we have:

$$
\begin{aligned}
y = \beta_0 &+ \beta_{1,A}x_{1,A} + \beta_{1,B}x_{1,B} + \dots + \beta_{1,S_1}x_{1,S_1} \\
&+ \beta_{2,A}x_{2,A} + \beta_{2,B}x_{2,B} + \dots + \beta_{2,S_2}x_{2,S_2} \\
&+ \dots \\
&+ \beta_{k,A}x_{k,A} + \beta_{k,B}x_{k,B} + \dots + \beta_{k,S_k}x_{k,S_k} \\
&+ \varepsilon
\end{aligned}
\qquad (A2.2a)
$$

The total number of coefficients of this model (and thus the number of effects of the different factors) is equal to $1 + \sum\limits_{i=1}^{k} s_i$

The variables associated with each factor are related. Thus, for the factor i:

$$\sum_{j=1}^{s_i} x_{i,j} = 1 \qquad\qquad\qquad\qquad (A2.2b)$$

So the values of the coefficients are also interdependent, being related to one another by k expressions of the kind:

$$\sum_{j=1}^{s_i} \beta_{i,j} = 0 \qquad\qquad\qquad\qquad (A2.2c)$$

for the ith variable summed over the s_i levels, because the mean effect of all levels is set arbitrarily at zero.

Thus the total number of independent effects is:

$$
\begin{aligned}
p &= 1 + (s_1 - 1) + (s_2 - 1) + \dots + (s_k - 1) \\
&= 1 + \sum_{i=1}^{k} (s_i - 1) \\
&= 1 + \sum_{i=1}^{k} s_i - k
\end{aligned}
\qquad (A2.3)
$$

2. Conversion to the reference state model

Two factors with 3 and 4 levels
The general model may be simplified by taking into account the interdependence of the (presence-absence) variables. Let F_1 be at 3 levels (called A, B, C) and the second factor F_2 at 4 levels (A, B, C, D). The model (from chapter 2, section II.C) is then:

$$y = \beta_0 + \beta_{1,A}x_{1,A} + \beta_{1,B}x_{1,B} + \beta_{1,C}x_{1,C}$$
$$+ \beta_{2,A}x_{2,A} + \beta_{2,B}x_{2,B} + \beta_{2,C}x_{2,C} + \beta_{2,D}x_{2,D}$$
$$+ \varepsilon$$

We select reference states arbitrarily, taking level C for F_1 and level D for F_2. Since the factor F_1 is forced to take one of the possible levels, A, B, or C, and similarly F_2 must take A, B, C or D, we have

$$x_{1,A} + x_{1,B} + x_{1,C} = 1 \quad \text{and} \quad x_{2,A} + x_{2,B} + x_{2,C} + x_{2,D} = 1$$

Substituting these relationships into the model, eliminating $x_{1,C}$ and $x_{2,D}$, we obtain:

$$y = \beta_0 + \beta_{1,C} + \beta_{2,D}$$
$$+ (\beta_{1,A} - \beta_{1,C})x_{1,A} + (\beta_{1,B} - \beta_{1,C})x_{1,B}$$
$$+ (\beta_{2,A} - \beta_{2,D})x_{2,A} + (\beta_{2,B} - \beta_{2,D})x_{2,B} + (\beta_{2,C} - \beta_{2,D})x_{2,C} + \varepsilon$$

or more simply:

$$y = \beta'_0 + \beta'_{1,A}x_{1,A} + \beta'_{1,B}x_{1,B} + \beta'_{2,A}x_{2,A} + \beta'_{2,B}x_{2,B} + \beta'_{2,C}x_{2,C} + \varepsilon \qquad (A2.4)$$

where we have replaced combinations of coefficients by new coefficients, distinguished as β'.

$$\beta'_0 = \beta_0 + \beta_{1,C} + \beta_{2,D}$$
$$\beta'_{1,A} = \beta_{1,A} - \beta_{1,C}$$
$$\beta'_{2,B} = \beta_{2,B} - \beta_{2,D} \quad \text{etc.}$$

Since:

$$\beta_{1,A} + \beta_{1,B} + \beta_{1,C} = 0$$

we can return to the presence-absence model by the following calculations:

$$\beta_{1,C} = -\tfrac{1}{3}(\beta'_{1,A} + \beta'_{1,B})$$
$$\beta_{1,A} = \beta'_{1,A} + \tfrac{1}{3}(\beta'_{1,A} + \beta'_{1,B})$$
$$\beta_{1,B} = \beta'_{1,B} + \tfrac{1}{3}(\beta'_{1,A} + \beta'_{1,B}) \quad \text{etc.} \qquad (A2.5)$$

General case

Each of the k factors has s_i associated variables of which $s_i - 1$ are independent. As before we choose one of these as the reference state. Let the reference state of the ith factor be S_i for this factor. We eliminate the variable corresponding to S_i from equation 2.3 and obtain a general equation equivalent to the specific case of equation 2.6:

$$
\begin{aligned}
y = \beta'_0 &+ \beta'_{1,A} x_{1,A} + \beta'_{1,B} x_{1,B} + \dots \\
&+ \dots \\
&+ \beta'_{i,A} x_{i,A} + \beta'_{i,B} x_{i,B} + \dots \\
&+ \dots \\
&+ \beta'_{k,A} x_{k,A} + \beta'_{k,B} x_{k,B} + \dots + \varepsilon
\end{aligned}
\tag{A2.6}
$$

where the coefficients β' are related to the β by expressions of the type:

$$
\beta'_{i,A} = \beta_{i,A} - \beta_{i,S_i}
$$

The reference-state coefficients are estimated by multi-linear regression. To return to the original coefficients (presence-absence model) we use the following relations:

$$
\beta_{i,S_i} = -\frac{1}{k} \sum_{i=1}^{s_i - 1} \beta_i
$$

$$
\beta_{i,A} = \beta'_{i,A} + \beta_{i,S_i}
\tag{A2.7}
$$

$$
\beta_0 = \beta'_0 - \sum_{i=1}^{k} \beta_{i,S_i}
$$

References

1. R. L. Plackett and J. P. Burman, The Design of Optimum Multifactorial Experiments, *Biometrica*, **33**, 305-325 (1946).
2. G. E. P. Box and R. D. Meyer, Finding the active factors in fractionated screening experiments, *J. Qual. Technol.*, **25**(2), 94-105 (1993).

APPENDIX III

DESIGNS FOR RESPONSE SURFACE MODELLING

A. Second-order designs for the spherical domain
B. Second-order designs for the cubic domain
  Central composite designs
  3^k factorial designs
  Special designs for the cubic domain

Certain designs for RSM were not described in the examples of chapter 5 and are not easily derived. They include the hybrid and other saturated or near-saturated designs, for 4 to 6 factors, in the spherical domain and also a number of special designs in the cubic domain. All references are to the bibliography of chapter 5.

A. Second-Order Designs for the Spherical Domain

Rules for setting up the central composite, Box-Behnken and Doehlert designs were given in chapter 5 and a number of them were set out in full in that chapter. Others, hybrid designs in particular, are given below. References to the tables for the various designs are as follows:

Design	Number of factors	Table
Central composite	2	5.1 & 5.2
Central composite	3	5.11
Central composite designs (summary)	2 - 6	5.16
Doehlert	2	5.18
Doehlert	3	5.18 & 5.19
Doehlert (summary)	2 - 5	5.22
Hybrid 311A (modified)	3	5.25
Hybrid	4	A3.1
D-optimal ("minimal")	5	A3.2
Hybrid	6	A3.3
Box-Behnken	3	5.26

474

Of the hybrid designs, only the one for 3 factors was given in chapter 5. There are no simple rules for deriving them, though they may be considered as a kind of (non-central) composite design. The designs given below, tables A3.1, A3.2, and A3.3, for 4, 5, and 6 factors respectively, are saturated, allow the model coefficients to be estimated with maximum precision, and show near isovariance of rotation. They are all normalised to lie within a sphere of radius 1. The centre-points of the hybrid designs (tables A3.1 and A3.3) may be replicated once or twice. Although not absolutely necessary, this is recommended in order to improve the estimations of certain coefficients, in particular those of the squared terms in the model.

There is no hybrid design for 5 factors, and the D-optimal design we give instead (table A3.2) is the only one existing for so few experiments. It has the disadvantage that each factor is set at as many levels as there are experiments.

Table A3.1 Hybrid Design for 4 Factors [after Franquart (11)]

No.	X_1	X_2	X_3	X_4
1	-0.354	-0.612	0.612	0.354
2	-0.354	0.612	-0.612	0.354
3	0.354	-0.612	-0.612	0.354
4	0.354	0.612	0.612	0.354
5	-0.75	-0.433	-0.433	0.250
6	-0.75	0.433	0.433	0.250
7	0.75	-0.433	0.433	0.250
8	0.75	0.433	-0.433	0.250
9	-0.707	0	0	-0.707
10	0.707	0	0	-0.707
11	0	-0.866	0	-0.5
12	0	0.866	0	-0.5
13	0	0	-0.866	-0.5
14	0	0	0.866	-0.5
15	0	0	0	1
16	0	0	0	0

Table A3.2 D-optimal ("Minimal") Design for 5 Factors [after Franquart (11)]

No.	X_1	X_2	X_3	X_4	X_5
1	-0.598	-0.354	-0.568	-0.434	-0.082
2	0.333	0.042	-0.197	0.918	0.080
3	0.340	0.537	0.184	0.168	0.731
4	-0.003	-0.003	0.003	-0.003	0.000
5	0.264	0.317	0.770	-0.487	0.002
6	0.245	0.319	-0.370	-0.757	0.258
7	0.122	-0.902	-0.152	0.254	-0.288
8	0.090	-0.417	0.704	0.375	0.426
9	-0.327	0.113	-0.335	0.380	-0.790
10	0.297	-0.163	-0.095	-0.604	-0.715
11	0.080	-0.656	0.233	-0.661	0.267
12	-0.495	0.609	0.042	-0.504	-0.360
13	0.404	-0.088	0.528	0.367	-0.636
14	-0.629	0.132	0.286	-0.311	0.639
15	-0.360	0.583	0.522	0.502	-0.075
16	0.069	-0.417	-0.430	0.108	0.791
17	0.387	-0.104	-0.900	0.030	-0.169
18	-0.576	-0.359	0.570	-0.162	-0.435
19	0.508	0.772	-0.140	0.056	-0.351
20	0.959	-0.221	0.063	-0.069	0.149
21	-0.716	-0.320	-0.077	0.592	0.168
22	-0.380	0.582	-0.619	0.230	0.285

Table A3.3 Hybrid Design for 6 Factors [after Roquemore (10)]

No.	X_1	X_2	X_3	X_4	X_5	X_6
1	0	0	0	0	0	0
2	-0.433	-0.433	-0.433	-0.433	-0.433	0.25
3	0.433	-0.433	-0.433	0.433	0.433	0.25
4	-0.433	0.433	-0.433	0.433	0.433	0.25
5	0.433	0.433	-0.433	-0.433	-0.433	0.25
6	-0.433	-0.433	0.433	0.433	0.433	0.25
7	0.433	-0.433	0.433	-0.433	-0.433	0.25
8	-0.433	0.433	0.433	-0.433	-0.433	0.25
9	0.433	0.433	0.433	0.433	0.433	0.25
10	-0.433	-0.433	-0.433	-0.433	0.433	0.25
11	0.433	-0.433	-0.433	-0.433	-0.433	0.25
12	-0.433	0.433	-0.433	-0.433	-0.433	0.25
13	0.433	0.433	-0.433	-0.433	0.433	0.25
14	-0.433	-0.433	0.433	0.433	-0.433	0.25
15	0.433	-0.433	0.433	0.433	0.433	0.25
16	-0.433	0.433	0.433	0.433	0.433	0.25
17	0.433	0.433	0.433	0.433	-0.433	0.25
18	-0.866	0	0	0	0	-0.5
19	0.866	0	0	0	0	-0.5
20	0	-0.866	0	0	0	-0.5
21	0	0.866	0	0	0	-0.5
22	0	0	-0.866	0	0	-0.5
23	0	0	0.866	0	0	-0.5
24	0	0	0	-0.866	0	-0.5
25	0	0	0	0.866	0	-0.5
26	0	0	0	0	-0.866	-0.5
27	0	0	0	0	0.866	-0.5
28	0	0	0	0	0	0

B. Second-Order Designs for the Cubic Domain

1. Central composite designs (1)

Sometimes known as Box-Wilson designs, these are the same as the central composite designs for the spherical domain, except for the positioning of the axial points. These are face-centred, with $\alpha = 1$. An example is given for 3 factors in table A3.4. They may be derived easily by applying the rules given in chapter 5.

Table A3.4 Central Composite Design (Box-Wilson) for 3 Factors

No.	X_1	X_2	X_3	No.	X_1	X_2	X_3
1	-1	-1	-1	9	-1	0	0
2	+1	-1	-1	10	+1	0	0
3	-1	+1	-1	11	0	-1	0
4	+1	+1	-1	12	0	+1	0
5	-1	-1	+1	13	0	0	-1
6	+1	-1	+1	14	0	0	+1
7	-1	+1	+1	15	0	0	0
8	+1	+1	+1				

2. 3^k factorial designs

The 3^2 *complete factorial design*, described in chapters 5 and 7, is identical to the square central composite design for 2 factors. The 3^{4-1} *reduced factorial design* for 4 factors is shown in table A3.5. Note that elimination of any of the columns gives the 3^3 full factorial design.

Table A3.5 3^{4-1} Reduced Factorial Design for 4 Factors

No.	X_1	X_2	X_3	X_4	No.	X_1	X_2	X_3	X_4	No.	X_1	X_2	X_3	X_4
1	-1	-1	-1	-1	10	0	-1	-1	+1	19	+1	-1	-1	0
2	-1	-1	0	0	11	0	-1	0	-1	20	+1	-1	0	+1
3	-1	-1	+1	+1	12	0	-1	+1	0	21	+1	-1	+1	-1
4	-1	0	-1	0	13	0	0	-1	-1	22	+1	0	-1	+1
5	-1	0	0	+1	14	0	0	0	0	23	+1	0	0	-1
6	-1	0	+1	-1	15	0	0	+1	+1	24	+1	0	+1	0
7	-1	+1	-1	+1	16	0	+1	-1	0	25	+1	+1	-1	-1
8	-1	+1	0	-1	17	0	+1	0	+1	26	+1	+1	0	0
9	-1	+1	+1	0	18	0	+1	+1	-1	27	+1	+1	+1	+1

3. Special designs for the cubic domain

Recommended designs for the cubic domain were listed in chapter 5 table 5.28. With the exception of the Box-Wilson designs (see section B.2) these are given in full in tables A3.6 to A3.10, as there are no simple rules for generating them.

Table A3.6 Designs for 2 Factors

(a) MB-207 [after Mitchell and Bayne (21)]			(b) BD-206 [after Box and Draper (22)]			(c) BD-208 [after Box and Draper (22)]		
No.	X_1	X_2	No.	X_1	X_2	No.	X_1	X_2
1	-1	-1	1	-1	-1	1	-1	-1
2	+1	-1	2	+1	-1	2	+1	-1
3	-1	+1	3	-1	+1	3	-1	+1
4	+1	+1	4	-0.132	-0.132	4	+1	+1
5	+1	0	5	+0.394	+1	5	+1	0
6	0	+1	6	+1	0.394	6	+0.082	+1
7	0	0				7	-0.082	-1
						8	-0.215	0

Table A3.7 Designs for 3 Factors

(a) L-311 after Lucas (23)]				(b) BD-310 after Box and Draper (22)]			
No.	X_1	X_2	X_3	No.	X_1	X_2	X_3
1	-1	-1	-1	1	-1	-1	-1
2	+1	-1	-1	2	+1	-1	-1
3	-1	+1	-1	3	-1	+1	-1
4	+1	+1	-1	4	-1	-1	+1
5	-1	-1	+1	5	+0.193	+0.193	-1
6	+1	-1	+1	6	+0.193	-1	+0.193
7	-1	+1	+1	7	-1	0.193	0.193
8	+1	+1	+1	8	-0.291	+1	+1
9	+1	0	0	9	+1	-0.291	+1
10	0	+1	0	10	+1	+1	-0.291
11	0	0	+1				

Table A3.8 Designs for 4 factors

(a) WH-415 [after Welch (24)]

No.	X_1	X_2	X_3	X_4
1	-1	-1	-1	-1
2	+1	-1	-1	-1
3	-1	+1	+1	-1
4	+1	+1	+1	-1
5	+1	+1	-1	+1
6	-1	+1	-1	+1
7	+1	-1	+1	+1
8	-1	-1	+1	+1
9	0	-1	-1	+1
10	0	-1	+1	-1
11	0	+1	-1	-1
12	0	+1	+1	+1
13	0	+1	0	0
14	0	0	+1	0
15	0	0	0	-1

(b) DF-415 after Dubrova and Federov (25)]

No.	X_1	X_2	X_3	X_4
1	+1	-0.25	-0.25	-0.25
2	-0.25	0	+1	-1
3	-0.25	-1	0	+1
4	-0.25	+1	-1	0
5	+1	-0.5	+1	+1
6	+1	+1	-0.5	+1
7	+1	+1	+1	-0.5
8	-1	0	-1	+1
9	-1	+1	0	-1
10	-1	-1	+1	0
11	-1	-1	-1	-1
12	+1	+1	-1	-1
13	+1	-1	+1	-1
14	+1	-1	-1	+1
15	-1	+1	+1	+1

(c) MB-421 [after Mitchell and Bayne (21)]

No.	X_1	X_2	X_3	X_4	No.	X_1	X_2	X_3	X_4
1	-1	-1	-1	-1	12	+1	+1	+1	0
2	-1	+1	-1	-1	13	+1	-1	0	-1
3	+1	+1	-1	-1	14	+1	+1	0	+1
4	-1	-1	+1	-1	15	+1	0	+1	-1
5	-1	+1	+1	-1	16	+1	0	-1	+1
6	-1	-1	-1	+1	17	0	+1	+1	+1
7	-1	+1	-1	+1	18	0	-1	+1	0
8	-1	-1	+1	+1	19	0	+1	0	-1
9	+1	-1	+1	+1	20	0	0	-1	-1
10	-1	+1	+1	+1	21	-1	0	0	0
11	+1	-1	-1	0					

Table A3.9 Designs for 5 Factors in the Cubic Domain

(a) NO-521 [after Notz (26)]

No.	X_1	X_2	X_3	X_4	X_5
1	+1	+1	+1	+1	+1
2	+1	+1	+1	-1	-1
3	+1	+1	-1	+1	-1
4	+1	+1	-1	-1	+1
5	+1	-1	+1	+1	-1
6	+1	-1	+1	-1	+1
7	+1	-1	-1	+1	+1
8	-1	+1	+1	+1	-1
9	-1	+1	+1	-1	+1
10	-1	+1	-1	+1	+1
11	-1	-1	+1	+1	+1
12	+1	-1	-1	-1	-1
13	-1	+1	-1	-1	-1
14	-1	-1	+1	-1	-1
15	-1	-1	-1	+1	-1
16	-1	-1	-1	-1	+1
17	+1	+1	0	0	+1
18	+1	+1	0	+1	0
19	0	0	0	+1	+1
20	0	1	+1	0	0
21	+1	0	+1	0	0

(b) MB-524 [after Mitchell and Bayne (21)]

No.	X_1	X_2	X_3	X_4	X_5
1	+1	-1	-1	-1	-1
2	-1	+1	-1	-1	-1
3	-1	-1	+1	-1	-1
4	+1	+1	+1	-1	-1
5	+1	+1	-1	+1	-1
6	+1	-1	+1	+1	-1
7	-1	+1	+1	+1	-1
8	-1	-1	-1	-1	+1
9	+1	+1	-1	-1	+1
10	+1	-1	+1	-1	+1
11	-1	+1	+1	-1	+1
12	+1	-1	-1	+1	+1
13	-1	+1	-1	+1	+1
14	-1	-1	+1	+1	+1
15	+1	+1	+1	+1	+1
16	-1	-1	-1	+1	0
17	-1	-1	-1	0	-1
18	0	-1	-1	+1	-1
19	-1	0	+1	-1	0
20	-1	0	0	+1	-1
21	0	+1	+1	+1	0
22	0	+1	0	-1	-1
23	+1	-1	0	0	0
24	0	0	-1	0	+1

(c) MB-530 [after Mitchell and Bayne (21)]

No.	X_1	X_2	X_3	X_4	X_5	No.	X_1	X_2	X_3	X_4	X_5
1	-1	+1	-1	-1	-1	16	-1	+1	+1	-1	0
2	-1	-1	+1	-1	-1	17	-1	+1	+1	0	+1
3	+1	+1	+1	-1	-1	18	+1	-1	0	-1	-1
4	-1	-1	-1	+1	-1	19	-1	0	+1	-1	+1
5	+1	-1	-1	+1	-1	20	0	-1	-1	-1	-1
6	+1	+1	-1	+1	-1	21	+1	-1	+1	0	0
7	+1	-1	+1	+1	-1	22	-1	-1	0	+1	0
8	-1	+1	+1	+1	-1	23	+1	0	-1	-1	0
9	-1	-1	-1	-1	+1	24	0	+1	-1	+1	0
10	+1	+1	-1	-1	+1	25	+1	+1	0	0	-1
11	+1	-1	+1	-1	+1	26	-1	0	-1	0	-1
12	+1	-1	-1	+1	+1	27	0	-1	-1	0	+1
13	-1	+1	-1	+1	+1	28	+1	0	0	+1	+1
14	-1	-1	+1	+1	+1	29	0	+1	0	-1	+1
15	+1	+1	+1	+1	+1	30	0	0	+1	+1	-1

Table A3.10   Designs for 6 Factors in the Cubic Domain

(a) HD628 [after Hoke (27)]

No.	X_1	X_2	X_3	X_4	X_5	X_6
1	0	0	0	0	0	0
2	-1	+1	+1	+1	+1	+1
3	+1	-1	+1	+1	+1	+1
4	+1	+1	-1	+1	+1	+1
5	+1	+1	+1	-1	+1	+1
6	+1	+1	+1	+1	-1	+1
7	+1	+1	+1	+1	+1	-1
8	-1	-1	-1	-1	-1	0
9	-1	-1	-1	-1	0	-1
10	-1	-1	-1	0	-1	-1
11	-1	-1	-0	-1	-1	-1
12	-1	0	-1	-1	-1	-1
13	0	-1	-1	-1	-1	-1
14	+1	+1	-1	-1	-1	-1
15	+1	-1	+1	-1	-1	-1
16	+1	-1	-1	+1	-1	-1
17	+1	-1	-1	-1	+1	-1
18	+1	-1	-1	-1	-1	+1
19	-1	+1	+1	-1	-1	-1
20	-1	+1	-1	+1	-1	-1
21	-1	+1	-1	-1	+1	-1
22	-1	+1	-1	-1	-1	+1
23	-1	-1	+1	+1	-1	-1
24	-1	-1	+1	-1	+1	-1
25	-1	-1	+1	-1	-1	+1
26	-1	-1	-1	+1	+1	-1
27	-1	-1	-1	+1	-1	+1
28	-1	-1	-1	-1	+1	+1

(b) HD-634 [after Hoke (27)]

No.	X_1	X_2	X_3	X_4	X_5	X_6
1	-1	-1	-1	-1	-1	-1
2	-1	+1	+1	+1	+1	+1
3	+1	-1	+1	+1	+1	+1
4	+1	+1	-1	+1	+1	+1
5	+1	+1	+1	-1	+1	+1
6	+1	+1	+1	+1	-1	+1
7	+1	+1	+1	+1	+1	-1
8	-1	0	0	0	0	0
9	0	-1	0	0	0	0
10	0	0	-1	0	0	0
11	0	0	0	-1	0	0
12	0	0	0	0	-1	0
13	0	0	0	0	0	-1
14	+1	+1	-1	-1	-1	-1
15	+1	-1	+1	-1	-1	-1
16	+1	-1	-1	+1	-1	-1
17	+1	-1	-1	-1	+1	-1
18	+1	-1	-1	-1	-1	+1
19	-1	+1	+1	-1	-1	-1
20	-1	+1	-1	+1	-1	-1
21	-1	+1	-1	-1	+1	-1
22	-1	+1	-1	-1	-1	+1
23	-1	-1	+1	+1	-1	-1
24	-1	-1	+1	-1	+1	-1
25	-1	-1	+1	-1	-1	+1
26	-1	-1	-1	+1	+1	-1
27	-1	-1	-1	+1	-1	+1
28	-1	-1	-1	-1	+1	+1
29	0	+1	+1	+1	+1	+1
30	+1	0	+1	+1	+1	+1
31	+1	+1	0	+1	+1	+1
32	+1	+1	+1	0	+1	+1
33	+1	+1	+1	+1	0	+1
34	+1	+1	+1	+1	+1	0

APPENDIX IV

CHOICE OF COMPUTER SOFTWARE

    A. Experimental domain and nature of factors
    B. Mathematical models
    C. Experimental designs
    D. Analysis of data

Programs for experimental design may be analysed and assessed according to the following criteria:

- The nature of the experimental domain and the factors being varied.
- The experimental designs proposed (whether standard, or specific).
- The mathematical models which may be set up and studied.
- The possibilities for analysis of experimental results (standard statistical analysis, specific treatments, numerical and graphical analysis).
- Various complementary tools and methods.

All the characteristics presented are needed to cover the designs discussed in this book, their degree of importance being indicated by the type: **essential**, *recommended*, or useful.

A. Experimental Domain and Nature of Factors

The tables given below show the different kinds of problem and the experimental domains which must be taken into account by a program for experimental design.

Table A4.1 Types of factors and variables

factors	qualitative	2 levels	
		> 2 levels	
or	quantitative	independent	
		mixture component	simple
variables			mixtures of mixtures
		mixed (independent variables *and* mixture components)	

Table A4.2 Types of experimental domain

experimental domain	independent variables	cubic
		spherical
		any (perhaps with constraints, some impossible experiments, etc...)
	mixture components	complete simplex
		simplex with constraints (individual or relational)
	mixed	independent variables *and* mixture components

B. Mathematical Models

An experimental design program must be able to treat two kinds of mathematical models:

- Standard polynomials.
- Polynomials containing only certain terms, for treating specific problems.

Standard polynomial models must be generated automatically. Table A4.3 summarises the different models used in the book and the conditions under which they are used.

We have not considered non-linear models in this work.

Table A.4.3 Mathematical Models and Types of Study

Model	Independent factors		Mixture components		Commments
	effects study	RSM	screen-ing	RSM	
screening	•				Additive model, qualitative or quantitiative factors at ≥ 2 levels, only first-order effects and constant term
2nd order synergistic		•			Additive model with first-order interactions
complete synergistic		•			Additive model with all possible interactions
1st order polynomial			•		
2nd order polynomial			•		
incomplete polynomial		•	•		
Scheffé canonical				•	•	* First-order Scheffé canonical polynomial model ** Scheffé canonical polynomial model
Scheffé reduced cubic					•	Reduced cubic canonical polynomial model
Scheffé + reciprocal					•	Canonical polynomial model + inverse terms as proposed by Box and Draper
Becker polynomial					•	Various additive blending models proposed by Becker
incomplete polynomial					•	Canonical polynomial model with some terms omitted
ratio model					•	
product polynomial			•		•	Polynomial model with both mixture component and independent variables

C. Experimental Designs

1. Standard and non-standard designs

The various possible designs and their conditions for use are summarised in the next three tables. Table A4.4 lists the standard designs and table A4.5 a number of designs that must be constructed specially. D-optimal designs are treated separately, in table A4.6, as they require detailed analysis of software capabilities.

Table A4.4 Standard designs and conditions for use

Name	Screen-ing	Effects study	RSM 1st order	RSM 2nd order	Mixture components Screen-ing	Mixture components RSM	
Hadamard	•		•				Also called Plackett-Burman designs
2^{k-r} **factorial**	•	•	•				Fractional design at 2 levels
s^{k-r} *fractional factorial*	•	•					Symmetrical designs at s levels
Asymmetric design	•	•					Designs at $s_1, s_2, ...s_n$ levels, (by collapsing the symmetrical designs)
¾ . $2^{k(-r)}$	•	•					Three-quarter factorial designs (2 levels)
2^k		•					Full factorial design (2 levels)
Simplex			•				
Rechtschaffner	•	•					
spherical composite				•			Central composite (Box-Wilson) rotatable design
Doehlert				•			Uniform shell design
Hybrid				•			Roquemore design or designs derived from these
Box-Behnken				•			
cubic composite				•			Central composite (Box-Wilson) design with $\alpha = 1$
axial design				•			
Scheffé simplex						•	Simplex lattice design
Scheffé centroid						•	Simplex centroid design

Table A4.5 Specific designs

Type	Objective
2^{k-r} minimal[1]	model consisting of principal effects plus certain interactions
particular designs	to satisfy the constraints of the problem
product designs	mixed problems (different types of factors: quantitative, qualitative, mixtures components)
product (inner and outer array) designs	study of quality (Taguchi approach: chapter 7)

Table A4.6 D-optimal designs

Step in design generation	Element	
candidate design	**automatic generation of the candidate design**	**lattice**
		product of two or more standard designs
		McLean-Anderson algorithm (for mixtures)
	manual generation: addition of any point	
	suppression of any generated point	
mathematical models	**treatment with any linear model, whether complete or incomplete**	
algorithm	**Fedorov's exchange algorithm, or other high-performance method**	
choice of experiments	**exhaustive**	without repetitions
	non-exhaustive	with repetitions
	protected points	
choice of the number N of experiments[2]	*trials of a range of values of N to choose the best value*	
	trials of a large number of solutions for each value of N, to choose the best solution at given N	
	use of several quality criteria ($tr(\mathbf{X'X})^{-1}$, d_{max} etc.)	
stop criteria	*possibility of modifying stop criteria (especially for mixtures)*	

[1] Fractional factorial designs of 2^{k-r} experiments for k factors at 2 levels.

[2] N = number of experiments in the design.

2. Analysis of experimental designs

It is often necessary to measure the quality of an experimental design, for example to compare several possible designs, or to determine the consequences of a forced or inadvertent modification of a standard design. Calculation of the following characteristics allows this to be done, before carrying out the experiments.

Table A4.7 List of Quality Criteria for Experimental Designs

dispersion matrix $(X'X)^{-1}$	determinant of matrix $(X'X)^{-1}$ or $(X'X)$
information matrix $(X'X)$	determinant of moments matrix (M)
correlation matrix	**variance inflation factor: VIF**
alias matrix	trace of matrix $(X'X)^{-1}$
	maximum variance function
	Khuri index (measure of rotatability)

D. Analysis of Data

1. Responses

A number of transformations of the response may be treated, as well as the raw (untransformed data). The transformation is often selected following Box-Cox analysis.

Table A4.8 List of Transformations

y	$\sqrt{y}$	*logit(y)*
$\log_{10}(y)$	$1/y$	arcsin(y)
$\log_e(y)$	$1/\sqrt{y}$	y^{λ} etc...

For investigations involving *quality* (chapter 7), we need to add the various different forms of *signal/noise* response.

2. Mathematical calculations

The coefficients are calculated by multi-linear regression, according to the least squares method. There are a very large number of different programs for doing these calculations. The use of properly structured experimental designs, which are usually quite close to orthogonality, has the result that the more sophisticated methods (partial least squares etc.) are not usually necessary.

Table A4.9 Calculations

Determination of model	**least squares multi-linear regression**
	weighted multi-linear regression
	robust regression
	ridge regression
	partial least squares
	generalised inverse
Determination of transformation	**Box-Cox analysis**
Analysis of coefficients	**analysis of variance (ANOVA)**
	significance of model and individual coefficients
	significance of going from one order to a higher order model
	significance of lack of fit
	R^2 **and** R^2_{adj}
	identifying active effects *
Analysis of residuals	**untreated (raw) or studentized residuals***
Prediction	**calculation of response at a given point in the factor space**
	calculation of upper and lower confidence limits

* See graphical analysis (section D.3.)

3. Graphical analysis

The graphical anaylsis requred for various types of studies is shown in table A4.11. Questions of the quality of the graphical display, and the compatibility of the graphics files with other programs (for reports or publications) may also need to be considered.

Table A4.10 Graphical analysis

regression analysis of residual	*normal and half-normal plots*
	plotted against experiment order (time-trend)
	plotted against each factor variable
	plotted against calculated response
analysis of effects	*normal and half-normal plots*
	first-order interaction effect diagram
	second-order interaction effect diagram
	Lenth approach
	Pareto plots
	bayesian analysis
	perturbation diagrams
RSM for independent variables and mixture components[3]	**2D contour plots**: *with overlay of contour lines for different responses, and limits[4] of the domain.*
	3D plot of response surface
optimization	**desirability** **RSM**
	linear and non linear functions
	weighting [5]
	optimal path method
	study of curvature of the response surface
design analysis	*plot of variance function d_A* (for rotatability)

[3] With conversion between coded and natural variables, or between component and pseudocomponent.

[4] Spherical, cubic, individual, or relational constraints for mixtures, irregular limits, etc., indicated (e.g. with unacceptable region shaded).

[5] Possibility of weighting each partial desirability function within the global desirability function.

4. Optimization

Table A4.11 Optimization methods

Optimization method	Summary of calculation
Canonical analysis	calculation of the stationary point
	change of axes
	canonical equation
	simultaneous optimization [by transformation of the fitted equations into the same set of axes as the "principal" response]
Graphical optimization	using RSM (see graphical analysis)
Desirability	see graphical analysis
Steepest ascent	*determination of line of steepest ascent*
	estimation of response along this line, with possible extrapolation [multiple stop criteria]
Optimum path method	*determination of optimum path*
	estimation of response along this curve, with possible extrapolation [multiple stop criteria]
Simplex	fixed step
	variable step
	with several responses, using the global desirability

APPENDIX V

SYMBOLS

β_i coefficients of a model equation: the true (unknown) values.

η_A true value of a response y at point A.

ν number of degrees of freedom.

σ true (unknown) standard deviation, of which s is an estimate.

$\overline{}$ experimental error.

ρ distance from the centre of the experimental domain, in terms of coded variables.

b_i estimated coefficients of a model equation, associated in the model with the response _ calculated using the fitted model.

c^{ij} an element of the matrix $(\mathbf{X'X})^{-1}$, row i and column j.

$E[y]$ the expectation of y.

G a generator.

i, j integers identifying factors, coefficients, experiments , etc.

k number of independent factors and independent variables in the problem.

L_i lower limit of the component X_i.

N number of experiments in the design.

p number of parameters (coefficients) in the model.

q number of components in a mixture.

r ratio of the proportions of 2 components.
       fraction of a fractional factorial design 2^{k-r}.

s estimate of the standard deviation σ.

s_i number of levels of the ith variable of an asymmetric factorial design

S/N "signal/noise" ratio as defined by G. Taguchi.

t Student's t.

U_i natural variable for the factor i.
       and upper limit of component i (for a mixture design).

X_i coded variable for the factor i. It is also used to describe the axis for the factor i in coded factor space, for both independent (process) variables and mixture components. In the case of a mixed problem (of mixture and process components), we use it for the mixture components only, and identify the process variables by Z_i.

x_i level of X_i in the model for an experiment.

$\mathbf{X}$ model matrix.

$\mathbf{X'}$ transpose of the model matrix.

$\mathbf{X'X}$ information matrix.

$(\mathbf{X'X})^{-1}$ dispersion matrix.

y_i measured response, either for the response i, or for the experiment number i.

$\_i$ response calculated by the fitted model, either for the response number i, or for the experiment i.

Z_i transformed set of axes by canonical analysis of the second order model, a process component in a mixed process-mixture model.

z_i level of the process component Z_i in a mixed process-mixture model.

Index